Sustainable and Digital Transformation of Road Infrastructures

Sustainable and Digital Transformation of Road Infrastructures

Guest Editors

Hugo M. R. D. Silva
Joel R. M. Oliveira

MDPI

Basel • Beijing • Wuhan • Barcelona • Belgrade • Novi Sad • Cluj • Manchester

Guest Editors
Hugo M. R. D. Silva
Institute for Sustainability and Innovation in Structural Engineering
University of Minho
Guimarães
Portugal

Joel R. M. Oliveira
Institute for Sustainability and Innovation in Structural Engineering
University of Minho
Guimarães
Portugal

Editorial Office
MDPI AG
Grosspeteranlage 5
4052 Basel, Switzerland

This is a reprint of the Special Issue, published open access by the journal *Infrastructures* (ISSN 2412-3811), freely accessible at: https://www.mdpi.com/journal/infrastructures/special_issues/53OG3E6042.

For citation purposes, cite each article independently as indicated on the article page online and as indicated below:

Lastname, A.A.; Lastname, B.B. Article Title. *Journal Name* **Year**, *Volume Number*, Page Range.

ISBN 978-3-7258-6466-9 (Hbk)
ISBN 978-3-7258-6467-6 (PDF)
https://doi.org/10.3390/books978-3-7258-6467-6

Cover image courtesy of Hugo M. R. D. Silva

Contents

About the Editors

Hugo M. R. D. Silva

Hugo M. R. D. Silva is an Assistant Professor of Civil Engineering at the University of Minho and a member of the Institute for Sustainability and Innovation in Structural Engineering (ISISE). Trained as a civil engineer, his research focuses on sustainable asphalt materials for road pavements, including the incorporation of waste and construction and demolition waste, high reclaimed asphalt pavement recycling, warm and half-warm mix technologies, foamed and bio-binders, polymer and rubber modification, self-healing materials, and pavement digitalisation using optical sensing technologies. He leads and participates in national and European research projects, serves as Director of the MSc in Civil Engineering, and teaches in the PhD programme. He has supervised several doctoral and master's theses, acts as an external evaluator for the European Commission, and serves on the Editorial Boards of peer-reviewed journals.

Joel R. M. Oliveira

Joel R. M. Oliveira is an Assistant Professor of Civil Engineering at the University of Minho and a member of the Institute for Sustainability and Innovation in Structural Engineering (ISISE). His work focuses on transport infrastructure, with particular emphasis on sustainable road pavements. His research interests include life-cycle assessment, pavement maintenance and preservation, recycling technologies for hot, warm, and cold mixtures, warm-mix asphalts, grouted macadam, and circular, low-carbon materials. He serves on the Scientific Council of the School of Engineering, has participated in scientific committees for international conferences, collaborated with Editorial Boards, co-edited Special Issues, and supervised doctoral and master's research. His publications include journal articles, books, and invited book chapters linking research, practice, and policy.

Preface

As Guest Editors, we approached this Reprint with deep appreciation for the diversity and quality of the contributions it brings together. The articles collected here reflect diverse perspectives, methods, and scales of analysis, yet they are united by a shared commitment to advancing road infrastructure engineering through sustainable approaches and digital innovation.

Taken together, the contributions address materials, monitoring practices, data-driven assessment, and decision-support tools, illustrating how sustainability and digitalization are increasingly intertwined in practical engineering contexts. This Reprint highlights not only individual research outcomes but also the added value that emerges when these works are read as part of a broader, coherent dialogue.

We hope this Reprint will be a valuable and accessible resource for researchers, engineers, and practitioners, supporting reflection, knowledge sharing, and informed decision making in the ongoing transformation of road infrastructures. We would also like to sincerely thank the authors for their high-quality contributions and the reviewers for their careful and constructive feedback.

Hugo M. R. D. Silva and Joel R. M. Oliveira
Guest Editors

infrastructures

Editorial

Editorial for "Sustainable and Digital Transformation of Road Infrastructures"

Hugo M. R. D. Silva * and Joel R. M. Oliveira

Department of Civil Engineering, Institute for Sustainability and Innovation in Structural Engineering, University of Minho, Campus de Azurem, 4800-058 Guimarães, Portugal; joliveira@civil.uminho.pt
* Correspondence: hugo@civil.uminho.pt

Road infrastructure is increasingly recognized as a critical asset for economic development, social cohesion, and territorial connectivity. However, it is under unprecedented pressure from climate change, resource scarcity, environmental constraints, and rapidly evolving societal expectations. Traditional construction materials, design approaches, and maintenance practices—mostly dependent on non-renewable resources and periodic inspections—are increasingly recognized as insufficient to ensure the long-term sustainability, resilience, and efficiency of road networks [1]. As a result, the transformation of road infrastructure is now widely understood to be a dual process in which sustainability and digitalization must evolve together rather than independently or sequentially [2].

Recent research has focused on reducing the environmental footprint of road construction and maintenance by adopting circular economy principles, using renewable or waste-derived materials, and optimizing life-cycle performance [3,4]. As a result, there is a growing amount of attention being paid to bio-based binders, alternative aggregates, recycled materials, and innovative mix designs to reduce dependence on petroleum-based products and minimize greenhouse gas emissions. These approaches are increasingly supported by experimental and field evidence, showing that environmental benefits can be achieved without compromising performance, durability, or safety [5,6]. The relevance of these developments is reinforced by broader concerns about bio-sustainability, biodiversity protection, and the responsible use of natural resources within transport infrastructure systems [7].

Climate change has emerged as a significant challenge with respect to the performance of road infrastructure. Increased temperature variability, more frequent extreme-weather events, and altered loading conditions require pavement structures and materials that can maintain functionality in more aggressive and uncertain environments. Recent studies have focused on performance-based design, advanced material characterization, and resilience-oriented assessment frameworks that explicitly account for climatic effects throughout the infrastructure life cycle [8,9]. These advances can help give rise to more robust design decisions, extend service life, and reduce long-term maintenance and adaptation costs.

Alongside material and structural innovation, the digital transformation of road infrastructure is rapidly reshaping how assets are monitored, managed, and maintained. Advances in sensing technologies, data acquisition systems, and communication networks enable the continuous collection of large volumes of data on traffic, environmental conditions, and structural responses [10]. When combined with artificial intelligence, machine learning, and advanced data analytics, these technologies support a shift from reactive to predictive and preventive maintenance strategies, improving resource efficiency and enhancing the long-term performance of road assets [11,12].

 https://doi.org/10.3390/infrastructures11010012

Recent advances in automated inspection, computer vision, and intelligent monitoring systems further illustrate this transition. Applications such as image-based distress detection, connected-vehicle data, fiber-optic sensing, and remote inspection platforms demonstrate how digital tools can enhance safety, reduce inspection costs, and increase temporal and spatial resolution in infrastructure condition assessment [13,14]. Notably, these approaches directly contribute to sustainability objectives by optimizing maintenance interventions, reducing unnecessary material consumption, and minimizing traffic disruptions throughout infrastructure's life cycle.

From a broader perspective, the transformation of road infrastructure is closely tied to ongoing shifts in mobility and energy. The electrification of transport, particularly in freight corridors, requires integrated planning to align road infrastructure development with energy supply, charging infrastructure deployment, and traffic demand [15]. This need highlights the growing interdisciplinary nature of road infrastructure research, integrating materials engineering, data science, transportation planning, and energy systems.

More recently, the concept of digital twins has gained prominence as a unifying framework for infrastructure management. By integrating sensing, modeling, simulation, and decision-support tools into a single digital environment, digital twins lead to more informed, transparent, and adaptive management strategies [16,17]. Within the broader vision of Construction 5.0, these approaches emphasize not only efficiency and automation but also resilience, sustainability, and human-centered decision-making.

Beyond technological and material innovation, the effective implementation of sustainable and digital road infrastructure increasingly depends on broader institutional and societal frameworks. Public procurement, standardization, and technical specifications, as key enablers of change, have recently been attracting an increasing amount of attention. Moving beyond decision-making processes that are predominantly based on initial cost or isolated performance indicators, there is a clear need for more holistic assessment frameworks that explicitly value durability, life-cycle performance, resilience to climate change, circularity, recyclability, and environmental protection, including biodiversity considerations. Such approaches also create opportunities for innovation-driven markets, skill development, and new forms of employment, reinforcing the societal dimension of infrastructure sustainability [18]. While these aspects were not the primary focus of this Special Issue, they are essential conditions for translating scientific advances in sustainability and digitalization into real-world impacts and long-term public value.

In parallel, emerging modeling approaches at the molecular scale are beginning to provide deeper insight into aging mechanisms, material interactions, and durability phenomena in paving materials. Although still in the early stages of practical application, these approaches offer promising pathways to explaining how processes at the microscale influence macroscopic performance, surface characteristics, and long-term behavior, thereby supporting the development of more durable, sustainable, and resilient materials [19,20]. Such perspectives are expected to become increasingly relevant in future research.

The themes addressed in this Special Issue align closely with current international research agendas and policy frameworks. European strategies, including the Green Deal, the Circular Economy Action Plan, and the Sustainable and Smart Mobility Strategy [21,22], along with global perspectives from organizations such as the World Bank, OECD, PIARC, and FHWA, consistently emphasize the need for sustainable, resilient, and digitally enabled transport infrastructure [23–27]. The recent launch of the HORIZON-CL5-2026-01-D6-07 call further confirms the strategic relevance and timeliness of research on the sustainable and digital transformation of road infrastructure [28].

The contributions collected in this Special Issue reflect these broader trends and offer concrete examples of how research is addressing pressing global challenges. The authors

 https://doi.org/10.3390/infrastructures11010012

of recent studies have advanced sustainable pavement materials by developing bio-based binders and incorporating alternative or recycled constituents, thereby reducing environmental impacts while maintaining functional performance [29,30]. In parallel, other contributions have demonstrated how digital technologies—such as embedded sensing systems, data-driven monitoring, and artificial intelligence—enable more efficient, scalable, and predictive approaches to infrastructure condition assessment and maintenance [31,32]. At a more integrative level, the inclusion of a review on cognitive digital twins within the Construction 5.0 paradigm highlights how sustainability, digitalization, and human-centered decision-making can be jointly addressed in road infrastructure management [33]. Collectively, these contributions reinforce the view that future progress in road infrastructure engineering will depend on integrating sustainability principles, digital technologies, life-cycle thinking, climate resilience, and societal considerations. Thus, this Special Issue offers a snapshot of an evolving research landscape rather than a closed body of work.

Given the pace of technological development and the ongoing urgency of sustainability and climate challenges, the editors anticipate that these themes will remain highly relevant in the near future. Preparing a second edition of this Special Issue is therefore timely, offering an opportunity to capture emerging advances, particularly in areas such as biosustainable materials, digital twins, resilience assessment, societal impact evaluation, and data-driven infrastructure management.

The editors are grateful to the authors for their contributions, the reviewers for their thorough and helpful feedback, and the *Infrastructures* editorial team for their support. We hope that this Special Issue will advance research, policy, and practice in road infrastructure toward greater sustainability, digital intelligence, and resilience.

Author Contributions: Conceptualization, all authors; writing—original draft preparation, all authors; writing—review and editing, all authors. All authors have read and agreed to the published version of the manuscript.

Conflicts of Interest: The authors declare that they have no conflicts of interest.

References

1. Kapatsa, C.; Kavishe, N.; Zulu, S.L.; Maro, G. Sustainability assessment indicators of road infrastructure projects: A systematic literature review. *J. Eng. Des. Technol.* **2024**. [CrossRef]
2. Lozano-Ramírez, N.E.; Sánchez, O.; Carrasco-Beltrán, D.; Vidal-Méndez, S.; Castañeda, K. Digitalization and Sustainability in Linear Projects Trends: A Bibliometric Analysis. *Sustainability* **2023**, *15*, 15962. [CrossRef]
3. Aryan, Y.; Dikshit, A.K.; Shinde, A.M. A critical review of the life cycle assessment studies on road pavements and road infrastructures. *J. Environ. Manag.* **2023**, *336*, 117697. [CrossRef]
4. Liu, Z.; Kringos, N. Transition from linear to circular economy in pavement engineering: A historical review. *J. Clean. Prod.* **2024**, *449*, 141809. [CrossRef]
5. Gu, F.; Ma, W.; West, R.C.; Taylor, A.J.; Zhang, Y. Structural performance and sustainability assessment of cold central-plant and in-place recycled asphalt pavements: A case study. *J. Clean. Prod.* **2019**, *208*, 1513–1523. [CrossRef]
6. Espinosa, L.V.; Gadler, F.; Mota, R.V.; Guatimosim, F.V.; Camargo, I.; Vasconcelos, K.; Barros, R.M.V.; Bernucci, L.L.B. Multi-scale study of bio-binder mixtures as surface layer: Laboratory evaluation and field application and monitoring. *Constr. Build. Mater.* **2021**, *287*, 122982. [CrossRef]
7. Hlatshwayo, T.I.; Zungu, M.M.; Collinson-Jonker, W.J.; Downs, C.T. Mainstreaming ecological connectivity and wildlife needs in green road transport infrastructure planning in South Africa. *J. Environ. Manag.* **2024**, *371*, 123062. [CrossRef]
8. Liu, H.J.; Love, P.E.D.; Sing, M.C.P.; Niu, B.; Zhao, J. Conceptual framework of life-cycle performance measurement: Ensuring the resilience of transport infrastructure assets. *Transp. Res. Part D Transp. Environ.* **2019**, *77*, 615–626. [CrossRef]
9. Qiao, Y.; Dawson, A.R.; Parry, T.; Flintsch, G.; Wang, W. Flexible pavements and climate change: A comprehensive review and implications. *Sustainability* **2020**, *12*, 1057. [CrossRef]
10. Senkans, U.; Silkans, N.; Spolitis, S.; Braunfelds, J. Comprehensive Analysis of FBG and Distributed Rayleigh, Brillouin, and Raman Optical Sensor-Based Solutions for Road Infrastructure Monitoring Applications. *Sensors* **2025**, *25*, 5283. [CrossRef] [PubMed]

https://doi.org/10.3390/infrastructures11010012

11. Shang, J.; Zhang, A.A.; Dong, Z.; Zhang, H.; He, A. Automated pavement detection and artificial intelligence pavement image data processing technology. *Autom. Constr.* **2024**, *168*, 105797. [CrossRef]
12. Fathy, A.E.A.E.R.; Abdel-Hamid, M.; Abdelhaleem, H.M. Developing a predictive maintenance model for the rehabilitation of roads' assets. *Int. J. Constr. Manag.* **2025**, *25*, 398–408. [CrossRef]
13. El Hakea, A.H.; Fakhr, M.W. Recent computer vision applications for pavement distress and condition assessment. *Autom. Constr.* **2023**, *146*, 104664. [CrossRef]
14. Chapeleau, X.; Blanc, J.; Hornych, P.; Gautier, J.L.; Carroget, J. Assessment of cracks detection in pavement by a distributed fiber optic sensing technology. *J. Civ. Struct. Health Monit.* **2017**, *7*, 459–470. [CrossRef]
15. Chen, F.; Taylor, N.; Kringos, N. Electrification of roads: Opportunities and challenges. *Appl. Energy* **2015**, *150*, 109–119. [CrossRef]
16. Vieira, J.; Martins, J.P.; Almeida, N.M.; Patrício, H.; Morgado, J.G. Towards Resilient and Sustainable Rail and Road Networks: A Systematic Literature Review on Digital Twins. *Sustainability* **2022**, *14*, 7060. [CrossRef]
17. Yan, Y.; Ni, L.; Sun, L.; Wang, Y.; Zhou, J. Digital Twin Enabling Technologies for Advancing Road Engineering and Lifecycle Applications. *Engineering* **2025**, *44*, 184–206. [CrossRef]
18. Karlsson, I.; Rootzén, J.; Johnsson, F. Reaching net-zero carbon emissions in construction supply chains—Analysis of a Swedish road construction project. *Renew. Sustain. Energy Rev.* **2020**, *120*, 109651. [CrossRef]
19. Luo, L.; Chu, L.; Fwa, T.F. Molecular dynamics analysis of oxidative aging effects on thermodynamic and interfacial bonding properties of asphalt mixtures. *Constr. Build Mater.* **2021**, *269*, 121299. [CrossRef]
20. Yin, P.; Pan, B.; Li, Z.; Jiao, B.; Liu, Y. Multiscale Analysis of the Function Mechanism of Aging Action on the Adhesion Properties between Asphalt and Aggregate. *J. Mater. Civ. Eng.* **2024**, *36*, 04024262. [CrossRef]
21. European Commission. *Sustainable and Smart Mobility Strategy—Putting European Transport on Track for the Future*; COM(2020) 789 Final; European Commission: Brussels, Belgium, 2020. Available online: https://eur-lex.europa.eu/legal-content/EN/TXT/HTML/?uri=CELEX:52020DC0789 (accessed on 18 December 2025).
22. European Commission. *The European Green Deal*; COM(2019) 640 final; European Commission: Brussels, Belgium, 2019. Available online: https://eur-lex.europa.eu/legal-content/EN/TXT/?uri=COM%3A2019%3A640%3AFIN (accessed on 18 December 2025).
23. Global Center on Adaptation. *Climate Resilient Infrastructure Handbook*; Global Center on Adaptation: Rotterdam, The Netherlands, 2025.
24. Hallegatte, S.; Rentschler, J.; Rozenberg, J. *Lifelines: The Resilient Infrastructure Opportunity*; World Bank: Washington, DC, USA, 2019. [CrossRef]
25. Organisation for Economic Co-operation and Development. *Infrastructure for a Climate-Resilient Future*; OECD Publishing: Paris, France, 2024. [CrossRef]
26. Federal Highway Administration. *The FHWA R&T Story: Innovating the Future of Transportation*; FHWA-HRT-22-035; Federal Highway Administration: Washington, DC, USA, 2022. Available online: https://highways.dot.gov/sites/fhwa.dot.gov/files/2021-12/HRT-22035.pdf (accessed on 19 December 2025).
27. PIARC. Envisioning the Transport Agency of the Future. In *State of Existing Knowledge*; World Road Association: Paris, France, 2024.
28. European Commission. *Innovative Construction and Maintenance, with the Use of New Materials and Techniques, for Resilient and Sustainable Transport Infrastructure*; HORIZON-CL5-2026-01-D6-07; Horizon Europe: Brussels, Belgium, 2025. Available online: https://cordis.europa.eu/programme/id/HORIZON_HORIZON-CL5-2026-01-D6-07 (accessed on 20 December 2025).
29. Frey, M.R.; Williams, S.L.; Srubar, W.V.; Torres-Machi, C. Characterization and Evaluation of Agar as a Bio-Based Asphalt Binder Alternative. *Infrastructures* **2025**, *10*, 223. [CrossRef]
30. Apaza, F.R.; Fernández Vázquez, V.; Paje, S.E.; Gulisano, F.; Gagliardi, V.; Saiz Rodríguez, L.; Gallego Medina, J. Towards Sustainable Road Pavements: Sound Absorption in Rubber-Modified Asphalt Mixtures. *Infrastructures* **2024**, *9*, 65. [CrossRef]
31. Rebelo, F.J.P.; Oliveira, J.R.M.; Silva, H.M.R.D.; Sá, J.O.E.; Marecos, V.; Afonso, J. Installation and Use of a Pavement Monitoring System Based on Fibre Bragg Grating Optical Sensors. *Infrastructures* **2023**, *8*, 149. [CrossRef]
32. Zhang, H.; Dong, Y.; Hou, Y.; Cheng, X.; Xie, P.; Di, K. Research on Asphalt Pavement Surface Distress Detection Technology Coupling Deep Learning and Object Detection Algorithms. *Infrastructures* **2025**, *10*, 72. [CrossRef]
33. Oditallah, M.; Alam, M.; Ekambaram, P.; Ranjha, S. Review and Insights Toward Cognitive Digital Twins in Pavement Assets for Construction 5.0. *Infrastructures* **2025**, *10*, 64. [CrossRef]

 https://doi.org/10.3390/infrastructures11010012

infrastructures

MDPI

Review

Review and Insights Toward Cognitive Digital Twins in Pavement Assets for Construction 5.0

Mohammad Oditallah, Morshed Alam *, Palaneeswaran Ekambaram and Sagheer Ranjha

Department of Civil and Construction Engineering, Swinburne University of Technology, Hawthorn, VIC 3122, Australia; moditallah@swin.edu.au (M.O.); pekambaram@swin.edu.au (P.E.); sranjha@swin.edu.au (S.R.)
* Correspondence: mmalam@swin.edu.au

Abstract: With the movement of the construction industry towards Construction 5.0, Digital Twin (DT) has emerged in recent years as a pivotal and comprehensive management tool for predictive strategies for infrastructure assets. However, its effective adoption and conceptual implementation remain limited in this domain. Current review works focused on applications and potentials of DT in general infrastructures. This review focuses on interpreting DT's conceptual foundation in the flexible pavement asset context, including core components, considerations, and methodologies. Existing pavement DT implementations are evaluated to uncover their strengths, limitations, and potential for improvement. Based on a systematic review, this study proposes a comprehensive cognitive DT framework for pavement management. It explores the extent of enhanced decision-making and a large-scale collaborative DT environment. This study also identifies current and emerging challenges and enablers, as well as highlights future research directions to advance DT implementation and support its alignment with the transformative goals of Construction 5.0.

Keywords: Digital Twins; pavement management; PMS; road pavement maintenance; Construction 5.0; asset management

1. Introduction

The construction industry exhibits lower productivity than other sectors, primarily due to factors such as the slow adoption of technologies [1,2]. According to a recent industrial survey report, construction was identified as one of the least digitised industries considering assets, usage, and workers [3]. Industry 4.0 technologies have contributed to the digital transformation of many sectors, including construction. Industries 2.0 and 3.0 focused on mechanisation and automation of manufacturing processes, respectively, whereas the fourth industrial revolution relied on digitalisation, automation, big data, and innovative processes [4]. Industry 4.0's recent focus is building interconnected environment applications to operate and maintain assets efficiently [5]. However, the proposed evolution to Industry 5.0 is based on combining collaborative human and machine intelligence. This aims for more informed decisions and integrated sustainable systems [4,6]. Furthermore, the connection between the physical and digital worlds is moving beyond current digital enablement and toward digital control [7]. This vision integrates the creativity of humans with intelligent machines and digital systems to collaborate productively [8,9]. For the construction industry, moving toward construction 5.0 involves incorporating advanced technologies to address the complexities of current infrastructure management, i.e., road

https://doi.org/10.3390/infrastructures10030064

pavement assets. This includes autonomous collection systems, realistic asset state prediction, and intelligent decision support systems. Currently, Construction 4.0 aims to enhance industry productivity and enable cooperation with assets by the help of technologies [10,11]. These technologies include Digital Twins (DTs), Artificial intelligence (AI), Internet of Things (IoT), Building Information Modelling (BIM), and Geographical Information Systems (GIS). Its applications expand across different phases of the construction assets, including operation and maintenance [6]. AI enables predictive analytics and recognition techniques, and IoT provides real-time data collection through facilitated communication between devices and cloud networks. Moreover, BIM involves modelling information that enriches 3D models for standardised lifecycle data management, and GIS supports spatial and temporal data analysis and visualisation. DTs are a virtual representation of a physical entity that enables real-time monitoring and lifecycle analysis of its mirrored entity [12–14]. However, DT leverages the combined capabilities of different technologies for comprehensive and enhanced data-informed management. Construction 4.0 and 5.0 within a broad industrial revolution timeline are presented in Figure 1 below.

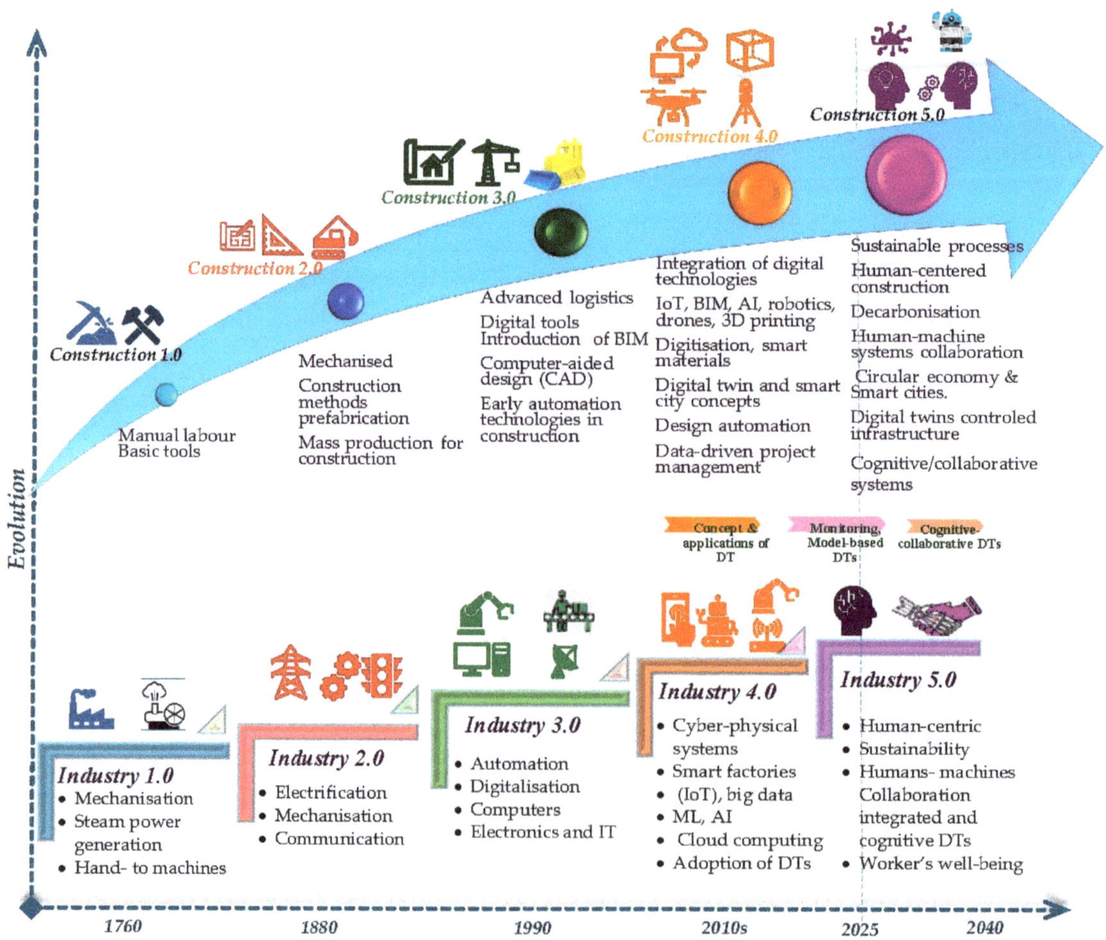

Figure 1. Industrial revolutions and Construction adoption timeline.

The global DT market is expected to have 50% compound annual growth, reaching 184.5 USD billion by 2030 [15]. In this market, built environments, including intelligent buildings, smart cities, energy, etc., form around 20% of the overall value [16]. Smart city DT solutions alone will reach $5.9 billion by 2029, and most IoT-based platforms will achieve DT capability [17]. Despite this significant growth in the DT market, the growth of the DT concept, and digitalisation in the construction industry, the development and implementation of mature DTs remain limited and in their early stages [18–21] Although examples are being developed in some applications, it is still in the prototype stage [20–23] and, in some areas, lacks the complete concept of DT adoption [21,24]. For instance, most examples of asset management are limited to information collection or virtual models, with less attention paid to road assets [25,26]. However, the barriers towards its efficient adoption in the construction industry include a lack and fragmented comprehension of DT concepts [24], absence of clear vision, and standardised DT implementation [27,28], which results in interoperability, data integration, and management issues [20]. Moreover, the limited prototypes and holistic frameworks, as well as the complexity of developing and maintaining DTs, have created more barriers. This is also due to a shortage of specialised digital skills in the construction industry [19,20,25].

For road pavement infrastructure management, DTs could enable predictive maintenance, proactive management of resources, and extended asset life [20,25,26]. Despite its numerous advantages, it is not yet well-developed and lacks full design considerations and implementation.

Existing road pavement infrastructure encounters several challenges, from adapting to growing demands and priorities of maintaining and improving service to expanding its lifespan. Pavement assets are the main contributors to energy consumption and emissions [29]. Its management is critical for the safe and efficient movement of users and the economy, and it must be managed efficiently [30]. Its operation and maintenance phase is usually a point of concern due to the cruciality of any possible improvement [31]. The following sections present an overview of flexible pavement management and the emerging solution of DTs.

1.1. Flexible Pavement Assets

Across industries, different maintenance strategies, planning, and implementation approaches are used. These approaches are corrective, reactive, preventive, condition-based, predictive, and prescriptive. Corrective approaches implement actions only when a part of the system fails. Preventive focuses on actions at a pre-defined set of times [32]. Condition-based assessments rely on evidence of asset deterioration or deviation from normal service levels [33]. The predictive approach predicts ahead of time whether the asset is going towards damage [34]. DTs can improve different interventional methods, from current preventive and reactive strategies to predictive and cognitive approaches, such as predictive and prescriptive maintenance. Figure 2 shows the developed industrial and technological evolution of maintenance strategies.

The flexible pavement asset consists of multiple structural layers. Its evaluation focuses on structural and functional aspects and is impacted by operational and environmental factors. Structural refers to the pavement's ability to withstand service conditions, such as resistance to deformation and cracking under continuous cycles of loads. This is assessed through field and laboratory testing and mechanistic modelling. Onsite structural evaluation is based on data derived from non-destructive testing (NDT) methods such as the Falling Weight Deflectometer (FWD) and traffic speed deflectometer [35]. In addition, inspections and measurements of structural cracking, rutting, or pothole distress are used as metrics for structural integrity, as they reflect subsurface failures. Ground penetration radar (GPR) is another onsite effective method for detecting pavement subsurface failures. The functional performance refers to the pavement's surface smoothness and conditions, presenting ride quality and roughness, i.e., surface cracks, potholes, and skid resistance. The severity and intensity of these conditions can be detected by visual inspection activities or through measuring methods, i.e., ride quality testing methods. On the other hand, structural and functional factors interact in terms of their development mechanisms, i.e., excessive rutting is a factor that causes crack formation. Moreover, environmental conditions, i.e., moisture and temperature, drive material interaction within pavement layer performance. This can cause or accelerate damage development under different operational loads, primarily due to the asphalt material's viscoelastic and plastic material mechanics. For example, pavement cracking and deformation are influenced by load patterns, loading speed, temperature differentials, material properties, aging, thermal stress, moisture, and wheel–surface contact stresses [36–38]. Performance and driving factors are crucial for effective pavement state prediction and management. However, in pavement maintenance, interventional decisions use predefined thresholds, such as quality metrics, to trigger maintenance. This is often after significant damage has occurred [30]. Maintenance decisions also depend on the broader context and factors, including network conditions, environmental factors, treatment costs, service levels, and user risks.

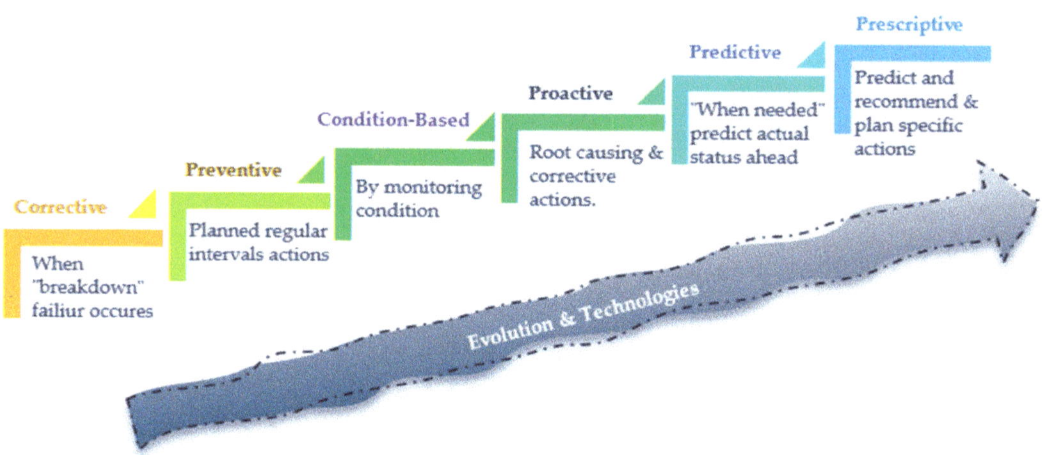

Figure 2. Maintenance and intervention approaches.

1.2. Pavement Management Systems

The pavement management system (PMS) was introduced in the early 1970s by the American Association of State Highway and Transportation Officials (AASHTO) [39]. PMS is "a set of defined procedures for collecting, analysing, maintaining, and reporting pavement data to assist the decision-makers in finding optimum strategies for maintaining pavements". In 1989, the Federal Highway Administration (FHWA) required related agencies to implement PMS by 1993 [40]. Over time, various organisations developed systems using different performance indicators and data standards. This results in PMS being a broad term for various systems. Some examples of these systems are PMS, PAVER, Pavement Information and Needs System (PINS), and Rural Infrastructure and Pavement Management System (RIPPS). For instance, road, bridge and culvert data were incorporated into a holistic system called the highway asset management system (HAMS) [41]. The U.S. Army Corps developed the PAVER system and adopted the pavement condition index (PCI) as a quality metric [42]. The RIPPS focuses on managing pavement for low-volume rural roads. Maintenance management systems (MMSs) are subsystems that focus on the function and effectiveness of the maintenance activities [43]. However, PMSs and MMSs are the most used systems in highway management [44].

Performance rating indices are used to rank the health of pavement, such as PCI, International Roughness Index (IRI), Present Serviceability Index (PSI), and Present Serviceability Rating (PSR) [45]. These rating indicators are measured or derived based on road segment evaluation processes using inspecting and testing survey data. However, IRI is the most employed within PMS [42]. The overall process of PMS involves data collection, evaluation and prediction to identify current or future deterioration trends.

1.2.1. Data Collection

Data collection is commonly based on manual pavement surface inspections, which are costly, labour-intensive, time-consuming, prone to errors, and reliant on expert evaluation [44,46,47]. Surveys involve onsite manual measurements and collect photos to report pavement conditions. This extends intervals of updating data and reduces the efficiency of maintenance planning and forecasting its status [42]. Therefore, some agencies have adopted specialised vehicles equipped with sensors, such as lasers, cameras, and deflectometers. However, this resulted in expensive processes and equipment [48]. Moreover, this process resulted in issues like data inconsistency and disruption of traffic flow [47] and, furthermore, the inability to cover the extended pavement area due to time and budget constraints [49], in addition to the limitation of using these expensive methods for relatively small municipalities due to insufficient budgets [50].

Recently, data collection has undergone a significant evolution with new tools and low-cost automated methods [51]. Image processing and AI detection methods for pavement images, such as using deep learning (DL) algorithms, have shown promising results [46,50,52,53]. These tools have improved the surface data collection performance of PMS [48]. Furthermore, the Machine Learning (ML) evolution has elevated processes to further severity measurements for various defect types, as in [54,55]. Moreover, there is a growing use of smartphones and unmanned aerial vehicles (UAVs) to collect pavement data. However, different subsystems and processes were developed separately, and PMSs also developed various subsystems used to support road management. Therefore, the data types and standards variability posed a limitation in the integration process, which resulted in a fragmented management approach [56]. PMSs have no centric dataset and lack smart capabilities in managing data [2]. Currently, to manage road asset conditions spatially and prioritise their interventions, most agencies have adopted global positioning systems (GPS) and GIS [57].

1.2.2. Pavement Performance Modelling

Prediction modelling is used to plan maintenance and optimise preventive maintenance schedules through existing statistical and empirical models. Models used are either developed for pavement design or using survey data to estimate the asset deterioration trends statically. Generally, prediction modelling can be classified as data-driven and physics-based models. Physics-based models, known as purely mechanistic models, rely on experimental structural response data such as stress and strain. Data-driven models include deterministic, empirical or statistical models, i.e., regression models. In addition, Mechanistic-empirical (ME) models incorporate some experimental response parameters in statistical form. Data-driven models rely on mathematical relationships between variables [58]. The modern form is ML models. On the other hand, Probabilistic models, such as Bayesian and Markov, are models that involve probability prediction to account for uncertainty [58]. They need extensive computational resources, work with limited data, and lack quantitative correlation [59].

Statistical models rely on time-independent data processing, which limits their prediction due to the incremental form of pavement deterioration [60]. These traditional models have different well-known limitations, such as data sensitivity, requiring structured and high-quality data, relying on assumptions for data preparation [61]. Moreover, they cannot account for complex and highly nonlinear relationships, i.e., pavement material interaction physics [62], in addition to randomness issues and limited prediction accuracy when a more extensive dataset or number of predictor variables is used [61,63,64].

The nonlinearity of ML models outperforms other methods in terms of accuracy and big datasets. It can predict patterns and deal effectively with multiple factors and randomness issues [51]. However, model knowledge is capped to a limited training dataset and requires high-quality training data. Moreover, it is prone to overfitting as it performs well on training data while poorly generalising to new data. It also lacks interpretability [62], which leads transport agencies to adopt other deterministic approaches due to ML model challenges for decision-makers [51,65]. For instance, Austroads, an Australasian head organisation of road transport agencies, in their last deterioration models reports, stated that most road agency members prefer to use deterministic models such as ME models over ML [65,66]. They noted that the ME model is more stable and reflects the explanatory deterioration of pavement. Moreover, it can be incorporated into the existing PMS [65]. Although ML provides robust models, overfitting and its limited transferability to other data, along with the black-box interpretability and complex implementation, are the main issues of its adoption in PMSs [59,67]. However, ME models have led to the advancement in pavement design [58], and they are more accurate than pure empirical models. The data-driven models often lack the physical interaction of the underlying pavement mechanics, which limits their accuracy in complex pavement systems [67]. However, ME models are data-driven models developed and validated through a one-time validation process and require continuous calibration [68]. Austroads deterioration models for thin pavements are based on a multi-variate nonlinear ME model [65]. The models were developed from data between 1994 to 2018. The fit of models (R^2) was only 0.23 for the roughness [67] and 0.5 for the rutting [65]. They indicated the need for calibration for local use conditions.

Using design ME models for analysing existing in-operation pavement involves critical assumptions, as these models were developed from pavement data in different environmental conditions and converted mixed vehicle classes as input [69]. This, in addition to relying on assumptions about the loads the asset handles, which can differ significantly from actual in-situ conditions [70], can lead to misrepresentation of asset performance. In pavement design, ME models use various traffic inputs for future scenario analysis. This can cause contradictions between existing and designed asset performance. Additionally, using these models for predictions during operation relies on estimated traffic loads and constant operation loads and patterns assumptions over the pavement's service life [70]. This often overlooks traffic patterns, operational changes, and various overloading scenarios. Traffic loading is a critical factor for pavement response, and it can change significantly over time. For instance, the FHWA technical report states that substantial traffic volume changes occurred over time at specific sites, often due to major road construction projects or events that alter traffic patterns [71].

1.3. Digital Twin for Pavement Management

Due to the limitations of data-driven models in capturing complex degradation and the lack of actual operational conditions, it is suggested that blending the strengths of these models could enhance accuracy [59]. The ME, as a simple blend, outperforms previous methods in pavement design [58]. The current PMS struggles with data integration, and limited predictive performance hinders predictive maintenance adoption [2,51,72]. Fragmented processes hinder efficient management and timely interventions, highlighting the need for a technology-driven PMS [51]. The DT concept uses multi-modelling with multi-sources lifecycle data, including live asset data, to improve modelling and integrate processes [2,73]. Predictive methods can enhance interventions, cut emissions, and improve operation and maintenance planning [26,74,75], necessitating frameworks for DT [20,21,24].

However, accelerated digitalisation and unstandardised DT adoption have led to misconceptions in some construction applications. Some technology integration works incorrectly labelled their efforts as DT, presenting it as a trendy term for traditional monitoring or modelling. This misrepresentation hinders the full implementation of true DTs, resulting in low maturity levels of DTs [76,77], along with unrealistic expectations of DT [78], and lack of focus on its purpose [19]. Opoku et al. [79] reviewed drivers of DT adoption in construction, highlighting enhanced predictive maintenance as a key success factor, demonstrating the expected added value in DT development.

Recent advancements in sensing and technologies for pavement have led to the emerging development of real-time monitoring, such as non-destructive methods, i.e., [35,80,81], and recent imagery-based AI defect detection, in addition to instrumented pavement through embedded sensing technologies, as in [82–84]. This allows dynamic evaluation of the underlying structural state of the asset for monitoring applications and capturing pavement response, i.e., stress, strain, moisture, and temperature. DTs and these real-time monitoring, i.e., Structural Health Monitoring (SHM), share various similarities. However, SHM provides periodical data collection or real-time monitoring for insights throughout parameter change reflection. Its advancement can enable mature implementation techniques for DTs [25]. DT integrates data processing and optimises employed sensing and monitoring capabilities for broader and more efficient data queries [20]. It expands SHM to simulate and predict changes for the long term, considering the entire life cycle and examining what-if scenario analysis [25,85]. These capabilities of DT are seen as an extension of the SHM and an expansion of its functions [19]. DT's uniqueness lies in interconnecting previously unconnected systems. This implementation required focus and holistic understanding for successful pavement DT that effectively incorporated suitable and mature monitoring methods.

Recent Reviews of DTs in Transportation

Conceptual DT architecture, the three-dimensional framework, was initially presented in three main components by Grieves [86]: a physical entity, a virtual entity, and a connection between them. However, with the expansion of DT adoption and the need to align with recent technology and applications, the five-dimensional (5D) conceptual DT model was proposed by Tao in 2018 [87]. The new conceptual model added data and service components for broader applicability. The 5D model was referenced in DT development for the mesostructure of asphalt mixture material [88], where virtual moulding of material samples was applied.

However, in transportation engineering, the previous reviews of DT focused on its general applications for maintenance [89], applications in road and railway networks [26], and broad technologies used for infrastructure life phases [90]. Other transportation reviews were conducted for BIM and DTs in the overall digitalisation process [91], autonomous vehicles and transportation operations [92], transportation safety, mobility, and environmental impact [93]. Furthermore, the very recent reviews still focus on applications of DTs for transportation infrastructure management [20], and evaluate general levels of adoption of all transport sectors [94].

For road pavement, in particular, a review focused entirely on data acquisition systems and only for bridge roads [95], general DT applications in road pavement with no framework provided [96], and pavement sensing technologies for DT development focused on sensing methods, including the asset construction phase, with no technical framework provided [25]. Reviews on the transportation asset DTs are summarised in Table 1 below.

Table 1. Recent reviews on DTs in transportation engineering.

Work	Year	Review Focus
[96]	2024	Explore DT applications in road pavement management.
[93]	2024	Transportation DTs for safety and mobility applications and architecture for a transportation DT (TDT).
[20]	2024	Overall Digital Twins of transportation infrastructure management.
[92]	2024	DT-ITS core considerations, progress, services development of DT-ITS concerning architectures.
[94]	2024	Transportation Operation and Maintenance adoption and general asset management framework.
[97]	2023	BIM applications in civil engineering using new Information and Communication Technologies (ICTs).
[24]	2023	Transportation infrastructure management DT concept definition, life cycle application, and technology.
[98]	2022	Predictive maintenance transportation motor vehicles.
[26]	2022	DT rail and road infrastructure networks.
[91]	2022	BIM and digital twins in the digitalisation of transportation.
[95]	2022	Digital twins of road bridges inspection.
[90]	2022	Underground infrastructure construction and Operation & Maintenance (O&M) on locating and mapping technology.
[89]	2020	General DT applications for maintenance.

Nevertheless, none of these works explored the structure and requirements of DT architecture, particularly for flexible pavement assets. Although one presented a framework [94], it remains a broad framework focused on general transportation assets, particularly regarding fleet and logistics aspects. Moreover, the work evaluated road transport DTs, focusing on applications such as autonomous vehicles, simulating vehicle operations, and operation safety. Furthermore, the developed framework is broad and generic. Therefore, it is unsuitable for pavement asset use, especially given flexible pavement assets' complex nature and management requirements. Thus, relying on this can pose challenges in implementation if asset-specific aspects are overlooked. Potential critical oversights and effects in pavement context are analysed in Table 2. A DT framework for practical adoption remains underdeveloped.

Current works neglect DT structure and design specific to asset needs for a comprehensive framework. Additionally, no enhanced cognitive DT framework or the potential of new construction 5.0 advancements have been proposed. Literature shows DT applications and potentials without attention to demonstrating their value for specific assets. This gap hinders pavement management in adopting DTs.

Therefore, in assessing the current proposed DT works for road pavement throughout the DT concept and structure, core considerations will uncover the current state of knowledge, outline the requirement for the actual implementation of cognitive pavement DT, and outline a framework that integrates possible solutions. This will also identify current gaps and direct future research. This research raised the following questions:

- What are the components and critical considerations of digital twins (DTs) in the pavement assets; their current maturity, and the gaps for cognitive DT capabilities?

- What are the required framework elements, processes, and methods for developing predictive cognitive DT for pavement that enable enhanced decision-making and support the Construction 5.0 paradigm?

Table 2. Flexible pavement critical aspects and potential challenges of the generalised framework adoption.

Aspect	General Framework Possible Assumptions	Asphalt Pavement Asset Criticality	Challenge/Effect
Modelling material behaviour	Not available (analytical only)/Elastic assumptions	Account for asphalt responses (i.e., viscoelastic/plastic response, time &temperature dependency, stress dependency response)	Unrealistic behaviour, inaccurate deterioration, overlooking temperature/load type or pattern)
Lifecycle/Simulation data input	Constant loads of operational/environmental data	Sensor data preparation (time-based loading, distributed temperature, overloading, etc.)	Inaccurate distress prediction (i.e., deformation), degradation rates, scenario analysis (i.e., event-impact scenario)
Integrated predictive model techniques	One modelling approach	Combine underlying physics of asset material response and facilitate	Enhanced accuracy and less computational cost allow faster and long-term predictions.
Continuous model updating and validation	One-off model development and training	Periodic data-driven update (simulation progress, survey data, sensor response for validation)	Updated knowledge of deterioration mechanisms, long-term prediction
Sensor data	Place on critical points/represent the structure	Rely on physics-based virtual sensors, validation study sections, incorporate users/mobile feeder	Unapplicable due to linear horizontal nature/ network level

The main objective of this research is to review the concept of DT in other industries, where it originated, and examine the literature on pavement asset DTs. This will clarify the current conceptual ambiguities of DT and the core requirements in this asset context. This research leverages the existing literature on DTs related to pavements. It discusses the context of flexible road pavements, mainly focusing on DT definition, the interconnected relationships among its required components, and the considerations needed for pavement assets. It will assess the developed pavement DTs to establish the foundation for the advanced adoption of cognitive DTs. Subsequently, this study will propose a holistic framework, outline the necessary considerations of potential decision systems for cognitive level DTs, propose conceptual collaborative DTs for the Construction 5.0 vision, and identify gaps and future needs for its comprehensive implementation.

1.4. Research Structure

The structure of this paper is divided into three sections. The first section presents the introduction, pavement management and gap, and methodology. Section 2 includes key features and considerations of the DT concept and discusses and analyses the developed pavement DT works. Section 3 addresses key points of cognitive pavement DT requirements, proposes a DT framework, highlights potential solutions and challenges, and summarises research gaps and future directions. The overall paper design is illustrated in Figure 3.

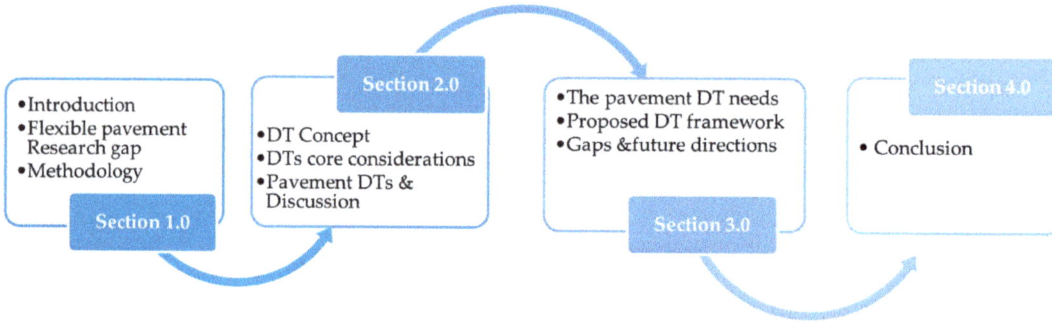

Figure 3. Summary of the paper's section structure.

1.5. Methodology

This narrative systematic review presents core research questions and specific objectives. The methodology leveraged multiple databases such as Scopus and Google Scholar to extract articles related to the subject areas. Additionally, this work focused on good-quality peer-reviewed articles, conferences, seminars, and books. The search was performed in the databases between January and November 2024. The review encompasses various papers in the last ten years that introduced and explored the core concept and implementation of DTs across different industries, where the DT concept initially developed. This broader perspective helped establish a foundational understanding of DTs before narrowing the focus to pavement DTs works.

For the pavement-related DT works search, the fields "title", "abstract", and "keywords" were used as "Digital Twin" AND "road pavement" OR "flexible pavement" OR "pavement maintenance". To avoid literature that focused on transportation operations such as mobility and automated vehicles, generic words alone, such as "road" and "infrastructure", were not used. The search was limited to works done between 2018 and November 2024, as DT applications in the civil and construction industry are relatively recent, and no DT or works related to road pavement were found in prior time.

The resulting papers passed a screening process considering the function of their titles. This was followed by an assessment of the abstract to filter works beyond the focus of this paper. The introduction section was reviewed for those not reflecting a clear scope in their work titles or abstracts. In this process, the focus of the filtering stage considered proposed DT works for road pavement, particularly flexible pavement assets only. In addition, this included works developed for various asset managing aspects such as data collection and sensing, evaluation, modelling, or decision-making, while proposed as DT-based. This allowed for a focused analysis of relevant information that addressed the research needs and aligned with the research objectives. The initial search resulted in 28 papers on DT for pavement management articles. The 28 papers were filtered out based on review criteria into 22 final papers, shown in Figure 4. The selection methodology is presented in Figure 5.

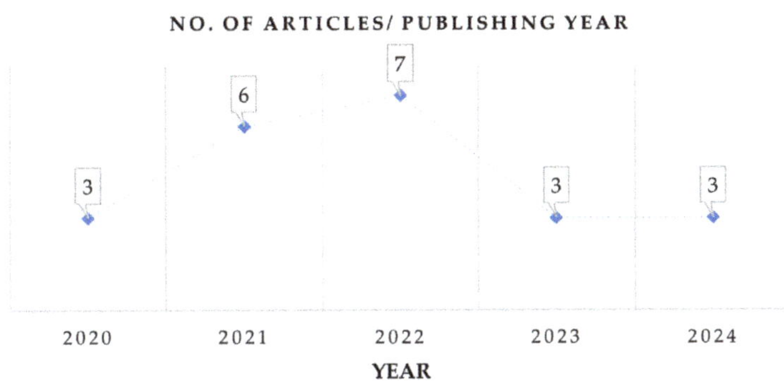

Figure 4. Selected pavement DT articles over publishing years.

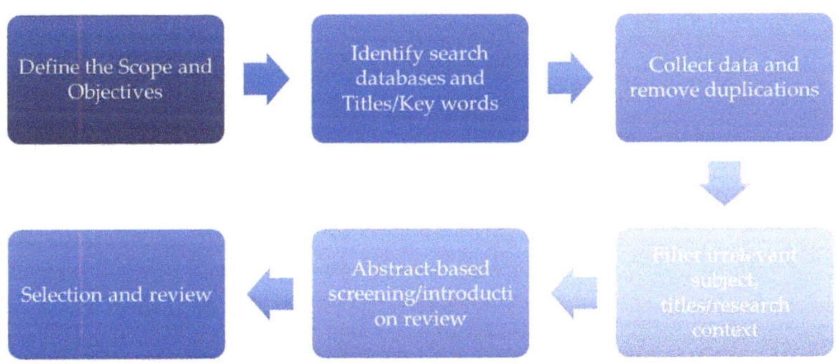

Figure 5. Paper selection and filtration procedures.

2. Digital Twins and Pavement Management Digital Twins

2.1. Digital Twins Concept

The DT concept was established in 2003 when the base for NASA product lifecycle management (PLM) was introduced in terms of virtual and physical spaces [86,99]. Afterwards, NASA, in 2010, published a technology roadmap [100], which provided a technical definition and revealed its key features [101]. Industries such as aerospace and manufacturing have adopted the concept and contributed to the design of the framework of DT in their sector. Therefore, the applications originated and were designed to meet their sector's standards and needs.

Different definitions were introduced of DTs. DTs are complex simulations built upon historical and real-time data designed to replicate the condition of a physical object [102]. They are also defined as a representation of a physical product that utilises data from the physical element or system to mimic real-world behaviour in the corresponding virtual counterpart [103]. The virtual space is reinforced by modelling and simulation tools, sensor data, and databases to enable more integrated and remote management. Another common definition indicates that DTs incorporate various physics, probabilistic analyses, and scales [2]. In most definitions, the simulations revealed the ability to analyse what-if scenarios of asset response under specific conditions. This is enabled by the connection and information exchange between components [101]. The benefit of a twin is achieved using actual operational data and integration of multiple data sources, as well as data exchange between the digital and physical spaces. The conceptual DT concept is illustrated in Figure 6.

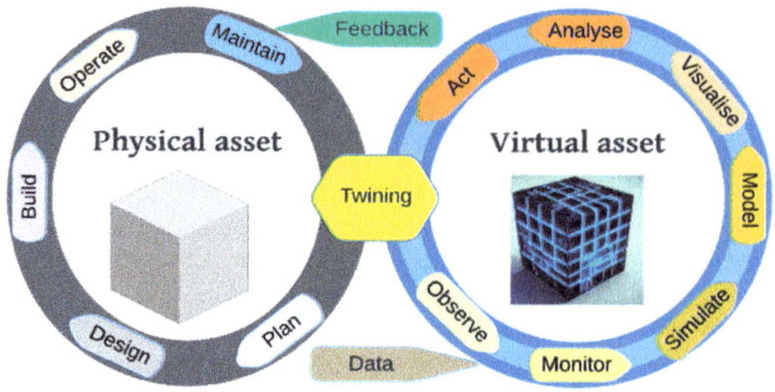

Figure 6. DT Conceptual model.

2.2. Civil Assets DT Core Considerations

DT core characteristics include the physical product lifecycle data of the asset, synchronisation of data [104], types of data connections between physical and virtual spaces (unidirectional or bidirectional flow) [105], visualisation, use of sensor data, and advanced analytical or simulation techniques [106]. In the context of construction industry assets, Oditallah et al. [74] defined the DT as "a virtual copy of a physical asset, process, or system of assets that comprise the twin purpose-related lifecycle data, reasonable data exchange and synchronisation, analytical or simulation models, and intelligent techniques to model real-world conditions and responses of the physical entity". This definition highlights the adoption of DT characteristics that fit the purpose of twinning. It emphasises purpose-related lifecycle data, applicable data exchange and synchronisation, and modelling real-world data or conditions. The key characteristics of DTs are summarised in Figure 7.

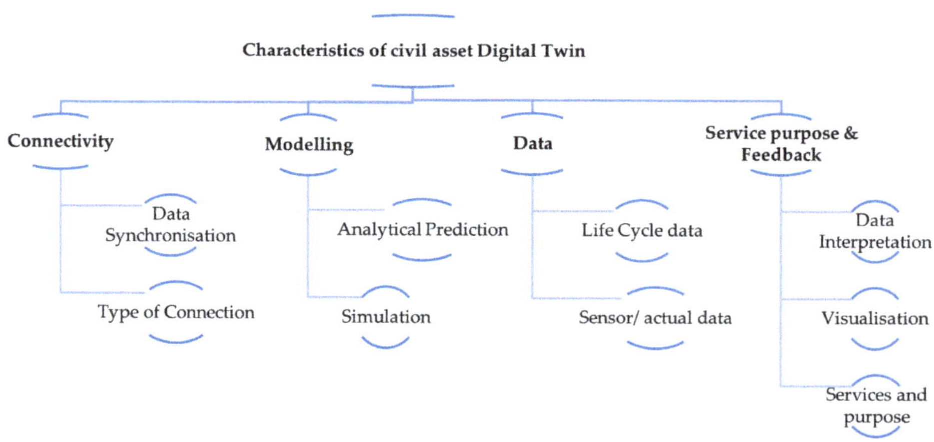

Figure 7. Characteristics of DT.

2.2.1. Connectivity

DT requires a data link to connect its digital and physical counterparts, facilitating data transfer between entities. The nature and frequency of connectivity depend on the specific application or service. A DT utilises the best available data, sensor readings, and other pertinent information to replicate its physical counterpart accurately. Connectivity encompasses the data transfer rate (e.g., real-time, near real-time, or periodic) and the connection type between the linked environments, whether unidirectional or bidirectional.

Connectivity requirements need to align with data updates to specific objectives of the developed twin [26]. For instance, in autonomous vehicle operations, DTs require real-time, bidirectional connectivity to provide immediate feedback, influencing vehicle performance in response to environmental and operational changes. In contrast, for pavement health monitoring, real-time data and bidirectional connectivity are not as important. In this context, DTs primarily evaluate asset conditions and provide maintenance recommendations based on periodic assessments, rather than real-time interventions. Therefore, the emphasis is on long term asset management and its responses instead of immediate operational changes, with actual data from the physical asset being more significant than real-time updates.

Bidirectional connectivity allows virtual actions, such as activating or deactivating physical components, to impact the physical asset. This is typically in mechanical or electrical systems such as production lines, where actuators can be controlled. For passive or non-mechanical assets, as in structural elements, the existence of bidirectional connections is often unnecessary. However, the feedback from the virtual space to the physical system may involve rescheduling maintenance or triggering manual interventions. Feedback and decision-making can affect the physical environment by either automated decisions or manual responses [107].

2.2.2. Modelling

Modelling and simulation are fundamental elements of DT systems. It can enable detailed visualisations, what-if scenario analysis, and solution validation. As a unique feature of DTs, accurate and continuously growing prediction models form the foundation for enhanced decision-making. Analytical modelling encompasses statistical models, including ML models, physics-based numerical models, and other intelligent techniques. However, the value lies in integrating multiple approaches and diverse data sources, particularly operational data from the physical asset and lifecycle data. Hybrid modelling, combining both approaches, is often used to leverage the advantages of each. Physics-based modelling can capture materials or internal structures in complex conditions to provide high predictive accuracy.

2.2.3. Data

Data representing the whole lifecycle of an asset are essential for linking assets and their operational activities for enhanced decision-making [108]. To support the DT's objectives, data such as geometrical and material properties related to the DT process can be incorporated [109]. In addition, operational and lifecycle data can be used for various processes within the DT. Material data can be used to model and understand the behaviour of asset response and affect informed decisions such as treatment type selection. Data on asset conditions, historical interventions, and design can also validate predicted conditions in the DT, and, furthermore, establish relationships between status change rates and operational data.

A DT typically consists of several data subsystems, such as asset inspection data, to process and reflect the asset's condition over time. Sensor data are used directly for monitoring or modelled to derive additional parameters for modelling, simulation, and visualisation. For example, raw vibration data may be used to derive internal forces that are more relevant for DT prediction purposes. The use and management of such data must align with the designed DT's functions.

2.2.4. Interpretation, Services, and Feedback

DTs can interpret the results in the context of the service required. This could include alerts and notifications triggered by predefined thresholds of predicted state. Current condition assessments using collected raw data or processed outputs can also be considered in the provided service. Decision support systems (DSS) can analyse the outcomes of related DT components and trigger recommendations based on monitored or predicted parameters. Some DSS and advanced knowledge-based systems incorporate advanced ML models or reinforcement learning to optimise planning decisions [110,111]. However, integrating DSS creates an intelligent layer and elevates the cognitive level of the DT framework. This enhances reliability and efficiency for managed assets and enables collaboration between machines and operators, reinforcing the vision of Construction 5.0 and elevating the maturity toward a truly cognitive DT. General Electric highlighted that a knowledge base system component incorporates data analysis and domain expertise and optimises DTs with industry best practices.

The visualisation rate of state updates is directly reflected by the data transfer rate and modelling or analysis practices. On the other hand, the required type and frequency of visualisations depend on the application's goal and purpose, as well as the needed system interactivity. Visualisation plays a role in data and output of modelling and simulation, such as investigating trends, identifying potential issues, and evaluating results in various scenarios. This is useful in complex systems where real-world activities are modelled over shorter timeframes. Real-time visualisation is essential when immediate responses are required. For instance, in traffic management or autonomous vehicle systems, high-frequency data updates require real-time responses and instant decisions. In contrast, health monitoring for pavement shows gradual increment changes and does not require continuous real-time updates. The periodic assessments can be adequate, allowing slower modelling and visualisation rates for long-term assessment systems. However, accuracy in visualising asset degradation can be more critical in this case. Precise feedback obtained from DTs not only benefits its performance but also reinforces the asset lifecycle outcomes for future asset development.

Overall, as connectivity determines the type of connection and the data rate, the visualisation and modelling requirements, along with feedback mechanisms, influence the connectivity design. Considering the core function and dependency in the DT structure. The role, relationships and criticality between its components are essential. The required service and the DT's primary purpose form the DT pyramid's foundation. This determines the needed modelling and simulation types, along with the necessary connectivity and visualisation. This structure specifies what data will be generated, how they will be utilised for the intended purpose, and how DT outcomes will be interpreted and integrated. Thus, Figure 8, shown below, describes the DT's base, which guides the dependency of other functional components.

2.3. *Road Pavement DTs Discussion*

The research on pavement DTs focuses on sensors for automating surface data collection and analysing defects. A group of works integrates these data into static 3D models using BIM and other spatial visualisations, while others apply real-time sensor data for monitoring and decision-making. Additionally, some studies aim to enhance virtual asset presentations and a few focus on updating 3D models to reflect surface condition changes for improved asset representation.

2.3.1. Pavement Surface Prediction DTs

Yu et al. [112] proposed DT to predict the highway pavement performance and improve its current preventive maintenance. Sensors were used to collect pavement surface current conditions. The prediction model learning process used ANN, random forest (RF), ridge regression (RR), and support vector regression (SVR). This involved utilising asset-related data, such as maintenance data and traffic flow. Data were visualised spatially using a BIM 3D model based on predicted data for support planning maintenance.

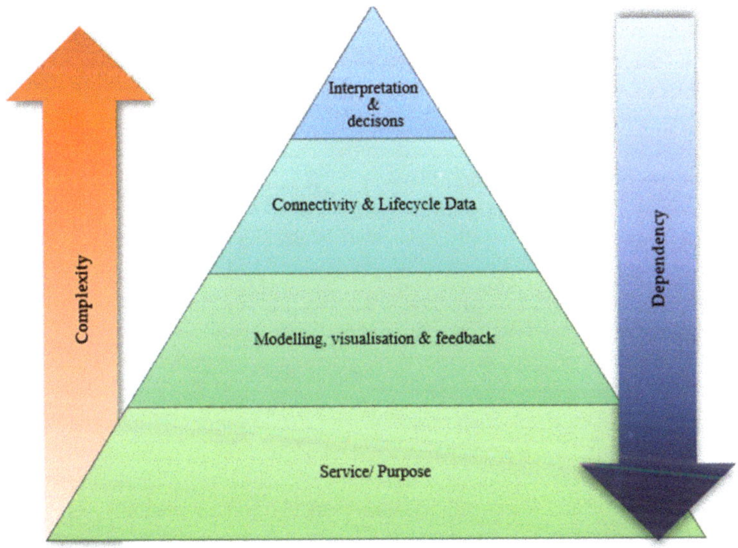

Figure 8. Digital Twin component's dependency pyramid.

Similarly, Consilvio et al. [113] provided architecture for the pavement DT, in which computer vision-based algorithms were utilised to evaluate pavement condition collection. Both Consilvio and Yu [112] involved other existing asset data, such as maintenance records, performance indicators, and BIM model visualising for decision support. Consilvio used an AI-based clustering method to filter road sections based on the severity of their condition. However, these works overlooked the accounting for the structural aspects of pavement or real operational data contribution, and the focus was confined to surface conditions only. Although ML predictive capabilities to forecast health metrics were used, this use has been long-standing; for instance, Marcelino et al. [60] employed a Boosting-based learning algorithm (TrAdaBoost), while other work compared five ML techniques [114]. Consequently, DT works could not address the added value of DT over the traditional ML models used. Nevertheless, reflecting predicted performance indices on a static 3D model does not constitute an entire DT. DTs should offer the expected enhanced precision and more cognitive service than traditional ones.

As a vision for futuristic road performance modelling, a proposed conceptual framework for DTs is presented by Kaliske et al. [115]. Their work proposed interactions of roads, tyres and vehicles as innovative technologies for vehicle–tyre interaction. The work introduced a potential contribution for tyre–road sensor data in monitoring road damage or changes in friction conditions. Although the work highlights the physics-based simulation data, their conceptual work proposal focused on the vehicle–tyre simulation perspectives only and road asset was not discussed. Ficara et al. [116] deployed a system that relies on edge cloud services to detect pavement anomalies. The work used statistical analysis based on the surface conditions and collected data from the crew and road users' vehicles. Road user video data were processed based on AI screening. The work suggests improvement in planning the manual collection of data based on the road user's preliminary data.

2.3.2. Sensing and Monitoring DTs

Barisic et al. [117] proposed a DT focusing solely on the thermal state of asphalt pavement. Specifically, temperature data were collected over the years using several instrumented highway sections and modelled for continuous condition monitoring. Another study used sensor data to propose a strain monitoring framework based on strip sensors under vehicle loads, which is used to assess the developed road pavement distress [118]. Data of loading variation were tested to assist, validate, and predict the measured strain scenarios using ML models. These works considered limited sensor data, neglecting other factors that involve pavement performance. Steyn et al. [119] used sensors for environmental monitoring data, asset imagery data, and camera data for vehicle counting. Point cloud data were derived from images of the demonstrated local road DT, in addition to road surface texture and temperature. The authors reported that acquiring the environmental parameters for monitoring supports management in the maintenance decisions of the roads. However, the role of collected real-time data and the use of environmental data were not discussed in their work. In addition, the work focused on collecting sensor data and no prediction techniques were used. Similarly, Meža et al. [120] proposed a DT of pavement for modified material structure and their work used point cloud scan data for road pavement model generation and visualisation. Data from temperature and moisture sensors, pressure pads, strain and deformation sensors were all mapped in a digital BIM model using the Common Data Environment (CDE) platform. The work reported that integrating the asset model with sensor data provided insight into the overall structural health of the monitored road. However, the data were mapped into the digital model for monitoring, and no modelling to predict structural measures was presented.

Digital Twin Box is a group of tools that was deployed in work to introduce DT for monitoring [121]. It comprises a GPS device, 360 camera, and IoT sensors for collecting data such as humidity and temperature. The work collected video data for the identification and tracking of objects in the road environment. The work considered DT development by linking real-time data to accessible cloud services for monitoring services. As in the previous works [117–120], these efforts focused on integrating sensor data into digital models as a support decision tool. Although sensor data integration is essential in DT, this part alone cannot comprehensively reflect the entire physical asset state. These works have not shown how these data contributed to the developed system. Instead, they solely presented IoT data on static digital models rather than establishing predictive analyses of asset deterioration based on actual leveraged data.

2.3.3. Surface Defect Detection, Visualisation, and As-Built Digitisation

For data collection automation, ML was frequently implemented for defect detection. Wang et al. [122] used imagery data for automated detection of distress and to assess the current condition of pavements. In their work, the ML training process involved multiple integrated data to build an enhanced detection model, which was described as a DT. Similarly, Sierra et al. [123] presented a road pavement twin based on the reality capture model. UAV images were used to detect current irregular surfaces and damages to the pavement and to detect surface defects in the captured model. In another work, laser scans were used to collect point cloud data and visualise the current pavement surface [124]. The collected 3D point cloud data were processed into a model and then used for surface-level flatness analysis and detection of defects. Collecting the current condition for pavement management is a promising aspect for feeding DT models, as in [122–124]. However, these data reflect the point-of-time conditions and are limited to surface status only. Therefore, it has contributed to automated defect classification. Its novelty lies in predicting current situations of surface conditions rather than predicting future asset status.

To manage the collected data for decision-making, a work implemented a DT for pavement based on BIM and GIS [125]. In their work, pavement surface data were collected using sensors, and the data were converted into a performance index. Road sections were modelled in BIM and georeferenced to GIS data. Similarly, Bosurgi et al. [126] used a BIM platform to manage the condition and quality information of pavement surveys. Another work proposed a framework for PMS in a systematic workflow containing data acquisition to support decision-making [72]. The framework shown in Figure 9 outlines automated road distress detection and 3D distress quantification by embedding data information into GIS. This work also aimed to use defect data on spatial platforms for collaboration as in previously mentioned works. Road agencies usually adopt spatial presentation and GIS for enhanced data management. These mentioned works proposed high-level frameworks that lack enhanced or advanced modelling and simulation for what-if scenarios.

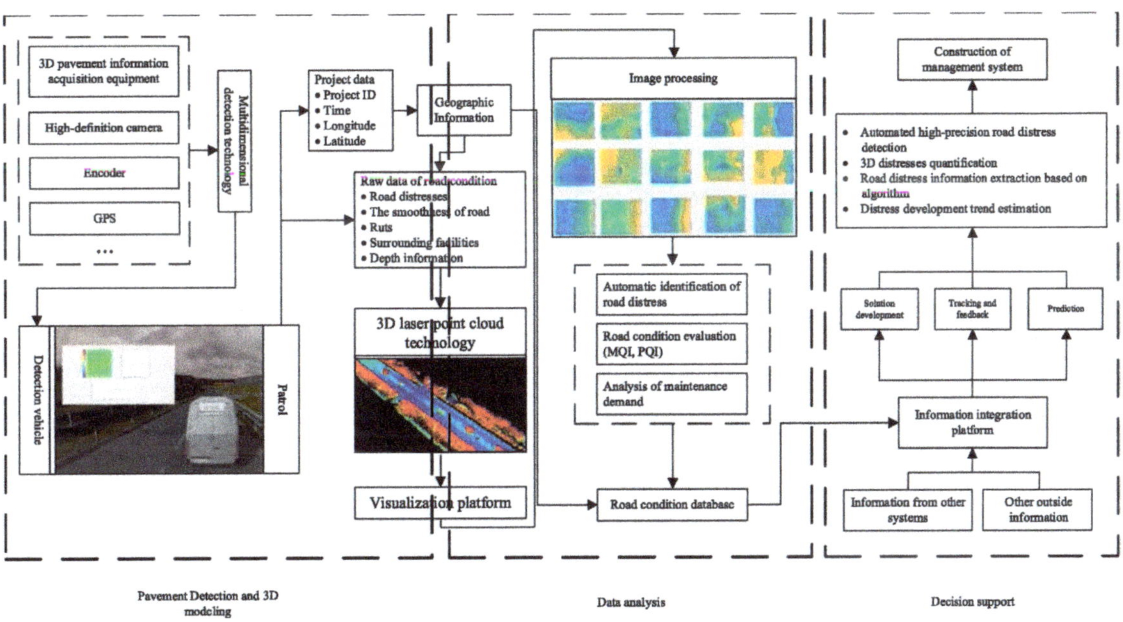

Figure 9. Systematic workflow for a Pavement Management System (PMS) [72].

Other works focused on the modelling of existing pavement assets. These works aimed to detect defects using ML, create 3D models, and reflect the conditions detected in the developed model. For instance, Cao et al. [127] developed an interactive system to present pavement cracks in 3D visualisation, where the crack boundary is detected and extracted based on a 3D crack edge feature algorithm. Data were acquired based on camera and laser scan sensors. The twin here is presented to replicate the existing shape and type of defects on the 3D model. This allowed the visualisation of pavement surface defects to reflect the conditions of road assets and help in decision-making and maintenance planning. Wang et al. [122] constructed a 3D model based on field-captured reality data and used deep-object detection algorithms for pavement distress detection. The authors integrated data by leveraging a lightweight engine used to graphically represent five distresses. Their proposed framework is shown in Figure 10. Cao's interactive system [127], along with this work, are promising enablers for visualisation aspects in DT, presenting updated visualisation capability of updated site data; however, it is not an entire DT system.

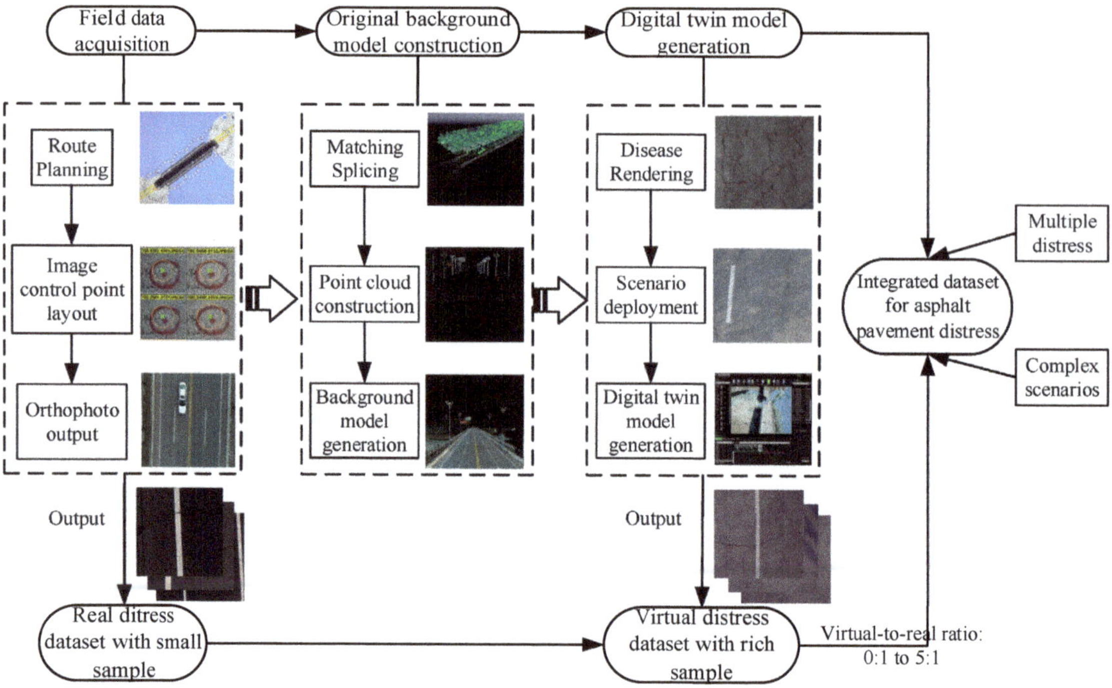

Figure 10. Pavement distress detection and rendering framework [122].

Furthermore, D'Amico et al. [128] used the existing road pavement condition database to develop a BIM model for that asset. The model was replicated based on two surveys carried out in their study. The collected data included point clouds, GPR scans and GPS records. The author used parametric elements in BIM that can adapt their model features to the information collected for road conditions. The processed data were integrated and adapted as 3D polylines extracted from the surveys, as shown in Figure 11. The proposed method used data to visualise the current road surface conditions. Bertolini et al. [129] introduced an approach that involves scanning the pavement surface to identify irregularities that may correspond to specific types of pavement distress. GPR surveys were utilised to generate multiple datasets, aiding in the detection of subsurface pavement failures. They followed a method based on a grid system for surface-level defects and 3D voxels for the subsurface. Although no prediction of the future status of pavement assets was achieved in these works [128,129], it is a promising method for visualising functionalities in DT. As DT is supposed to replicate the defect on the 3D model to reflect its status and future status, these approaches potentially benefit DT.

Other works on pavement DTs focused on creating the digital model of pavement assets. This also allowed the building of the pavement model surface, including its deformations and irregularities in terms of pavement layers. Forming the 3D model of an existing road can establish a DT base. This can indeed, along with previously mentioned works, be useful for visualisation in DTs. For instance, Fox-Ivey et al. [130] proposed a DT model using a 3D scanning and positioning system for inspection and damage detection. To develop a digital model of an existing asset, Jiang et al. [131] mapped collected data to build a digital surface model (DSM). Then, fitting processes for horizontal and vertical alignment and cross-section generation were implemented (Figure 12). Similarly, Pan et al. [132] utilised a 3D point cloud segmentation process to acquire semantic information using 3D deep learning models. The process involved clustering the point cloud and alignments fitting based on polynomial approximation. The points extracted and segmentation of road surfaces were converted to detailed components using two effective deep learning models, KPConv and Superpoint Transformer. The fitted curves were then used to separate road surface points into lanes, shoulders, and central medians. The authors reported that the work offers the base model of physical assets, which potentially can be expanded to complete DTs. An illustration of the process is shown in Figure 13.

Figure 11. Proposed visualisation of the current road surface conditions [128].

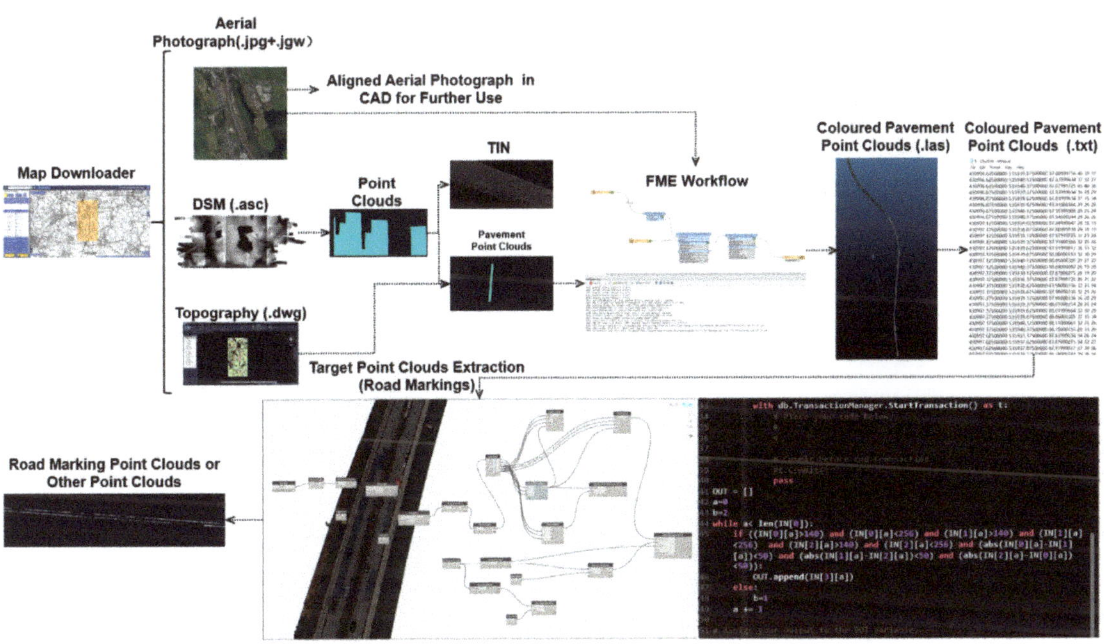

Figure 12. Generating a digital model for existing infrastructure highway workflow [131].

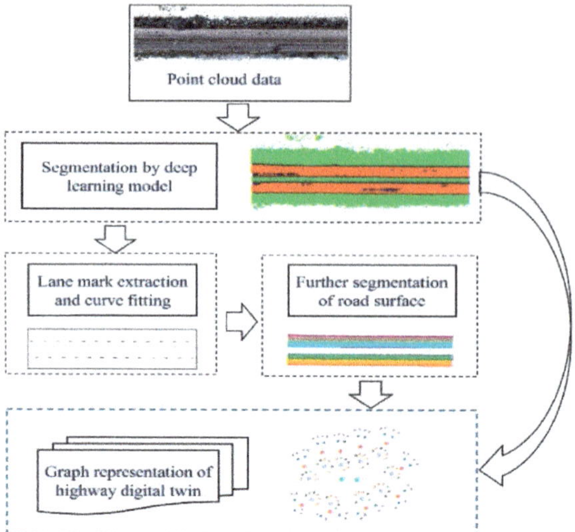

Figure 13. Hierarchical relationships among various components [102].

2.3.4. Developed Pavement DTs Summary

The discussed works have used various tools and approaches in their studies. Table 3 summarises and presents the data acquisition methods, approaches used, platforms or integration strategies utilised, and types of output feedback.

Table 3. Developed DT methodologies.

Work	Methodologies			
	Data Acquisition	**Model Type/ Development/Approaches**	**Platform/Integration Strategies/Tools**	**Interaction Type/ Output/Feedback**
[112]	Vehicle-line scan, Accelerometer	ML-Prediction: multi-ANN, XGBoost, etc.	BIM software (Autodesk Revit)/CDE, Visual programming	3D model-Roughness IRI index visualisation
[117]	Set-temperature sensors (profile), weather station sensors	Statistical prediction, numerical analysis/ validation	Analytical/numerical software	Real-time temperature prediction
[118]	Strain sensor	Physics-based model	Numerical analysis software (LS-DYNA)	Strain, load prediction
[119]	Smartphone, UAV camera, CVV camera, infrared, LiDAR, temperature	Depth map modelling, multi-view stereo (MVSNet)	Cloud IoT platform	3D model/real-time data streaming
[120]	Embedded (temperature, humidity, displacement, pressure, stress, strain)	3D BIM model	Centralised data management, visualisation and collaboration platform	3D model/real-time data streaming
[121]	Camera/Video, GPS, Gyroscope	Object detection model (SSD MobileNet V2)	Open-source cloud computing	Real-time data streaming, object detection
[127]	LiDAR, Camera	3D Object crack segmentation (CNN), 3D feature extraction	Python-ML platform (TensorFlow software)	As-is 3D crack visualisation
[122]	UAV-Camera	Object Detection (cracks, potholes) based on (YOLOv5)	Photogrammetry/mapping software, Game engine, real-time 3D platform	As-is distress visualisation
[123]	UAV-Camera	Object segmentation crack (U-Net, VGG-16)	Reality modelling (ContextCapture software)	Distress detection
[124]	3D LiDAR, GIS	Point cloud processing, parametric modelling	Bridge/road software (Civil 3D), plugin-integrated visualisation (Leica's toolset for Revit), 3D Surveying software (Cyclone 3DR)	As-built 3D model/virtual assessment (flatness or distortion)

Table 3. *Cont.*

Work	Methodologies			
	Data Acquisition	**Model Type/ Development/Approaches**	**Platform/Integration Strategies/Tools**	**Interaction Type/ Output/Feedback**
[125]	Camera, GPS camera	Spatial data analytics	BIM modelling, GIS software	3D model GIS integration
[126]	NA	3D object -inspection, lifecycle data integration	BIM Infra software	3D Model-based data
[116]	Smartphone sensors/camera/GPS, edge computing	NA	Cloud-based storage/services/ query platform	Condition indicators visualisation
[72]	Camera, LiDAR, GIS	Image processing, data analysis	GIS, point cloud computing software	Distress detection, quality index visualisation
[113]	UAV camera, GPS, Light Detection and Ranging, survey data	Parametric modelling, computer vision-based algorithms, AI clustering decision optimisation	Cloud computing platform. Game engine, real-time 3D platform	Mapped/visualised data support decision system
[128]	GPR, Light Detection and Ranging, Laser Profiler	3D data processing, Parametric modelling	Design/visualisation software (civil, infra)	As-is 3D pavement model/distress visualisation
[129]	Equip vehicle-laser Measuring System, GPR surveys	Point cloud processing/reasoning/Cells and voxels classification, visualisation	Specialised point cloud computing software	As-is Surface reproducing/subsurface distress visualisation
[130]	GPS/GNSS, 3D scan-integrated inertial measurement units (IMUs)-	Geo-referencing, 3D points cloud processing	LDTM software, Cad viewer	As-is 3D surface modelling
[131]	Existing Map database, aerial images, derived point cloud	3D data processing	3D modelling, point cloud processing software, visual programming	3D model element reconstruction/digitising
[132]	3D Point cloud data	Deep learning-segmentation, elements fitting	Point cloud processing software, Algorithm-Python	3D Design element reconstruction/digitising

The analysed literature reveals that current implementations often fail to fully adopt the DT concept. This is seen in the limited inclusion of in-service behaviour of modelled assets and the absence of enhanced predictive techniques over already existing ones. This hinders the realisation of DT's full benefits. The proposed frameworks mainly focus on the automated detection and extraction of defects, considering the functional aspects based on surface conditions. Although the proposed work is not the entire DT, it has valuable contributions toward its full development.

The literature has different viewpoints proposing the conceptual structure of DTs for pavement management. Nevertheless, the predictive and analytical capabilities of the DT concept remain key components. This includes advanced decision support system (DSS) models, which have not been fully integrated into the DT framework to propose cognitive and reliable decisions and continuously learn and share knowledge for asset management. The developed DT works in the literature have been classified into different categories based on their focus and contribution area in Table 4. In this table, 'X' is used to indicate which focus category is associated with each work.

However, cognitive DT implementation is intended not only to facilitate services related to current and future asset status assessment; it further optimises the accuracy of the overall reflected performance. Figure 14 illustrates the expected services of Pavement DT.

Table 4. Developed DT's focus and contribution areas.

Work	Category			
	Current Condition Detection	3D Model Creation/ Visualisation	Support Monitoring Data	Analytical Prediction/Decision Support
[116]				X
[112]	X			
[72]				X
[128]	X	X		
[115]			X	
[117]				
[124]		X	X	
[119]			X	
[113]	X			X
[122]	X			
[127]	X	X		
[131]		X		
[122]	X	X		
[130]		X		
[123]	X			
[120]		X	X	
[126]				X
[118]			X	
[125]				X
[129]	X	X		

X denotes that the feature is applicable to the corresponding work.

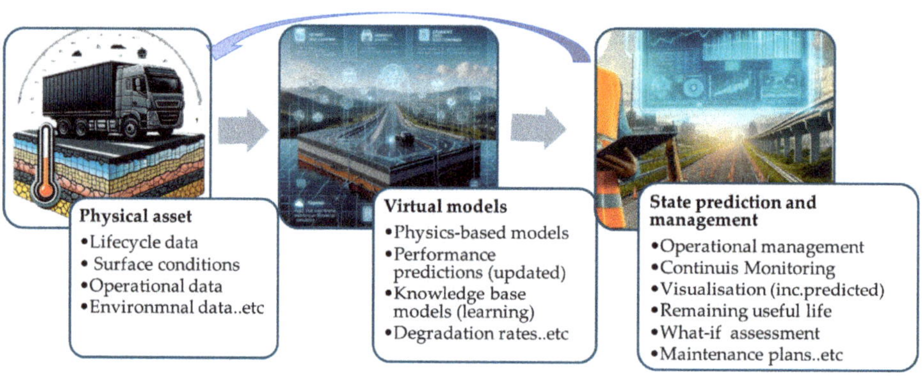

Figure 14. Conceptual services of Pavement DT.

3. Advancing Toward Cognitive Pavement DT

The framework design aims to bridge the asset needs from an operational point of view to provide a starting point for effective pavement DT development. The following discussion includes considering actual operational conditions, integrating multi-modelling methods, and defining visualising aspects in the context of DT structure and flexible pavement needs. This would advance the developed framework toward full DT implementation for construction 5.0 environments.

3.1. Operational Data, Asset Databases and Performance Modelling

Actual operational data and incorporated material modelling in DT are key for enhancing the prediction of pavement actual performance. Service condition data, such as temperature and operational data, i.e., load frequencies, speed, and overloading, are essential in performance degradation. However, sensors and weather station data can be leveraged and calibrated based on formulas and models, such as models developed in [133–135]. This will allow DT to account for in-service assets' actual temperature and moisture levels. Moreover, other existing pavement lifecycle data can be acquired using asset databases, such as LTPP, which can facilitate actual data for analysis and prediction. The existing databases, i.e., PMS data, involve the asset structure data, i.e., layer thicknesses and material test data under various environmental conditions. This would yield insightful contributions to the enhanced prediction of DTs. Regarding operational loading data, Weigh-In-Motion (WIM) data, which is sensor data, classifies stream traffic loads and presents the weight of vehicles, loading speed, actual repetition of each type, and other detailed traffic. These data can be modelled in time series form and leveraged to present specific site operations characteristics. When sensors are absent, they can also undergo metamodeling approaches to estimate other network pavement loads. This connection between the actual site environment and the operational scenario in the virtual space can enable predictive DT. Sensor systems can rely on WIM or similar camera-based systems and embedded sensors for temperature or leveraging weather station data within the asset zone. These data can be employed in DT modelling, prediction, and decision selection processes.

Other data, such as existing maintenance records and detected surface condition data, can be integrated into DT using data inquiry and management methods, i.e., SQL databases. Surface condition automated methods, i.e., cameras and laser sensors, can be used for frequent inspections, where acquired data can be classified based on AI detection and stored for further prediction operations. Recent works proposed systems adaptable to vehicles to be used as collectors. These potentials are supported by edge computing technologies that use cloud models and access various data and computing resources. It is possible that future integration of autonomous vehicles can help to capture road surface data over time and analyse them based on edge computing. This includes sensing technologies developed that can be leveraged for structural testing through advanced non-distractive methods. These data, i.e., deflection responses, can be acquired from road vehicle users to indicate the structural health of pavement and used in calibration and validation processes in the DT modelling component.

One of DT's challenges is determining data updating and connectivity aspects. The structural health of infrastructure assets changes slowly. Data collection is periodically based on the rate of deterioration and frequency of decisions needed. Therefore, periodical use of the continuously collected data can be used to update the periodical modelling. This can be done by segregating and redistributing the newly available operational data to fit the modelling frequency.

In terms of predictive modelling, data-driven models within DT, such as ML models, can predict pavement temperature at various depths using other environmental factors and weather sensors. Moreover, at the pavement performance modelling stage, data such as traffic, axle load, etc., which are collected from WIM and integrated asset databases, require data modelling in DT functionalities. For instance, predicting traffic flow at a given time of the day or under extraordinary event changes for what-if scenario analyses. For performance prediction, the physics-based models can contribute to forming enhanced hybrid models such as surrogate models. Physics-based models can incorporate material characteristics for response prediction. This considers operational and environmental conditions leveraged from the physical pavement to forecast degradation based on the physics law of the pavement, i.e., viscoelastic response. Continuous prediction results can be incorporated into ML models to facilitate faster and generalised predictions of pavement for decisions. This overall process involves the lifecycle data for accurate prediction and predictive DT. Furthermore, NDT devices are used to gather structural performance data and can be adapted to validate and optimise the performance models in DT. For instance, various works, as in [35,136–141], used ML to establish a model based on simulation data for structural analysis using NDT data and optimise its performance. These surrogate models open a window for further innovative modelling in DTs.

3.2. Intelligent Decision-Making Support Systems

Enhanced performance prediction in the DT approach allows a precise decision-making process. Furthermore, integrated DSS can automate plan prediction based on advanced knowledge models and optimise decisions based on incorporated industry best practices. This decision model can leverage performance modelling outputs, other system and asset inputs, asset operation and maintenance policies, and best practice considerations to inform optimised decisions. Its cognitive feature can adopt budget limitations, intervention factors, treatment costs, etc., [142]. This is essential to close the DT feedback loop and achieve a cognitive level. ML models in DSS can guide, support and optimise decisions and elevate the practice to a prescriptive maintenance approach [143], which predicts and prescribes maintenance actions. For instance, reinforcement learning is frequently used to optimise planning decisions [110,111,144]. This allows the cognitive DT to be optimised for continuous learning over time. In this context, various works developed decision models: Georgios et al. [145] used a decision tree to provide repair strategy, defect cause and treatment based on distress type, severity, and budget; Philip and AlJassmi [146] adopted Bayesian Belief Networks (BBN) to produce optimal sustainable decisions; and Abu Dabous et al. [147] proposed multi-criteria decision analysis (MCDA) based on multi-quantitative evaluation, while another work adopted a risk assessment BBN model trained to expand its knowledge domain over time [148].

Considering Construction 5.0 principles and focus, DSS integration is a significant step in aligning the future cognitive DTs with its goals. In addition to advancing DT maturity, it combines automation with skilled human oversight. This can enhance the intelligence, connectivity, and collaborative aspects for the next evolution in construction. DSS in a DT can interpret and present complex data in a more accessible way to support teamwork in digital systems for optimised outcomes. Furthermore, the continuous learning ability of reinforcement learning helps knowledge-based systems adapt. These systems can plan, facilitate and automate routine decisions, which allows expert humans to focus more on critical and strategic decisions. In addition, the recent generative AI has a promising role in enabling workers to gain real-time insights. This facilitates complex DT functionalities and translates insights into plain language, bringing human creativity and machine cognition closer.

3.3. Visualisation & Data Integration

The DT visualisation interface can help to clearly understand the actual state as visual information about defects. This plays a fundamental role in the route causing the deterioration mechanisms. Visualisation in the pavement asset context can be divided into macro and micro levels. The macro level refers to the visualised conditions in terms of performance indices. This data visualisation helps spatial-based investigation for planning interventions. Micro visualisation can involve pavement structure geometry changes, surface, and appearance degradation, reflecting the nature and severity of defects. It can be used to determine deterioration mechanisms and their interaction to support decisions.

Although some developed DT works presented in the previous discussions are promising methods for micro-scale visualisation, further visualisation methods for DTs are required.

Operable platforms and tools that handle Application Programming Interfaces (API) can facilitate data exchange between different systems, leveraging various data sources and model results among multiple applications and platforms, and GIS adds a spatial context to asset visualisation. BIM models also provide 3D adaptable models rich with information. Simulation model visualisation is also essential for generating and updating visualised assets. Designed DT could involve multiple types of visualisations for short and long-term decisions.

3.4. DT Platforms

Industrial platforms are enablers for DTs and serve as ecosystems for various industries' applications. Examples include the Siemens Xcelerator Platform, Microsoft Azure DTs, and IBM DT Exchange. These cloud-based platforms provide tools to create comprehensive, integrated digital environments. However, other DT-enabled platforms were explicitly developed for the construction industry, such as platforms or CDEs. For instance, Bentley's iTwin platform supports reality capture to create a digital context and integrates iTwin IoT for analysing sensor data. Autodesk Tandem focuses on the construction and operational phases of assets. Similarly, Trimble Connect CDE provides DT capabilities by integrating with field data and offering real-time project analytics.

These platforms are centralised environments to store and manage all relevant asset data. This includes the development of dashboard visualisation, date information, cloud-based simulations, predictive models, and other data sources. These technologies and processes enable DT to control the integration of multi-modelling and analytics approaches based on various cloud-based services and data storage throughout APIs. Thus, a well-designed visualisation helps understand current and projected asset conditions. Furthermore, it allows integrated workflows with GIS and DSS systems to ensure comprehensive decision-making.

3.5. Integrated & Collaborative DTs

An integrated collaborative DT is a system of interconnected DTs communicating and sharing data to support more comprehensive decisions. In this concept, each DT holds its functions and purposes, such as pavement management DT, traffic management DT, road users service DT, and autonomous vehicle DT. However, they also share essential data across the network as needed. When requested through data protocols, data from one twin can directly benefit or inform the processes of another. For instance, a DT of road traffic can share data about traffic flow and density, helping pavement DT with periodical asset assessment. This includes transferring data from one twin, i.e., traffic DT, for what-if analysis in pavement DT. Similarly, pavement DT can feed road traffic DT with any predicted hazards that could affect the traffic in the future, helping adjust routes based on current or future conditions. Furthermore, an autonomous vehicle twin system can provide pavement DT with surface-related data, tyre contact data, and defect location. Combining such structural data-sharing and protocols in a collaborative environment can be systematically achieved. However, sharing data protocols and security issues can be an issue in its adoption. Figure 15 illustrates conceptual data-sharing actions between several DT systems, and Figure 16 illustrates pavement DT within an integrated DT collaborative environment.

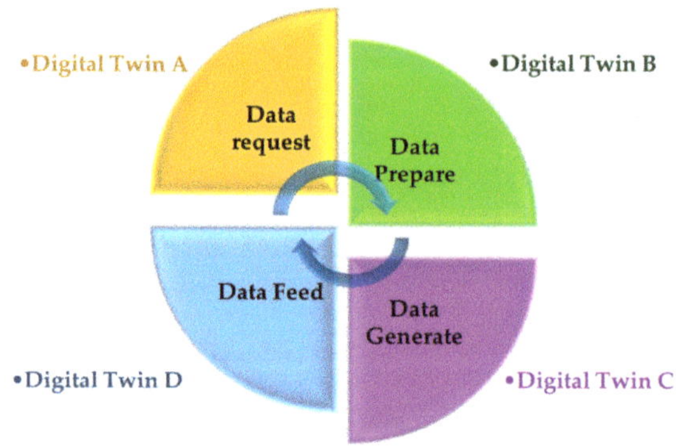

Figure 15. DT data sharing environment.

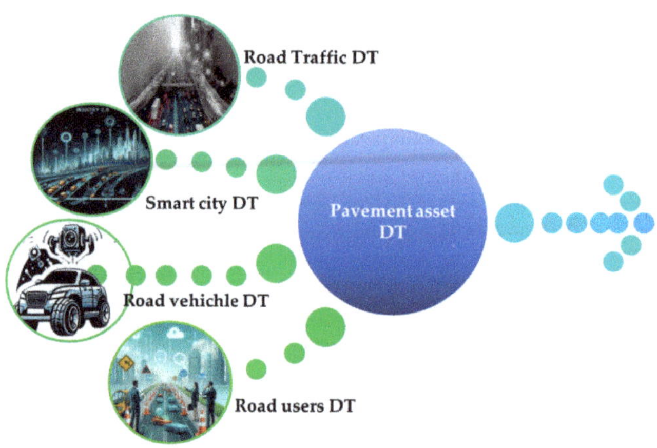

Figure 16. Illustrated pavement DT within an integrated DT collaborative environment.

The proposed federated DT system, Twins Chain, a possible term, can also improve resource efficiency by reducing the redundancy of duplicating data collection or analysis efforts by an isolated DT system. This would save computing power, data storage, and cloud service interruptions. This collaborative approach also aligns with the Construction 5.0 vision for sustainable, intelligent systems for adaptive management of construction resources. It also provides a conceptual step toward standardisation for centric transport DTs, allowing shared access and collaboration across disciplines to realise the concept of smart city DTs.

3.6. Proposed Cognitive Pavement DT Framework

Prior analysis and discussion sections put forward a map considering all required data, possible models, and support components of a cognitive DT for pavement management. This outlines the most effective architecture framework for road pavement based on DT structure and pavement-developed systems. The conceptual framework in Figure 17 consists of a skeleton equipped with different tools and technologies.

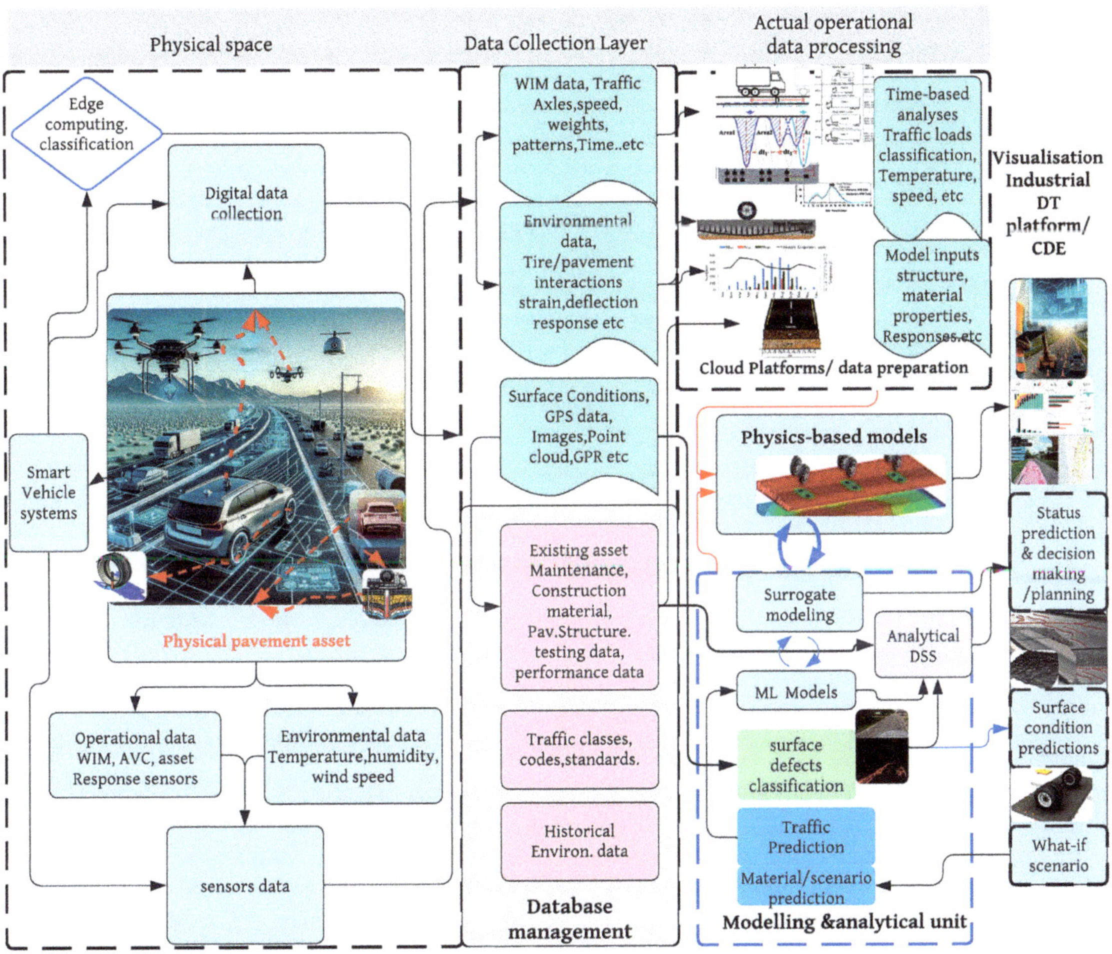

Figure 17. Proposed Cognitive Pavement DT Framework.

The system utilises data from multiple sources, including sensors on user vehicles, existing databases, and surveying technologies, to collect physical asset condition data. It employs cloud-edge services for data detection, analysis, and classification, which are used in different functional units within the proposed system. The framework also incorporates data preparation and management processes within dedicated data collection and management layers.

The proposed DT framework integrates multisource data, including processed data on operational, environmental, and surface conditions. A processing unit organises operational data into time-series sets (e.g., traffic load, asset temperature, and material properties) for physics-based and ML model predictions. These models use a multi-modelling approach with surrogate models, enabling advanced analysis, what-if scenarios, and visualisation. This framework can potentially bridge the existing gap in the literature concerning discussed works. It incorporates the DT concept's multi-modelling feature to improve pavement systems' predictive modelling. This enhancement facilitates superior forecasting and enables more informed decision-making. Thus, it highlights the additional value of the DT approach in comparison to traditional practices.

A DSS combines predicted and existing asset data to inform and optimise planning, incorporating classified surface conditions from processed field data. The framework relies on cloud-based dashboards, CDEs, or DT platforms for seamless integration and user interaction. This dynamic, cognitive DT evolves by updating surrogate models and DSS knowledge. This can be leveraged based on reinforced learning DSS, improving predictions as new data become available, and enhancing decisions as more properties and practices are provided throughout the asset's service life. In contrast to the discussed foundational level of previously developed DTs, comprehending processes within the presented pavement framework encompasses all required management components. This includes data collection, detection of current conditions, and an enhanced approach to future state predictions. This framework integrates actual asset operational data and proposes sensing pathways, in addition to intelligent decision-making systems with continuous learning features inherent in the DT concept, allowing prediction models and decision experts to grow for ongoing improvements in DT performance. Such advancements will effectively bridge the current adoption gap and align with future industry 5.0 requirements during implementation.

However, the proposed architecture has potential challenges. Data integration (e.g., multi-sensor data) and managing diverse data types (e.g., sensors, databases, surveys) impose technical difficulties. In addition, the suggested continuous real-time physics-based and ML models require significant processing power. This could demand excessive computational power and cap its implementation. Furthermore, model calibration for ensuring surrogate models and DSS adapt to changes can be a challenge. These, in addition to high digital skills, are required in its full implementation.

Implementing the proposed DT framework in real-world projects will validate model accuracy and potentially refine system integration. It will also identify further practical challenges and assess how well the DT framework adapts to real-world conditions.

3.7. Research Gaps and Future Recommendations

This work comprehensively reviewed, analysed and discussed possible solutions and technologies for a holistic framework of pavement DTs. To advance to practical implementation and operate the proposed frameworks, real-world case implementations to validate and refine system architecture and focus on facilitating the technical challenges are needed in the future. Some of the key research gaps and future research recommendations are as follows:

- To advance the current underdevelopment practice of DTs, implementation should align with the proposed DT concept. The structural aspect of the asset and actual operational conditions need to be incorporated. Furthermore, the predictions and decision processes of pavement DTs need improvements. Simulation is a core feature of the DT fundamentals. However, it is not mentioned in most developed works. Methods for integrating modelling and simulation for advanced analytical modelling and what-if scenario functionalities require further research. Physics-based simulations, such as the finite element method, and other model updating and optimisation techniques outlined in the proposed framework require further investigations. In addition, real case implementation as proof of concept is required for physics-based methods and ML hybrid integration. This considers real asset operational and lifecycle data to overcome current limitations and achieve improved prediction in growing DT. This work recommends further exploring and assessing these hybrid simulation methods incorporating real-life data.
- Visualisation tools are critical in predicting complex systems for user-friendly use, interpreting findings and making informed decisions through intelligent systems. It is a conjunction of all data types, from simulation, modelling, and decisions to asset state

visualisation as proposed in the proposed framework. A few works have developed virtual models that replicate assets and visualise current defects and surface conditions, which are proposed in the literature. However, there is a need for methods to reflect further predicted data and conditions for the future state of the asset, in addition to developing visualisation components for users within DT platforms to align with full DT capabilities.

- DT depends on leveraging remote actual data from the physical asset. Recent NDT methods, i.e., [80], used 3D-digital image correlation to the in situ testing based on vehicle tyre load deflections, or a Laser dynamic deflectometer (LDD) to capture the deflection as the vehicle moves [81]. A work presented tyre–pavement interaction to provide data potentially helping road damage analysis [115]. However, these approaches imply collecting actual data and asset responses based on on-road vehicles, as highlighted in the asset response sensor data in the proposed framework. This significant research direction on the future of non-destructive sensing enables the enhanced asset status prediction of future DT implementation.
- Cognitive DT implementation involves a complex and wide range of data parameters and information. This can also challenge the traditional decision-making processes due to their complexity and numerous dependencies. Therefore, optimising these decisions would require advanced self-learning DSS for managing, interpreting, and optimising detailed and complex datasets. Consequently, more sophisticated DSSs are required to facilitate the effective implementation of DTs and adapt for Construction 5.0.
- The reliance of DTs on existing asset data highlights the critical importance of structured databases, such as LTPP. However, this presents other challenges due to the lack of digitised data records in integrating data within DTs. The solution for data operability, security, and existing database use requires developing innovative methods to facilitate their implementation. Therefore, interoperability and data security solutions must be addressed, and cloud service solutions in integrated diverse data for modelling, storing, and visualising need to be explored. This will help move toward a sustainable digital built-in shared environment across multiple sectors.
- In the current literature, there is an absence of works investigating the cost and environmental considerations of DT implementation for asset management. This includes human–machine relations, trust, connection, and sustainability factors within infrastructure asset DTs for Construction 5.0 readiness. In addition, it is essential to understand the factors that influence the adaptation and adoption of DT systems. These required contributions will impact the successful deployment of DTs.

4. Conclusions

This research provides a crucial step in developing a comprehensive DT framework capable of meeting DT requirements, enhancing status prediction, and incorporating growing knowledge DSS to achieve cognitive systems aligned with the Construction 5.0 principles. This paper reviewed the concept of DT structure, outlined critical considerations, and assessed existing DT implementations for pavement assets. The review spanned various aspects, highlighted current gaps, and proposed asset-specific needs and required components with recent advancements that could facilitate DT implementation. The redeveloped cognitive DT framework will foster its implementation in the management of pavement infrastructure, including integration of decision systems and prediction improvement from advanced technologies and profound DT concept realisation to enhance the overall DT efficiency. Research gaps to fully implement the proposed framework were also presented.

The review revealed that the existing literature on pavement DTs lacks a full DT concept presentation, which includes enhanced predictive capabilities, efficient use of actual operational data, intelligent DSSs, and the incorporation of what-if scenario modelling. Most studies emphasise surface condition data with a significant focus on digital models or assessing present conditions rather than extending to fully predictive DTs.

The proposed framework in this work addresses the gap between current practice and the true concept of DTs. This was achieved by addressing the added value of DTs over the previously developed practices. It adopted the full DT concept's features, incorporating a multi-modelling approach and actual physical asset-handled data to improve predictive modelling, and comprised comprehensive components, including specific asset needs and growing predictive capabilities. This was through advanced DSS incorporation to facilitate data interpretation and growing models to expand knowledge and practices of actions, adding the cognition layer to the current developed DT maturity. In addition, this work proposed collaborative DTs for optimising and standardising smart city-scale DT implementation potentials. This enhances decision-making and optimises planning strategies while aligning with Industry 5.0. This research contributes to the design of DTs for pavement infrastructure, enabling predictive and cognitive systems. While existing systems do not fully address the concept of true DTs, the proposed cognitive DT implementation needs to overcome some identified gaps. However, implementing DTs remains challenging due to their complexity and lack of standardised processes. Moreover, data interoperability related to multi-sensor and source data fusion and security poses significant challenges, particularly for large-scale DT applications.

Author Contributions: Conceptualisation and methodology, M.O. and M.A.; investigation, M.O.; resources, M.O.; data curation, M.O.; writing—original draft preparation, M.O.; writing—review and editing, M.O. and M.A.; visualisation, M.O.; supervision, M.A., P.E. and S.R. project administration, M.A. All authors have read and agreed to the published version of the manuscript.

Funding: This research received no external funding.

Data Availability Statement: The original contributions presented in this study are included in the article. Further inquiries can be directed to the corresponding authors.

Conflicts of Interest: The authors declare no conflicts of interest.

References

1. Dunhoft, M.-C. *"This Is a Paradigm Shift for the Entire Industry": Employee Resistance to Technological Innovation in the German AEC Industry*; University of Twente: Enschede, The Netherlands, 2022.
2. Singh, M.; Fuenmayor, E.; Hinchy, E.; Qiao, Y.; Murray, N.; Devine, D. Digital Twin: Origin to Future. *Appl. Syst. Innov.* **2021**, *4*, 36. [CrossRef]
3. Index, D. *Digital America: A Tale of the Haves and Have-Mores*; McKinsey: Chicago, IL, USA, 2015.
4. Moshood, T.D.; Rotimi, J.O.B.; Shahzad, W.; Bamgbade, J.A. Infrastructure digital twin technology: A new paradigm for future construction industry. *Technol. Soc.* **2024**, *77*, 102519. [CrossRef]
5. Sawhney, A.; Riley, M.; Irizarry, J.; Riley, M. (Eds.) *Construction 4.0.*; Routledge: London, UK, 2020.
6. Musarat, M.A.; Irfan, M.; Alaloul, W.S.; Maqsoom, A.; Ghufran, M. A Review on the Way Forward in Construction through Industrial Revolution 5.0. *Sustainability* **2023**, *15*, 13862. [CrossRef]
7. Qi, Q.; Tao, F.; Hu, T.; Anwer, N.; Liu, A.; Wei, Y.; Wang, L.; Nee, A.Y.C. Enabling technologies and tools for digital twin. *J. Manuf. Syst.* **2021**, *58*, 3–21. [CrossRef]
8. Sindhwani, R.; Afridi, S.; Kumar, A.; Banaitis, A.; Luthra, S.; Singh, P.L. Can industry 5.0 revolutionize the wave of resilience and social value creation? A multi-criteria framework to analyze enablers. *Technol. Soc.* **2022**, *68*, 101887. [CrossRef]
9. Tyagi, A.K.; Lakshmi Priya, R.; Mishra, A.K.; Balamurugan, G. Industry 5.0. In *Privacy Preservation of Genomic and Medical Data*; Wiley: Hoboken, NJ, USA, 2023; pp. 409–432.
10. Forcael, E.; Ferrari, I.; Opazo-Vega, A.; Pulido-Arcas, J.A. Construction 4.0: A literature review. *Sustainability* **2020**, *12*, 9755. [CrossRef]
11. Vu, T.L.A.; Le, T.Q. Development orientation for higher education training programme of mechanical engineering in industrial revolution 4.0: A perspective in Vietnam. *J. Mech. Eng. Res. Dev.* **2019**, *42*, 68–70.

12. Tao, F.; Sui, F.; Liu, A.; Qi, Q.; Zhang, M.; Song, B.; Guo, Z.; Lu, S.C.-Y.; Nee, A.Y. Digital twin-driven product design framework. *Int. J. Prod. Res.* **2019**, *57*, 3935–3953. [CrossRef]
13. Bao, J.; Guo, D.; Li, J.; Zhang, J. The modelling and operations for the digital twin in the context of manufacturing. *Enterp. Inf. Syst.* **2019**, *13*, 534–556. [CrossRef]
14. Lee, H.; Kim, T. Smart factory use case model based on digital twin. *ICIC Express Lett. Part B Appl.* **2018**, *9*, 931–936.
15. Page, G.; Yorke-Biggs, B.; De-Guido, S. *IDE Digital Twin White Paper: Harnessing the Digital Twin for Real Competitive Advantage*; Loughborough University: Loughborough, UK, 2023.
16. Seaton, H.; Savian, C.; Sepasgozar, S.; Sawhney, A. *Digital Twins from Design to Handover of Constructed Assets*; Royal Institute of Chartered Surveyors: London, UK, 2022.
17. Research and Markets. Digital Twins Market by Technology, Twinning Type, Cyber-to-Physical Solutions, Use Cases and Applications in Industry Verticals 2024–2029. Available online: https://www.researchandmarkets.com/reports/5308850/digital-twins-market-by-technology-twinning?utm_source=dynamic&utm_medium=CI&utm_code=6q68tb&utm_campaign=1366076+-The+Future+of+the+Digital+Twins+Industry+to+2025+in+Manufacturing,+Smart+Cities,+Auto (accessed on 3 February 2024).
18. Pregnolato, M.; Gunner, S.; Voyagaki, E.; De Risi, R.; Carhart, N.; Gavriel, G.; Tully, P.; Tryfonas, T.; Macdonald, J.; Taylor, C. Towards Civil Engineering 4.0: Concept, workflow and application of Digital Twins for existing infrastructure. *Autom. Constr.* **2022**, *141*, 104421. [CrossRef]
19. Saback, V.; Popescu, C.; Blanksvärd, T.; Täljsten, B. Analysis of Digital Twins in the Construction Industry: Practical Applications, Purpose, and Parallel with other Industries. *Buildings* **2024**, *14*, 1361. [CrossRef]
20. Wang, W.; Zaheer, Q.; Qiu, S.; Wang, W.; Ai, C.; Wang, J.; Wang, S.; Hu, W. *Digital Twin Technologies in Transportation Infrastructure Management*; Springer: Berlin/Heidelberg, Germany, 2024.
21. Zhong, D.; Xia, Z.; Zhu, Y.; Duan, J. Overview of predictive maintenance based on digital twin technology. *Heliyon* **2023**, *9*, e14534. [CrossRef] [PubMed]
22. Broo, D.G.; Schooling, J. Digital twins in infrastructure: Definitions, current practices, challenges and strategies. *Int. J. Constr. Manag.* **2021**, *23*, 1254–1263. [CrossRef]
23. Jones, D.; Snider, C.; Nassehi, A.; Yon, J.; Hicks, B. Characterising the Digital Twin: A systematic literature review. *CIRP J. Manuf. Sci. Technol.* **2020**, *29*, 36–52. [CrossRef]
24. Yan, B.; Yang, F.; Qiu, S.; Wang, J.; Cai, B.; Wang, S.; Zaheer, Q.; Wang, W.; Chen, Y.; Hu, W. Digital twin in transportation infrastructure management: A systematic review. *Intell. Transp. Infrastruct.* **2023**, *2*, liad024. [CrossRef]
25. Wang, X.; Zhang, Y.; Li, H.; Wang, C.; Feng, P. Applications and challenges of digital twin intelligent sensing technologies for asphalt pavements. *Autom. Constr.* **2024**, *164*, 105480. [CrossRef]
26. Vieira, J.; Poças Martins, J.; Marques de Almeida, N.; Patrício, H.; Gomes Morgado, J. Towards Resilient and Sustainable Rail and Road Networks: A Systematic Literature Review on Digital Twins. *Sustainability* **2022**, *14*, 7060. [CrossRef]
27. Autodesk. What is a Digital Twin? Available online: https://www.autodesk.com/design-make/articles/what-is-a-digital-twin (accessed on 1 February 2025).
28. Chen, Z.-S.; Chen, K.-D.; Xu, Y.-Q.; Pedrycz, W.; Skibniewski, M.J. Multiobjective optimization-based decision support for building digital twin maturity measurement. *Adv. Eng. Inform.* **2024**, *59*, 102245. [CrossRef]
29. Mantalovas, K.; Di Mino, G.; Jimenez Del Barco Carrion, A.; Keijzer, E.; Kalman, B.; Parry, T.; Lo Presti, D. European National Road Authorities and Circular Economy: An Insight into Their Approaches. *Sustainability* **2020**, *12*, 7160. [CrossRef]
30. Liu, Z.; Balieu, R.; Kringos, N. Integrating sustainability into pavement maintenance effectiveness evaluation: A systematic review. *Transp. Res. Part D Transp. Environ.* **2022**, *104*, 103187. [CrossRef]
31. Lu, R.; Brilakis, I. *Generating Bridge Geometric Digital Twins from Point Clouds*; European Council on Computing in Construction: St-Niklaas, Belgium, 2019; pp. 367–376.
32. van Dinter, R.; Tekinerdogan, B.; Catal, C. Predictive maintenance using digital twins: A systematic literature review. *Inf. Softw. Technol.* **2022**, *151*, 107008. [CrossRef]
33. Nikolaev, S.; Belov, S.; Gusev, M.; Uzhinsky, I. Hybrid Data-Driven and Physics-Based Modelling for Prescriptive Maintenance of Gas-Turbine Power Plant. In Proceedings of the Product Lifecycle Management in the Digital Twin Era, Moscow, Russia, 8–12 July 2019; Springer: Cham, Switzerland, 2019; pp. 379–388.
34. Camposano, J.C.; Smolander, K.; Ruippo, T. Seven metaphors to understand digital twins of built assets. *IEEE Access* **2021**, *9*, 27167–27181. [CrossRef]
35. Adigopula, V.K. A Simplified Empirical Approach for Prediction of Pavement Layer Moduli Values Using Lightweight Deflectometer Data. *Int. J. Pavement Res. Technol.* **2022**, *15*, 751–763. [CrossRef]
36. Canestrari, F.; Ingrassia, L.P.; Virgili, A. A semi-empirical model for top-down cracking depth evolution in thick asphalt pavements with open-graded friction courses. *J. Traffic Transp. Eng. (Engl. Ed.)* **2022**, *9*, 244–260. [CrossRef]
37. Zhang, Y.; Gu, F.; Birgisson, B.; Lytton, R.L. Viscoelasticplastic–fracture modeling of asphalt mixtures under monotonic and repeated loads. *Transp. Res. Rec.* **2017**, *2631*, 20–29. [CrossRef]

38. Wang, H.; Al-Qadi, I.L. Combined effect of moving wheel loading and three-dimensional contact stresses on perpetual pavement responses. *Transp. Res. Rec.* **2009**, *2095*, 53–61. [CrossRef]
39. Brewer, K.A. *AASHTO Maintenance Manual for Roadways and Bridges*; Aashto: Washington, DC, USA, 2007.
40. Irrgang, F.C.; Maze, T. *Status of Pavement Management Systems and Data Analysis Models at State Highway Agencies. Transportation Research Record No. 1397*; National Academy Press: Washington, DC, USA, 1993; pp. 1–6.
41. Gharaibeh, N.G.; Darter, M.I.; Uzarski, D.R. Development of prototype highway asset management system. *J. Infrastruct. Syst.* **1999**, *5*, 61–68. [CrossRef]
42. Amândio, M.; Parente, M.; Neves, J.; Fonseca, P. Integration of Smart Pavement Data with Decision Support Systems: A Systematic Review. *Buildings* **2021**, *11*, 579. [CrossRef]
43. Chen, W.; Zheng, M. Multi-objective optimization for pavement maintenance and rehabilitation decision-making: A critical review and future directions. *Autom. Constr.* **2021**, *130*, 103840. [CrossRef]
44. Haas, R.; Hudson, W.R. *Pavement Asset Management*; John Wiley & Sons: Hoboken, NJ, USA, 2015.
45. Brown, D.; Liu, W.; Henning, T. *Identifying Pavement Deterioration by Enhancing the Definition of Road Roughness*; NZ Transport Agency: Auckland, New Zealand, 2010.
46. Coenen, T.B.; Golroo, A. A review on automated pavement distress detection methods. *Cogent Eng.* **2017**, *4*, 1374822. [CrossRef]
47. Zhang, S.; Lippitt, C.D.; Bogus, S.M.; Neville, P.R. Characterizing pavement surface distress conditions with hyper-spatial resolution natural color aerial photography. *Remote Sens.* **2016**, *8*, 392. [CrossRef]
48. Aleadelat, W.; Ksaibati, K. Estimation of pavement serviceability index through android-based smartphone application for local roads. *Transp. Res. Rec.* **2017**, *2639*, 129–135. [CrossRef]
49. Pan, Y.; Zhang, X.; Sun, M.; Zhao, Q. Object-based and supervised detection of potholes and cracks from the pavement images acquired by UAV. *Int. Arch. Photogramm. Remote Sens. Spat. Inf. Sci.* **2017**, *42*, 209–217. [CrossRef]
50. Maeda, H.; Sekimoto, Y.; Seto, T.; Kashiyama, T.; Omata, H. Road damage detection using deep neural networks with images captured through a smartphone. *arXiv* **2018**, arXiv:1801.09454.
51. Peraka, N.S.P.; Biligiri, K.P. Pavement asset management systems and technologies: A review. *Autom. Constr.* **2020**, *119*, 103336. [CrossRef]
52. Gopalakrishnan, K. Deep learning in data-driven pavement image analysis and automated distress detection: A review. *Data* **2018**, *3*, 28. [CrossRef]
53. Zhu, J.; Zhong, J.; Ma, T.; Huang, X.; Zhang, W.; Zhou, Y. Pavement distress detection using convolutional neural networks with images captured via UAV. *Autom. Constr.* **2022**, *133*, 103991. [CrossRef]
54. Du, Y.; Pan, N.; Xu, Z.; Deng, F.; Shen, Y.; Kang, H. Pavement distress detection and classification based on YOLO network. *Int. J. Pavement Eng.* **2021**, *22*, 1659–1672. [CrossRef]
55. Zakeri, H.; Nejad, F.M.; Fahimifar, A. Image based techniques for crack detection, classification and quantification in asphalt pavement: A review. *Arch. Comput. Methods Eng.* **2017**, *24*, 935–977. [CrossRef]
56. Li, Z. *Transportation Asset Management: Methodology and Applications*; CRC Press: Boca Raton, FL, USA, 2018.
57. Al-Mansoori, T.; Abdalkadhum, A.; Al-Husainy, A.S. A GIS-enhanced pavement management system: A case study in Iraq. *J. Eng. Sci. Technol.* **2020**, *15*, 2639–2648.
58. Justo-Silva, R.; Ferreira, A.; Flintsch, G. Review on Machine Learning Techniques for Developing Pavement Performance Prediction Models. *Sustainability* **2021**, *13*, 5248. [CrossRef]
59. Wang, W.; Qin, Y.; Li, X.; Wang, D.; Chen, H. Comparisons of Faulting-Based Pavement Performance Prediction Models. *Adv. Mater. Sci. Eng.* **2017**, *2017*, 6845215. [CrossRef]
60. Marcelino, P.; de Lurdes Antunes, M.; Fortunato, E.; Gomes, M.C. Machine learning approach for pavement performance prediction. *Int. J. Pavement Eng.* **2021**, *22*, 341–354. [CrossRef]
61. Abdualaziz Ali, A.; Heneash, U.; Hussein, A.; Khan, S. Application of Artificial neural network technique for prediction of pavement roughness as a performance indicator. *J. King Saud Univ. Eng. Sci.* **2023**, *36*, 128–139. [CrossRef]
62. Wang, X.; Zhao, J.; Li, Q.; Fang, N.; Wang, P.; Ding, L.; Li, S. A Hybrid Model for Prediction in Asphalt Pavement Performance Based on Support Vector Machine and Grey Relation Analysis. *J. Adv. Transp.* **2020**, *2020*, 7534970. [CrossRef]
63. Barros, R.; Yasarer, H.; Uddin, W.; Sultana, S. Roughness Modeling for Composite Pavements using Machine Learning. *IOP Conf. Ser. Mater. Sci. Eng.* **2021**, *1203*, 032035. [CrossRef]
64. Sandamal, K.; Shashiprabha, S.; Muttil, N.; Rathnayake, U. Pavement Roughness Prediction Using Explainable and Supervised Machine Learning Technique for Long-Term Performance. *Sustainability* **2023**, *15*, 9617. [CrossRef]
65. Martin, T.; Sen, R. *Austroads Road Deterioration Model Update: Rutting*; Austroads: Sydney, Australia, 2023; ISBN 978-1-922700-70-4.
66. Martin, T.; Sen, R. *Austroads Road Deterioration Model Update: Roughness*; Austroads: Sydney, Australia, 2023; ISBN 978-1-922700-69-8.
67. Basnet, K.S.; Shrestha, J.K.; Shrestha, R.N. Pavement performance model for road maintenance and repair planning: A review of predictive techniques. *Digit. Transp. Saf.* **2023**, *2*, 253–267. [CrossRef]

68. Gunner, S.; Voyagaki, E.; Gavriel, G.; Carhart, N.; MacDonald, J.; Tryfonas, T.; Taylor, C.; Pregnolato, M. Digital Twins for civil engineering: The Clifton Suspension Bridge (UK). In Proceedings of the 10th International Conference on Structural Health Monitoring of Intelligent Infrastructure, SHMII, Porto, Portugal, 30 June–2 July 2021; pp. 1–6.
69. American Association of State Highway and Transportation Officials. *Mechanistic-Empirical Pavement Design Guide: A Manual of Practice*; AASHTO: Washington, DC, USA, 2008.
70. Wang, H.; Al-Qadi, I.L.; Stanciulescu, I. Simulation of tyre–pavement interaction for predicting contact stresses at static and various rolling conditions. *Int. J. Pavement Eng.* **2012**, *13*, 310–321. [CrossRef]
71. Selezneva, O.I.; Hallenbeck, M.E. *Facilitating Analysts' Use of Traffic Data from the Long-Term Pavement Performance (LTPP) Program*; United States Federal Highway Administration, Office of Infrastructure: Washington, DC, USA, 2022.
72. Dong, J.; Meng, W.; Liu, Y.; Ti, J. A framework of pavement management system based on IoT and big data. *Adv. Eng. Inform.* **2021**, *47*, 101226. [CrossRef]
73. Piryonesi, S.M. *The Application of Data Analytics to Asset Management: Deterioration and Climate Change Adaptation in Ontario Roads*; University of Toronto (Canada): Toronto, ON, Canada, 2019.
74. Oditallah, M.; Alam, M.M.; Ekambaram, P. Digital Twins in AEC Infrastructures and Buildings Management Systems. In Proceedings of the 2nd International Conference on Civil Infrastructure and Construction (CIC 2023), Doha, Qatar, 5–8 February 2023.
75. Kun, C.; Mehran, E.T.; Mingjie, C.; Long, Z.; Alvaro, G.-H. Identifying the most suitable machine learning approach for a road digital twin. *Proc. Inst. Civ. Eng. Smart Infrastruct. Constr.* **2021**, *174*, 88–101.
76. Arup. *Digital Twin: Towards a Meaningful Framework*; Technical Report; Arup: London, UK, 2019.
77. MEED. *Digital Evolution—The Critical Impact of Digital Twins on Middle East Construction*; MEED: Dubai, United Arab Emirates, 2021.
78. Love, P.E.; Matthews, J.; Zhou, J. Is it just too good to be true? Unearthing the benefits of disruptive technology. *Int. J. Inf. Manag.* **2020**, *52*, 102096. [CrossRef]
79. Opoku, D.-G.J.; Perera, S.; Osei-Kyei, R.; Rashidi, M.; Famakinwa, T.; Bamdad, K. Drivers for Digital Twin Adoption in the Construction Industry: A Systematic Literature Review. *Buildings* **2022**, *12*, 113. [CrossRef]
80. Núñez-Temes, C.; Bastos, G.; Arza-García, M.; Castro, A.; Lorenzana Fernández, J.A.; Ortiz-Sanz, J.; Portela, M.; Gil-Docampo, M.; Prego, F.J. Assessment of pavement deflection under vehicle loads using a 3D-DIC system in the field. *Sci. Rep.* **2022**, *12*, 9491. [CrossRef]
81. Liao, J.; Lin, H.; Li, Q.; Zhang, D. A correction model for the continuous deflection measurement of pavements under dynamic loads. *IEEE Access* **2019**, *7*, 154770–154785. [CrossRef]
82. Wang, X.; Wang, H.; Wang, C.; Feng, P.; Irfan, M.; Gao, J.; Leng, Z.; Zhang, C. Real-time monitoring of asphalt pavement structure fatigue response based on tri-axis accelerometer. *Int. J. Pavement Eng.* **2024**, *25*, 2308158. [CrossRef]
83. Braunfelds, J.; Senkans, U.; Skels, P.; Porins, J.; Haritonovs, V.; Spolitis, S.; Bobrovs, V. Development of the strain measurement calibration technique for road pavement structural health monitoring applications using optical FBG sensors. In Proceedings of the 2023 Photonics & Electromagnetics Research Symposium (PIERS), Prague, Czech Republic, 3–6 July 2023; pp. 1060–1065.
84. Wang, D.; Lv, H.; Tang, F.; Ye, C.; Liu, P. Intelligent detection technology of contact state between asphalt pavement layers based on 3D ground penetrating radar technology. *J. Beijing Univ. Technol* **2022**, *48*, 573.
85. Khajavi, S.H.; Motlagh, N.H.; Jaribion, A.; Werner, L.C.; Holmström, J. Digital twin: Vision, benefits, boundaries, and creation for buildings. *IEEE Access* **2019**, *7*, 147406–147419. [CrossRef]
86. Grieves, M. Digital twin: Manufacturing excellence through virtual factory replication. *White Pap.* **2014**, *1*, 1–7.
87. Tao, F.; Liu, W.; Zhang, M.; Hu, T.-L.; Qi, Q.; Zhang, H.; Sui, F.; Wang, T.; Xu, H.; Huang, Z. Five-dimension digital twin model and its ten applications. *Comput. Integr. Manuf. Syst* **2019**, *25*, 1–18.
88. Yu, H.; Yao, D.; Qian, G.; Zhu, X.; Shi, Z.; Zhang, C.; Li, P. Review on Digital Twin Model of Asphalt Mixture Performance Based on Mesostructure Characteristics. *China J. Highw. Transp* **2023**, *36*, 20–44.
89. Errandonea, I.; Beltrán, S.; Arrizabalaga, S. Digital Twin for maintenance: A literature review. *Comput. Ind.* **2020**, *123*, 103316. [CrossRef]
90. Wang, M.; Yin, X. Construction and maintenance of urban underground infrastructure with digital technologies. *Autom. Constr.* **2022**, *141*, 104464. [CrossRef]
91. Gao, C.; Wang, J.; Dong, S.; Liu, Z.; Cui, Z.; Ma, N.; Zhao, X. Application of Digital Twins and Building Information Modeling in the Digitization of Transportation: A Bibliometric Review. *Appl. Sci.* **2022**, *12*, 11203. [CrossRef]
92. Ge, C.; Qin, S. Digital twin intelligent transportation system (DT-ITS)—A systematic review. *IET Intell. Transp. Syst.* **2024**, *18*, 2325–2358. [CrossRef]
93. Irfan, M.S.; Dasgupta, S.; Rahman, M. Toward Transportation Digital Twin Systems for Traffic Safety and Mobility: A Review. *IEEE Internet Things J.* **2024**, *11*, 24581–24603. [CrossRef]

94. Werbińska-Wojciechowska, S.; Giel, R.; Winiarska, K. Digital Twin Approach for Operation and Maintenance of Transportation System—Systematic Review. *Sensors* **2024**, *24*, 6069. [CrossRef]
95. Sanfilippo, F.; Thorstensen, R.T.; Jha, A.; Jiang, Z.; Robbersmyr, K.G. A Perspective Review on Digital Twins for Roads, Bridges, and Civil Infrastructures. In Proceedings of the 2022 International Conference on Electrical, Computer, Communications and Mechatronics Engineering (ICECCME), Male, Maldives, 16–18 November 2022; pp. 1–6.
96. Talaghat, M.A.; Golroo, A.; Kharbouch, A.; Rasti, M.; Heikkilä, R.; Jurva, R. Digital twin technology for road pavement. *Autom. Constr.* **2024**, *168*, 105826. [CrossRef]
97. Cepa, J.J.; Pavón, R.M.; Alberti, M.G.; Ciccone, A.; Asprone, D. A Review on the Implementation of the BIM Methodology in the Operation Maintenance and Transport Infrastructure. *Appl. Sci.* **2023**, *13*, 3176. [CrossRef]
98. Ersöz, O.Ö.; İnal, A.F.; Aktepe, A.; Türker, A.K.; Ersöz, S. A Systematic Literature Review of the Predictive Maintenance from Transportation Systems Aspect. *Sustainability* **2022**, *14*, 14536. [CrossRef]
99. Grieves, M.W. Product lifecycle management: The new paradigm for enterprises. *Int. J. Prod. Dev.* **2005**, *2*, 71–84. [CrossRef]
100. Shafto, M.; Conroy, M.; Doyle, R.; Glaessgen, E.; Kemp, C.; LeMoigne, J.; Wang, L. *Modeling, Simulation, Information Technology and Processing Roadmap*; NASA: Washington, DC, USA, 2010.
101. Bado, M.F.; Tonelli, D.; Poli, F.; Zonta, D.; Casas, J.R. Digital Twin for Civil Engineering Systems: An Exploratory Review for Distributed Sensing Updating. *Sensors* **2022**, *22*, 3168. [CrossRef]
102. Glaessgen, E.; Stargel, D. The digital twin paradigm for future NASA and US Air Force vehicles. In Proceedings of the 53rd AIAA/ASME/ASCE/AHS/ASC Structures, Structural Dynamics and Materials Conference 20th AIAA/ASME/AHS Adaptive Structures Conference 14th AIAA, Honolulu, HI, USA, 23–26 April 2012; p. 1818.
103. Tao, F.; Zhang, M. Digital Twin Shop-Floor: A New Shop-Floor Paradigm Towards Smart Manufacturing. *IEEE Access* **2017**, *5*, 20418–20427. [CrossRef]
104. Söderberg, R.; Wärmefjord, K.; Carlson, J.S.; Lindkvist, L. Toward a Digital Twin for real-time geometry assurance in individualized production. *CIRP Ann.* **2017**, *66*, 137–140. [CrossRef]
105. Kritzinger, W.; Karner, M.; Traar, G.; Henjes, J.; Sihn, W. Digital Twin in manufacturing: A categorical literature review and classification. *IFAC-PapersOnLine* **2018**, *51*, 1016–1022. [CrossRef]
106. Zhang, H.; Liu, Q.; Chen, X.; Zhang, D.; Leng, J. A digital twin-based approach for designing and multi-objective optimization of hollow glass production line. *IEEE Access* **2017**, *5*, 26901–26911. [CrossRef]
107. Stark, R.; Damerau, T. Digital Twin. In *CIRP Encyclopedia of Production Engineering*; Chatti, S., Tolio, T., Eds.; Springer: Berlin/Heidelberg, Germany, 2019; pp. 1–8.
108. Zhang, L.; Zhou, L.; Ren, L.; Laili, Y. Modeling and simulation in intelligent manufacturing. *Comput. Ind.* **2019**, *112*, 103123. [CrossRef]
109. Defraeye, T.; Shrivastava, C.; Berry, T.; Verboven, P.; Onwude, D.; Schudel, S.; Bühlmann, A.; Cronje, P.; Rossi, R.M. Digital twins are coming: Will we need them in supply chains of fresh horticultural produce? *Trends Food Sci. Technol.* **2021**, *109*, 245–258. [CrossRef]
110. Li, J.; Yin, G.; Wang, X.; Yan, W. Automated decision making in highway pavement preventive maintenance based on deep learning. *Autom. Constr.* **2022**, *135*, 104111. [CrossRef]
111. Gao, L.; Yu, Y.; Ren, Y.H.; Lu, P. Detection of Pavement Maintenance Treatments using Deep-Learning Network. *Transp. Res. Rec.* **2021**, *2675*, 1434–1443. [CrossRef]
112. Yu, G.; Zhang, S.; Hu, M.; Wang, Y.K. Prediction of Highway Tunnel Pavement Performance Based on Digital Twin and Multiple Time Series Stacking. *Adv. Civ. Eng.* **2020**, *2020*, 8824135. [CrossRef]
113. Consilvio, A.; Hernández, J.S.; Chen, W.; Brilakis, I.; Bartoccini, L.; Gennaro, F.D.; Welie, M.V. Towards a digital twin-based intelligent decision support for road maintenance. *Transp. Res. Procedia* **2023**, *69*, 791–798. [CrossRef]
114. Zeiada, W.; Dabous, S.A.; Hamad, K.; Al-Ruzouq, R.; Khalil, M.A. Machine Learning for Pavement Performance Modelling in Warm Climate Regions. *Arab. J. Sci. Eng.* **2020**, *45*, 4091–4109. [CrossRef]
115. Kaliske, M.; Behnke, R.; Wollny, I. Vision on a Digital Twin of the Road-Tire-Vehicle System for Future Mobility. *Tire Sci. Technol.* **2021**, *49*, 2–18. [CrossRef]
116. Ficara, A.; Fazio, M.; Pellegrino, O.; Ruggeri, A.; Ruggeri, A.; Sollazzo, G.; Bosurgi, G. Distributed Services for Pavement Anomaly Detection in Smart Roads. In Proceedings of the 2024 IEEE International Symposium on Measurements & Networking (M&N), Rome, Italy, 2–5 July 2024; pp. 1–6.
117. Barisic, L.; Levenberg, E.; Skar, A.; Boyd, A.; Zoulis, P. *A Thermal Digital Twin for Condition Monitoring of Asphalt Roads*; CRC Press: Boca Raton, FL, USA, 2021; pp. 709–713.
118. Sotiriadis, G.; Kotzakolios, T.; Kostopoulos, V.; Gemou, M. A Digital Twin Assisted and Embedded Strain Gauge Monitoring System. *Transp. Res. Procedia* **2023**, *72*, 2038–2045. [CrossRef]
119. Steyn, W.J.; Broekman, A. Development of a digital twin of a local road network: A case study. *J. Test. Eval.* **2022**, *50*, 2901–2915. [CrossRef]

120. Meža, S.; Mauko Pranjic, A.; Vezočnik, R.; Osmokrović, I.; Lenart, S. Digital Twins and Road Construction Using Secondary Raw Materials. *J. Adv. Transp.* **2021**, *2021*, 8833058. [CrossRef]
121. Marai, O.E.; Taleb, T.; Song, J. Roads Infrastructure Digital Twin: A Step Toward Smarter Cities Realization. *IEEE Netw.* **2021**, *35*, 136–143. [CrossRef]
122. Wang, W.; Xu, X.; Peng, J.; Hu, W.; Wu, D. Fine-Grained Detection of Pavement Distress Based on Integrated Data Using Digital Twin. *Appl. Sci.* **2023**, *13*, 4549. [CrossRef]
123. Sierra, C.; Paul, S.; Rahman, A.; Kulkarni, A. Development of a cognitive digital twin for pavement infrastructure health monitoring. *Infrastructures* **2022**, *7*, 113. [CrossRef]
124. Sofia, H.; Anas, E.; Faïz, O. Mobile Mapping, Machine Learning and Digital Twin for Road Infrastructure Monitoring and Maintenance: Case Study of Mohammed VI Bridge in Morocco. In Proceedings of the 2020 IEEE International conference of Moroccan Geomatics (Morgeo), Casablanca, Morocco, 11–13 May 2020; pp. 1–6.
125. Hidayat, F.; Supangkat, S.H.; Hanafi, K. Digital Twin of Road and Bridge Construction Monitoring and Maintenance. In Proceedings of the 2022 IEEE International Smart Cities Conference (ISC2), Pafos, Cyprus, 26–29 September 2022; pp. 1–7.
126. Bosurgi, G.; Celauro, C.; Pellegrino, O.; Rustica, N.; Giuseppe, S. The BIM (Building Information Modeling)-Based Approach for Road Pavement Maintenance. In Proceedings of the 5th International Symposium on Asphalt Pavements & Environment (APE), Padua, Italy, 11–13 September 2019; Springer: Cham, Switzerland, 2020; pp. 480–490.
127. Cao, T.; Wang, Y.; Liu, S. Pavement Crack Detection Based on 3D Edge Representation and Data Communication with Digital Twins. *IEEE Trans. Intell. Transp. Syst.* **2022**, *24*, 7697–7706. [CrossRef]
128. D'Amico, F.; Bianchini Ciampoli, L.; Di Benedetto, A.; Bertolini, L.; Napolitano, A. Integrating Non-Destructive Surveys into a Preliminary BIM-Oriented Digital Model for Possible Future Application in Road Pavements Management. *Infrastructures* **2022**, *7*, 10. [CrossRef]
129. Bertolini, L.; Giorgadze, I.M.; Vahdatikhaki, F.; D'Amico, F. A Semantic Digital Twinning Approach for the Management of Road Distress Data. *ISPRS Ann. Photogramm. Remote Sens. Spat. Inf. Sci.* **2024**, *X-4/W5-2024*, 49–54. [CrossRef]
130. Fox-Ivey, R.; Laurent, J.; Petitclerc, B. Using 3D Pavement Surveys to Create a Digital Twin of Your Runway or Highway. In *Airfield and Highway Pavements 2021*; Proceedings; American Society of Civil Engineers (ASCE): Reston, VA, USA, 2021; pp. 180–192.
131. Jiang, F.; Ma, L.; Broyd, T.; Chen, W.; Luo, H. Building digital twins of existing highways using map data based on engineering expertise. *Autom. Constr.* **2022**, *134*, 104081. [CrossRef]
132. Pan, Y.; Wang, M.; Lu, L.; Wei, R.; Cavazzi, S.; Peck, M.; Brilakis, I. Scan-to-graph: Automatic generation and representation of highway geometric digital twins from point cloud data. *Autom. Constr.* **2024**, *166*, 105654. [CrossRef]
133. Lukanen, E.O.; Stubstad, R.; Briggs, R.C.; Intertec, B. *Temperature Predictions and Adjustment Factors for Asphalt Pavement*; Turner-Fairbank Highway Research Center: McLean, VA, USA, 2000.
134. Rajapaksha, M.M.C.; Athukorallage, B.; Senadheera, S.; James, D. Temporal and spatial temperature predictions for flexible pavement layers using numerical thermal analysis and verified with large datasets. *Case Stud. Constr. Mater.* **2023**, *18*, e02008. [CrossRef]
135. Xu, B.; Dan, H.-C.; Li, L. Temperature prediction model of asphalt pavement in cold regions based on an improved BP neural network. *Appl. Therm. Eng.* **2017**, *120*, 568–580. [CrossRef]
136. You, L.; Yan, K.; Liu, N. Assessing artificial neural network performance for predicting interlayer conditions and layer modulus of multi-layered flexible pavement. *Front. Struct. Civ. Eng.* **2020**, *14*, 487–500. [CrossRef]
137. Shafabakhsh, G.; Talebsafa, M.; Motamedi, M.; Badroodi, S.K. Analytical evaluation of load movement on flexible pavement and selection of optimum neural network algorithm. *KSCE J. Civ. Eng.* **2015**, *19*, 1738–1746. [CrossRef]
138. Ziyadi, M.; Al-Qadi, I.L. Efficient surrogate method for predicting pavement response to various tire configurations. *Neural Comput. Appl.* **2017**, *28*, 1355–1367. [CrossRef]
139. Okte, E.; Al-Qadi, I.L. Prediction of flexible pavement 3-D finite element responses using Bayesian neural networks. *Int. J. Pavement Eng.* **2022**, *23*, 5066–5076. [CrossRef]
140. Al-Qadi, I.L.; Hernandez, J.A.; Gamez, A.; Ziyadi, M.; Gungor, O.E.; Kang, S. Impact of Wide-Base Tires on Pavements: A National Study. *Transp. Res. Rec.* **2018**, *2672*, 186–196. [CrossRef]
141. Li, M.; Wang, H. Development of ANN-GA program for backcalculation of pavement moduli under FWD testing with viscoelastic and nonlinear parameters. *Int. J. Pavement Eng.* **2019**, *20*, 490–498. [CrossRef]
142. Matchett, R. *Digital Twins for Road Infrastructure*; University of Pretoria: Pretoria, South Africa, 2022.
143. Ansari, F.; Glawar, R.; Nemeth, T. PriMa: A prescriptive maintenance model for cyber-physical production systems. *Int. J. Comput. Integr. Manuf.* **2019**, *32*, 482–503. [CrossRef]
144. Vasiliev, Y.E.; Fineeva, M.A.; Belyakov, A.B.; Varshavsky, S.V.; Caesar, A.A. Decision support system for street-road network objects repair. *IOP Conf. Ser. Mater. Sci. Eng.* **2021**, *1159*, 012025. [CrossRef]

145. Hadjidemetriou, G.M.; Masino, J.; Christodoulou, S.E.; Gauterin, F.; Brilakis, I. Comprehensive Decision Support System for Managing Asphalt Pavements. *J. Transp. Eng. Part B: Pavements* **2020**, *146*, 06020001. [CrossRef]
146. Philip, B.; AlJassmi, H. A Bayesian decision support system for optimizing pavement management programs. *Heliyon* **2024**, *10*, e25625. [CrossRef]
147. Abu Dabous, S.; Zeiada, W.; Zayed, T.; Al-Ruzouq, R. Sustainability-informed multi-criteria decision support framework for ranking and prioritising pavement sections. *J. Clean. Prod.* **2020**, *244*, 118755. [CrossRef]
148. Pour, S.M.; Masoumi, A.M.; Dujardin, N.S. Towards a Technology-Driven Adaptive Decision Support System for Integrated Pavement and Maintenance strategies (TDADSS-IPM): Focus on risk assessment framework for climate change adaptation. *Transp. Res. Procedia* **2023**, *72*, 4468–4475. [CrossRef]

infrastructures

MDPI

Article

A Big Data System Architecture to Support the Monitoring of Paved Roads

Jorge Oliveira e Sá [1,*], Francisco Rebelo [2], Diogo Silva [2], Gabriel Teles [2], Diogo Ramos [2] and José Romeu [2]

1 Algoritmi Center, University of Minho, 4710-057 Braga, Portugal
2 Institute for Sustainability and Innovation in Structural Engineering (ISISE), University of Minho, 4800-058 Guimarães, Portugal; id9826@alunos.uminho.pt (F.R.); a82927@alunos.uminho.pt (D.S.); a80418@alunos.uminho.pt (G.T.); a74399@alunos.uminho.pt (D.R.); a89267@alunos.uminho.pt (J.R.)
* Correspondence: jos@dsi.uminho.pt; Tel.: +351-253510315

Abstract: Today, everything is connected, including the exchange of data and the generation of new information. As a result, large amounts of data are being collected at an ever-increasing rate and in a variety of forms, a phenomenon now known as Big Data. Recent developments in information and communication technologies are driving the generation of significant amounts of data from multiple sources, namely sensors. In response to these technological advances and data challenges, this paper proposes a Big Data system architecture for paved road monitoring and implements part of this architecture on a section of road in Portugal as a case study. The challenge in the case study architecture is to collect and process sensor data in real time, at a rate of 500 records per second, producing 15 GBytes of data per day, using a real-time data stream for real-time monitoring and a batch data stream for deeper analysis. This allows users to obtain instant updates on road conditions such as the number of vehicles, loads, weather, and pavement temperatures on the road. They can monitor what is happening on the road in real time, receive alerts, and even gain insight into historical data, such as analysing the condition of structures or identifying traffic patterns.

Keywords: Big Data; Big Data architecture; fibre-optic sensors; monitor road health; real-time monitoring; batch monitoring

1. Introduction

The demonstration approach has been around for a long time, but recent developments in computing have allowed huge amounts of data to be processed to anticipate, understand, and predict behaviours through data patterns [1].

Collecting data from sensors embedded in structures such as roads allows us to demonstrate their behaviour when they undergo deformation caused by constant strain. This usage of Big Data falls under the domain of the Internet of Things (IoT). The use of computer technology makes it possible to collect a huge volume of data generated by sensors, transform it, and store it in such a way that it can be analysed, visualised, and fed into intelligent algorithms that can predict road behaviour over time. IoT is seen as a network of sensors embedded in roads, which is a significant source of Big Data.

Monitoring of transport infrastructure is very important in predicting ageing and loss of performance [2]. It is imperative that this prevention be as effective as possible to ensure the safety and comfort of those who use them every day.

Based on the ontology proposed by [3], it is assumed that there are currently two types of monitoring systems: dynamic and static, regardless of the type of road and pavement used. It is difficult to place static embedded systems on unpaved roads, although dynamic systems can be used. On paved roads, regardless of the type of pavement, both monitoring systems are viable. It is important to note that the most traditionally used monitoring system is visual inspection, which is a dynamic system performed by humans. Dynamic road monitoring systems are mobile systems capable of monitoring multiple locations and roads, whereas static monitoring systems are limited to a single monitoring location, often embedded within the road.

A literature review revealed several data collection technologies that are applicable to road monitoring systems:

- Digital cameras (dynamic): To capture images of roads and analyse the surface road conditions, the cameras should be mounted on a vehicle [4–6].
- Smartphones (dynamic): Use of smartphones with an accelerometer and Global Position System (GPS) attached to a vehicle, such as a car or bicycle, and comparing the vibration with the vehicle's speed to detect irregular road surfaces [7–17].
- Wireless and high-precision sensors (dynamic): To capture the roughness of the road surface, for example, use noise sensors and vibration sensors [18,19].
- Radio Frequency (RF) sensors (static): Based on wireless sensors embedded on the road, this could include an RF transmitter and an RF reader installed in vehicles to capture the data [20].
- Fibre optic sensors (static): To accurately measure road stresses, capture the vibrations felt on the road, allowing the identification of the vehicle's weight and the damping effect [21,22].
- Temperature sensors (static): Capturing temperature variations provides insights about road surface stress [23].
- Strain gauge sensors (static): To measure the deformations and stress on the road [24,25].
- Temperature, strain, and inclinometer sensors (static): To measure the thermal curling on the concrete pavement slab [26].
- Temperature, Strain, and Pressure Cells (Static): To measure the conditions of airport pavements [27].
- Geophones (static). To measure road deflection [27].
- Ground-penetrating radar (static): The radars were installed at different depths, with their antennas suspended above the road surface to monitor road conditions [28–31].
- Mechanical systems sensors (static). Integrates mechanical, sensor, and electronic elements on a small chip: It allows you to collect and analyse temperature, humidity, and deformation [32].
- Smart Rocks (static): It is a device that integrates different sensors, including piezoresistive stress and strain sensors, as well as a triaxial accelerometer for direct measurement of 3D stress, strain, and acceleration on the SmartRock surface [33].
- Distributed Acoustic Sensing (static): It is used to measure seismic signals propagating over long distances using fibre-optic cables. The technology was capable of measuring small deformations when a sensor fibre is coupled to the system and road noise over long distances [34].

Our study uses static Fibre Optic Sensors not to analyse the road surface but rather its internal structural layers and the deformation within them. To achieve this, it is necessary to design an architecture for collecting, preparing, storing, analysing, and utilising (big) data to support road monitoring while also taking into account issues such as data security and privacy.

In general, the aforementioned case studies either lack descriptions of the architecture used to deploy the proposed solutions or provide very limited detail. Data collection, processing, and subsequent analysis are considered essential for pavement monitoring systems, but most studies do not describe this component [35]. However, one study stands out by introducing a cloud-based architecture for monitoring cyber-physical systems. Its primary aim is to automate the transmission of data generated by IoT devices to the cloud. Within the cloud environment, the data are analysed and stored using blockchain technology to ensure authenticity and integrity [36]. The authors argue that this architecture constitutes a suitable cloud-based solution for modern monitoring systems and can serve as a reference model for future research.

Therefore, this paper presents a more versatile architecture capable of accommodating the requirements of road monitoring systems, whether they are cloud-based or not. A comprehensive Big Data Systems Architecture (BDSA) is proposed, integrating several layers and components for data collection, storage, processing, analysis, and distribution. This integrated environment facilitates informed decision-making and predictions regarding structural behaviour.

The aim of this paper is to describe the use of the BDSA in a case where it was installed on a real road section to validate a specific set of components within the BDSA. This study covers the use and testing of different layers of the BDSA and some of its specific components, covering the entire data lifecycle from data collection to analysis and visualisation. It also introduces and models the necessary data structures to support storage and processing within a (Big) Data Warehouse.

Methodologically, this work was carried out following the research methodology of Design Science Research [37]. The validation of the proposed system was achieved by developing a prototype of a Big Data System designed for monitoring a road section.

The remainder of this paper is structured as follows:

- Section 2 explores the evolution of Business Intelligence and Big Data Analytics, considering the concept of Big Data.
- Section 3 introduces the BDSA and outlines its architecture for road monitoring. It details the various layers and emphasises the selected layers and components for implementing the proof-of-concept.
- Section 4 presents a comprehensive account of the implementation process, encompassing (Big) Data Warehouse modelling, along with associated tasks such as data integration, cleansing, transformation, loading, and data analysis and visualisation.
- The results, which show the importance of using a BDSA, are presented in Section 5.
- Finally, in Section 6, we offer concluding remarks and provide insights into future directions.

2. Business Intelligence and Big Data Analytics Evolution

In recent years, successive advances in digital technologies have triggered important changes in organisations, their strategies, and business models. In this way, digital has become an integral part of companies' daily lives, becoming an unavoidable reality that increasingly contributes to their success [38].

Therefore, over the last years, the interest in Analytics and Big Data has increased considerably [39], particularly after 2012, as can be seen in Figure 1. This may be due to the consequent skyrocketing growth in the volume of data generated by companies, requiring new tools and technologies able to process and analyse large datasets. Now it becomes relevant to carry out a retrospective analysis of the evolution of Business Intelligence (BI) and (big) Data Analytics to reach the current reality, the Big Data era.

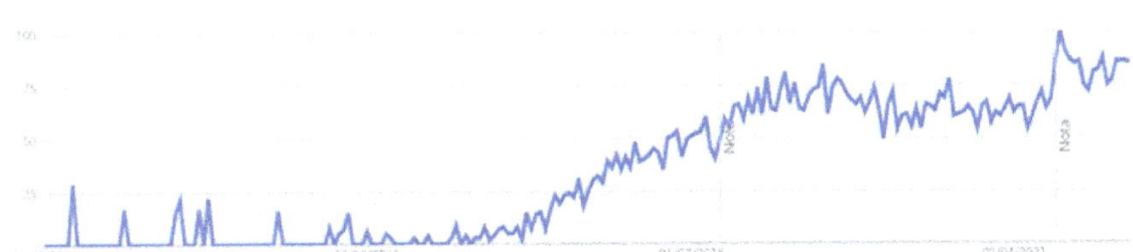

Figure 1. Interest in Analytics and Big Data.

According to IBM, BI is software that considers business data and presents it in user-friendly visualisations such as reports, dashboards, charts, and graphs, allowing business users to access different types of data and enabling analysis of this information to gain a better insight into company performance. Organisations can use the knowledge gained through BI and data analysis to improve business decisions, identify problems or issues, spot market trends, and find new revenues or business opportunities [40]. Although this is a broader definition, the emphasis is maintained on data processing capabilities to provide useful information and knowledge to improve the business, i.e., BI is a set of strategies and technologies that companies may use to analyse business information and transform it into practical knowledge that serves as a basis for commercial, strategic, and tactical decisions.

(Big) Data Analytics can be explained as the use of advanced analytical techniques on diverse (big) data sets that include structured, semi-structured, and unstructured data from different sources and of different sizes and whose size or type exceeds the ability of traditional relational databases to capture, manage, and process the (big) data with low latency [40]. A more detailed definition of Big Data can be achieved, such as high-volume, high-velocity, and/or high-variety information assets.

Given that sources of Big Data are becoming more complex than traditional data because they are being driven by Artificial Intelligence (AI), mobile devices, social media, and the IoT, and much of that Big Data are generated in real-time at a very large scale (for example, data originates from sensors), it is extremely necessary to adopt Big Data Analytic tools capable of providing insights to perform better and faster decisions, to model and predict future outcomes, and to improve business intelligence. The key asset is still information and data processing for supporting the decision-making process and enhancing the business.

In this context, an evolution from BI and Analytics (BI&A) to Big Data Analytics can be observed in terms of support technologies and development frameworks. This evolution is addressed in the work of [41], which makes a retrospective characterisation of the BI&A itself and shows what changes for Big Data. According to these authors, Business Analytics (BA) is understood as the main analytical component of BI; some authors refer to BA as a system that offers advanced techniques for the reporting of data, an evolution of BI [42], while Big Data are used to describe data sets so large and complex that they require advanced technologies for data storage, management, analysis, and visualisation. Therefore, Big Data Analytics offers new research directions for BI&A.

The BI&A evolution over time is divided into four eras: BI&A 1.0, BI&A 2.0, BI&A 3.0, and BI&A 4.0, with applications and emerging research areas with different data sources, as can be seen in Figure 2 [41–43]:

- BI&A 1.0 refers to the first era of Business Intelligence and Analytics, in which data were essentially structured and came from multiple data sources. Data warehouses (DW) are a creation that dates back to this era and play a crucial role in it, as they serve as the basis for integrating and consolidating company data. DW schemas also prove essential to organising and structuring data effectively by developing Extract, Transform, and Load (ETL) mechanisms to extract data from various sources, transform it into a suitable format, and load it into the DW.
- Online Analytical Processing (OLAP) and reporting tools are used in BI&A 1.0 to explore and analyse data, providing interactive environments with intuitive graphs and enabling ad-hoc querying processing, complemented by statistical methods and data mining algorithms for advanced data analytics.
- BI&A 2.0 represents the next phase, where data collection expands beyond structured sources. In this era, data are collected from websites using cookies and server logs, allowing companies to gain insight into user needs and preferences. The focus shifts to text and web analytics, involving techniques such as web intelligence, web analytics, text mining, web mining, social network analysis, and spatio-temporal data analysis. These methods help extract meaningful information from unstructured data, enabling businesses to identify new opportunities and make data-driven decisions [41].
- In BI&A 3.0, the emergence of mobile devices takes centre stage. Due to the proliferation of mobile phones, tablets, sensor-based devices, Internet-enabled devices, barcodes, and radio tags, the IoT has become a significant source of data. And consequently, this era generates a vast amount of detailed data from various devices and sensors, which can be leveraged to gain new insights. The mobile device and IoT ecosystem create opportunities for businesses to understand user behavior, pref-

erences, and real-time interactions, enabling personalised experiences and targeted marketing efforts.

- BI&A 4.0 is the AI era, in which the focus will shift from descriptive analysis to prescriptive and predictive analysis. Given the exponential development of AI and the improved predictive analytics—smarter; faster; and more actionable—enabled by AI techniques and their consequent capabilities to handle large amounts of data in real-time; delivering results with high accuracy [44]; due to the high potential of AI to transform business; the impact of Analytics 4.0 is likely to be significantly larger and more disruptive than previous technology transitions [43].

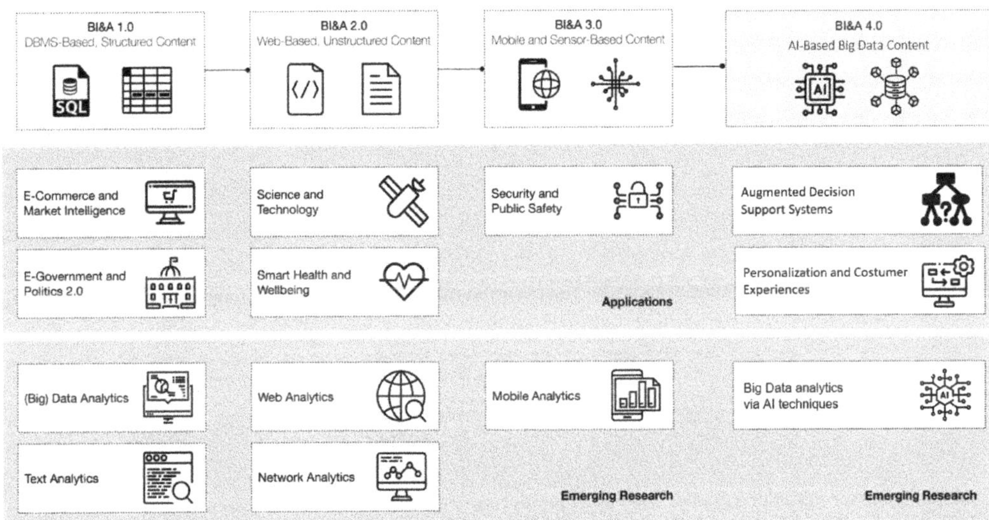

Figure 2. Interest in Analytics and Big Data, adapted from [41].

3. Big Data System Architecture

The emergence of IoT devices, namely sensors, and their availability to connect with each other to generate an intelligent network for the constant exchange of information has generated a new concept for dealing with data called Big Data. This required the adoption of proper technologies to fulfil the data collection, storage, processing, and analysis needs.

Thus, in order to monitor road pavements, Figure 3 shows the proposed BDSA as well as its main layers and components. The components tested in the case study presented in this paper are highlighted in the figure.

The BDSA is divided into seven layers, each of which includes components. Each layer is represented by a grey rectangle, while white rectangles are used to specify the components. Data flows between layers are also represented in this figure and go from data collection to data visualisation.

The first layer of the BDSA is the Data Sources, which can include road sensors, monitoring cameras, and traffic control system data, among others. These data sources are responsible for collecting relevant information about traffic and roads. The variety of data sources implies that the data can be collected in various formats, namely structured, unstructured, or semi-structured data.

The second layer is Data Processing, where raw data are analysed and transformed into more useful and meaningful information. In this stage, techniques such as data cleansing, standardisation, and enrichment can be applied, as well as Machine Learning algorithms for more advanced data processing and transformation.

After processing, the data are directed to the storage layer, where it is stored in a structured manner and made accessible for future queries. This layer may involve the use of databases or distributed storage systems like Hadoop's Distributed File System (HDFS), Google's Big Table, or Amazon's S3 in a Big Data context. The main challenge at this layer is to efficiently store massive volumes of data in a way that allows for fast retrieval and analysis.

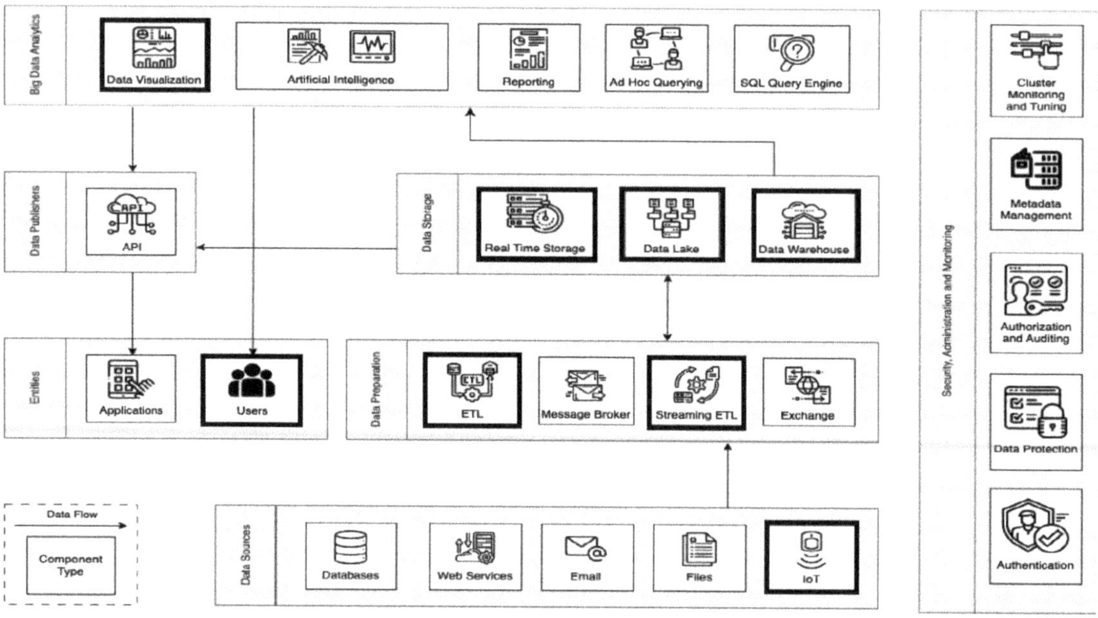

Figure 3. BDSA for Roads Monitoring, adapted from [45].

Once the data are stored, the Big Data Analytics layer involves the analysis of this data. This usually involves Data Visualisation, AI to model predictions about health roads, Reporting, Ad-Hoc Querying (ideal for exploratory interactive data analysis), and a Structured Query Language (SQL) Query Engine (used for routine reporting and transaction processing). The goal here is to derive insights, trends, and patterns from the stored data presented through graphs, tables, or other forms of visual representation. This allows external apps or users, such as road engineers or traffic authorities, to easily access and interpret the information obtained from road monitoring.

The fifth layer is the Data Publisher, which is responsible for publishing the data and the analysis results to external systems, i.e., applications. These could be databases, other Big Data systems, or Application Program Interfaces (APIs). This layer ensures that the results of the Big Data processing system are widely available to be used by other components of the broader technology ecosystem.

The layer entities represent the consumers of the insights derived from the analytics layer and can be humans or machines. Humans, like road engineers or traffic authorities, are looking for important insights in dashboards or reports. Machines can consume this data to drive decision-making or automation.

Finally, the Security, Administration, and Monitoring layer cuts across all the previous layers. It involves ensuring the security of the data and the Big Data System itself, administering the system (e.g., managing user access, allocating resources), and monitoring the system (e.g., tracking performance, identifying, and handling failures). These aspects are critical for the Big Data System's overall integrity, reliability, and performance.

The BDSA is a layered architecture that provides a logical separation of concerns, where each layer has a specific role and interacts with the layers above and below it.

This helps build a scalable, robust, and capable architecture capable of handling Big Data processing and storage demands.

In the next section, we will use some layers and components to instantiate a case study to monitor a real road section.

4. A Case Study

4.1. Case Study Contextualisation

Having proposed a Big Data System and its BDSA for road monitoring, it is now important to implement it, a task that will validate some layers and components of the architecture.

The case study was carried out on a section of road in northern Portugal called IC5, where the sensors used to collect data are optical sensors that allow strain and temperature data to be collected, as it is a static installation. This location was chosen for its proximity to a power source and internet infrastructure.

These sensors generate data with a measurement frequency of 500 records per second, which is equivalent to 15 Gigabytes per day. This huge volume of data has to be collected, prepared, and analysed and it is impossible to use conventional technologies. Therefore, the Big Data System for Road Monitoring was adopted. In this installation, not all the layers and components proposed in Figure 3 were used, but the layers and components used are justified, serving to evaluate the proposed architecture.

The first step was to install the sensors embedded in the pavement in the IC5 road section. The installation took place on two one-way lanes, the right lane for slow, heavy traffic and the left lane for fast, light traffic. Two channels were installed in each lane: the right lane has Channel A with seven strain sensors and one for temperature measurement, and Channel B with fifteen strain sensors and one for temperature measurement. The left lane is a replica of the right lane, and the channels are designated D and E. Each channel was embedded perpendicularly into the two-lane road with a distance of three metres between channels A and B, the same distance for channels D and E (see Figure 4).

One of the system's requirements is to be able to monitor traffic in real-time, called real-time processing. It is responsible for collecting and processing data at high speeds while also making it available in a visualisation tool, i.e., a dashboard. In addition to the dashboard, it is also able to show a table containing logs of alarming events for registry vehicles with very low speed, excessive speed, or excessive weight, as well as low or high temperatures. The other system requirement is to support batch processing; it is responsible for collecting and transforming data, but the storage phases are more complex and apply to a larger volume of historic data (see Figure 5).

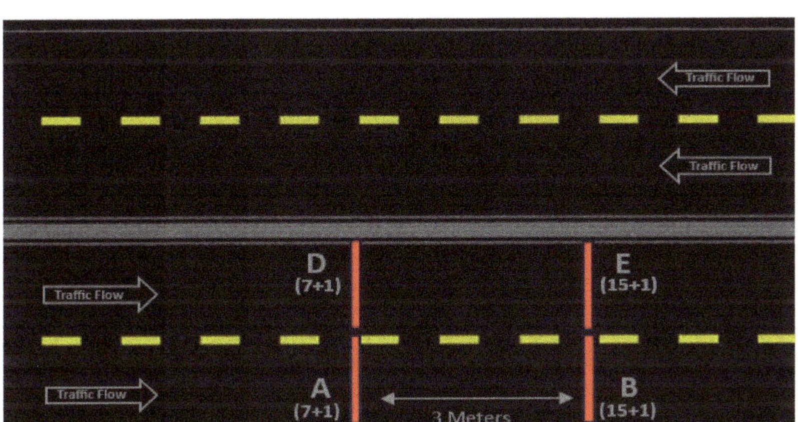

Figure 4. Installation of IC5 road section sensors.

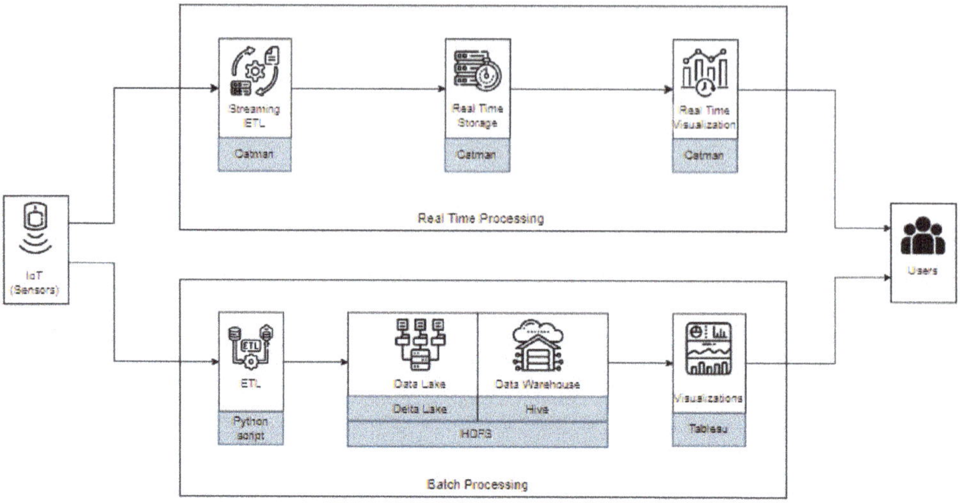

Figure 5. Architecture data flow.

Therefore, some layers and components of the BDSA proposed in Figure 3 were used, as shown in Table 1.

Table 1. Layers, components, and technologies from BDSA are used.

Layer	Component	Technology
Data Source	IoT	Sensor Optical FBG (Strain and Temperature static sensors with Optical Interrogator and Catman®Easy software)
Data Preparation	ETL	Python script (Z-score algorithm and others)
	Streaming ETL	Catman®Easy software (filters, functions and scripts)
Data Storage	Real Time Storage	Catman®Easy software (internal database to store real time data and alerts)
	Data Lake	Delta Lake (HDFS)
	Data Warehouse	Hive (HDFS)
Big Data Analytics	Data Visualisation	Catman®Easy software (Real time monitor) Tableau (Batch time monitor)
Entities	Users	Road engineers and/or Traffic authorities

4.2. Real-Time Processing

The first step in real-time processing is to collect the data directly, seamlessly, and with minimal delay from the sensors. The data acquisition software needs to identify and categorise the different channels and sensors so the collected data are correctly organised for further analysis and interpretation.

Following the data collection step, it enters the data processing phase. Data filtering tasks are important in order to exclude outliers, anomalies, and other values that fall outside user-defined checks. Ensuring the time synchronisation of readings from different sensors is also essential to maintaining a logical stream of values. Furthermore, some form of event detection and subsequent application of mathematical functions are needed in order to convert the signal data into valuable information (e.g., to calculate vehicle speed and weight, to obtain the time of occurrences, etc.).

This data are stored in a real-time way to keep the transformation conducted and calculations made in the data, as well as to register the events.

Finally, the data reaches the data visualisation phase, where it is presented in a variety of visual representations, including graphs, tables, and digital indicators, in a way that a regular user can easily and quickly interpret the information. Moreover, in case an event has exceptional parameters that break user-defined limits, it has to be registered along with its information, and a system response may be triggered (e.g., sending a push notification or communicating the incident via email).

Catman® Easy software ensures all these requirements are covered. It identifies the channels and sensors as well as their information while also allowing the user to configure some settings (e.g., sensor name, sampling rate, sensor type/role, etc.) and associates the streams of collected data to each sensor while quickly processing them and saving them in real-time storage. The data processing capabilities include filters, functions, and scripts executed in near real-time in order to derive insights. For instance, it is able to count the daily number of vehicles that have passed and calculate the approximate speed and weight they were going while encountering low to no performance issues. It also offers a vast variety of data visualisation tools and a panel that allows free use of them to create real-time dashboards, enabling dynamic data visualisation and analysis.

In order to analyse this data in real-time, a dashboard for each lane was developed. In Figure 6, the left lane's dashboard is presented, displaying the maximum values measured in channels D (indicated by the yellow line) and E (indicated by the blue line) over a two-minute interval in the top graph. The bottom section features two graphs: on the left, the D channel (depicted by yellow bars) with seven sensors, and on the right, the E channel (represented by blue bars) with fifteen sensors. Additionally, on the right side of the dashboard, you can find information about road and air temperatures, a list of vehicles that have passed, and the speed of the last vehicle.

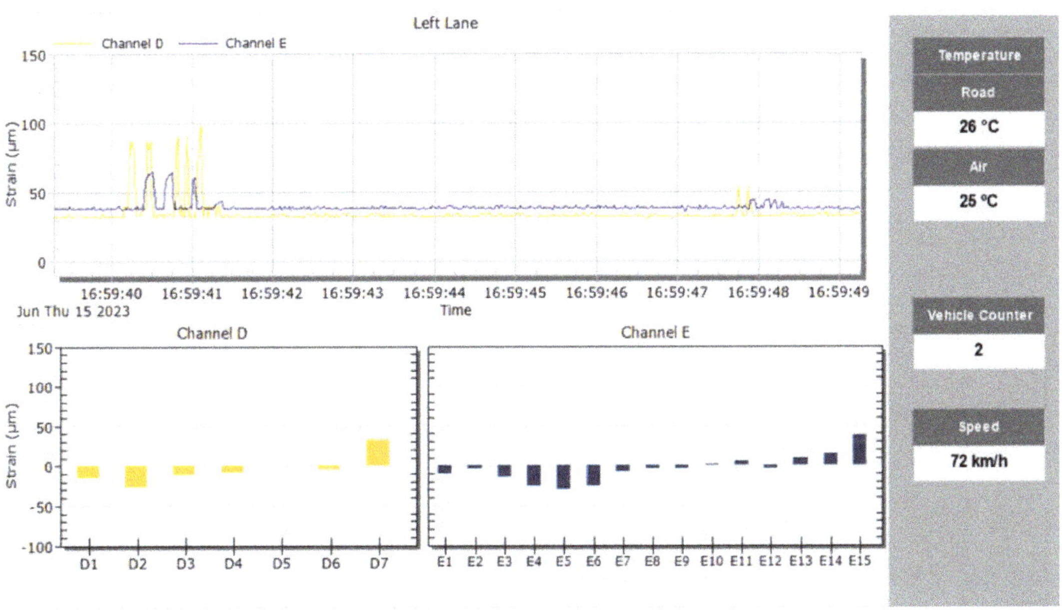

Figure 6. Real-time dashboard.

In the main graph of the dashboard, the peaks of the lines show the strains measured when the vehicle wheel axles pass over the sensors. In this example, we can see the passage of a heavy vehicle with five axles and the passage of a light vehicle with two axles.

Several calculations are executed in the monitoring process. For example, when a vehicle axle traverses the D channel, the time $t1$ of this occurrence is recorded. Subsequently, when the same axle traverses the E channel, located three metres ahead of the D channel, the time $t2$ is recorded. In this way, it is possible to calculate the speed of a vehicle and to determine the number of vehicles that have passed through by detecting the axles that pass through the two channels at the same speed and with a certain cadence, regardless of the number of axles in the vehicle. Since we know the distance d between channels is three meters, we can successfully calculate the speed and multiply it by 3.6 to convert the value from m/s to km/h; see Formula (1).

$$s = \frac{d}{t2 - t1} \times 3.6 \tag{1}$$

The real-time monitor includes alerts to facilitate the detection of specific events, e.g., a vehicle exceeding the speed limit, which then trigger appropriate actions. These actions serve as effective measures to respond to these specific events in a timely manner, ranging from drawing attention to the dashboard with sound effects, pushing notifications, or sending emails to remote users (see Figure 7).

Date	Time	Type of occurrence	Value
15/06/2023	16:30:27	Excessive Speed	124 km/h
15/06/2023	17:42:12	Excessive Strain	957 µm
15/06/2023	17:58:44	Low Speed	42 km/h

Figure 7. Real-time alerts.

It is worth noting that the sensor values are reset to zero on a weekly basis to enhance data visualisation's user-friendliness and simplicity. This reset is particularly beneficial as the stabilised values of the sensors, even within the same channel, can initially exhibit significant dispersion. However, it is important to emphasise that the original sensor values are stored in real-time, and they are indispensable for conducting further studies on the road structure's health.

4.3. Batch Processing

We started by defining the technological infrastructure needed to support the BDSA. To perform this, we created a technological cluster consisting of four computers, one master, and three workers. These machines run on the Ubuntu operating system, Hadoop to manage the distribution of computing in the cluster, and HDFS to manage the data files and ensure that no information are lost during the storage process.

Hadoop was selected because it facilitates the creation of this type of machine cluster using less expensive and sophisticated hardware, as well as the existence of a wide range of programmes and technologies compatible with the Hadoop platform and, consequently, with HDFS.

One example is Spark, a distributed computing framework that we use to process the data because it is faster than Hadoop's data processing component, MapReduce, for certain types of processing when performing in-memory operations and reusing data in multiple processing steps. It provides a fast and general-purpose cluster computing system, and it is well-suited for processing large volumes of data, including sensor data, in parallel across a cluster of machines.

Spark can distribute data processing tasks across a cluster of machines, enabling efficient processing of large datasets like what can be found in this project, and cache data in memory, leading to faster iterative processing and reduced data access latency. What more is, it offers a wide range of data transformation operations (e.g., filtering, mapping, and aggregating) through its RDD (Resilient Distributed Dataset) and DataFrame APIs and supports machine learning, graph processing, and SQL queries, enabling complex analytical tasks on sensor data.

We used a data lake to ingest and store the data, specifically the Delta Lake solution with its three layers, Bronze, Silver, and Gold [46]. Delta Lake is an open-source storage layer that brings ACID (atomicity, consistency, isolation, and durability) transactions to Apache Spark and big data workloads. It is built on top of existing storage systems like HDFS and provides data reliability, consistency, and performance improvements for data lakes. Other features include allowing for schema evolution over time, making it easier to adapt to changes in sensor data formats without disrupting downstream processes; tracking historical versions of data, enabling audit trails and the ability to revert to previous data states; employing optimisations like compaction and indexing to improve query performance on large datasets; and the time travel feature permitting the query of previous data versions, aiding historical analysis and debugging.

When used with HDFS to process and store sensor data, Spark and Delta Lake together provide a robust and scalable solution for managing sensor data at scale. Spark's distributed processing capabilities enable efficient data analysis and transformation, while Delta Lake's ACID transactions and versioning features ensure data reliability and consistency, making it well-suited for managing sensor data in dynamic and evolving environments.

In the Gold layer of Delta Lake, we use the Hive database to implement a (Big) Data Warehouse to store data ready for analysis and visualisation. Apache Hive is a data warehousing and SQL-like query language tool built on top of Hadoop that allows the management and querying of structured data stored in HDFS using familiar SQL syntax and defining schemas for the data using the Data Definition Language (DDL), making it possible to impose structure on sensor data. Other capabilities include support for various data processing operations like filtering, aggregation, and transformation using SQL-like syntax, query optimisation mechanisms to improve query performance, making it suitable for querying large volumes of data, and the possibility of integrating with Tableau, a popular data visualisation tool that enables users to create interactive dashboards.

The data flow used is shown in Figure 5. This is explained in more detail below, and it is an ETLT (combination of ETL and ELT), i.e., extraction, transformation, loading, and transformation phases. The data are generated by sensors at a frequency of 500 readings per second, although it should be noted that this frequency can be set to a higher (max. 1000) or lower value. The value of 500 has been set in order to be able to measure the passage of the wheels over the sensors, but it is a value that can still be adjusted. The main transformation at this stage is to eliminate data that does not register vehicle passages, as over time there are periods when there are no vehicles on the road and the sensors are constantly generating data.

To detect vehicle passages, in this case called peaks, the *z-score* peak detection algorithm is used, with the definition of a threshold that defines the value from which the *z-score* must correspond to a peak. This algorithm is based on the principle of dispersion, according to which if a new data point is above the threshold, i.e., within a certain number of standard deviations from an average value, the algorithm classifies this point as a peak [47]. The *z-score* is given by the Formula (2).

$$z - score = \frac{(x - mean)}{standard\ deviation} \tag{2}$$

where x is the value measured by one sensor, i.e., A1, channel A sensor 1, the *mean* and *standard deviation* are calculated over all the values obtained by one sensor, e.g., sensor A1. The calculated *z-score*, together with the threshold, makes it possible to identify peaks and then the passage of a vehicle wheel over the sensor. This is calculated for all 44 sensors. The data are reduced in the time interval of 2 s before the peak and 2 s after the peak, which corresponds to 2000 lines of data per axle pass; the data that does not represent peaks is discarded.

These data are ingested in the Bronze Layer of Delta Lake as raw data.

The data in the Bronze layer is transformed and stored in the Silver layer by transforming:

- Column names—the column names of the generated data are very complex; making them difficult to handle in processing scripts. The solution is to rename the column names to a simple name.
- Numerical values—numerical values had a comma as a decimal separator; and this caused problems in handling these values in the processing scripts. The solution is to replace the commas with dots.
- Dates—the dates (timestamps) of the generated data were in Unix TimeStamp format and have been converted to DateTime format.

The next step is to transform the data from the Silver layer to the Gold layer. The Gold layer contains highly structured data that has undergone additional processing and aggregation from the Silver layer to answer specific business questions. From the Silver layer to the Gold layer, the data goes through a process of identifying vehicle classes, using machine learning algorithms to classify vehicles by axle pass and registered weight, as well as calculating vehicle speed and aggregating external data on weather in relation to road conditions (dry, wet, with a probability of ice or snow). The Gold layer implements a (Big) Data Warehouse with a dimensional data model, i.e., a constellation of three stars. This model makes it possible to answer questions relevant to the perception of road monitoring and to use algorithms to model the state of the road. The dimensional model developed for this project, see Figure 8, has three star schemas with different levels of aggregation:

- Passages Management—records vehicle passes; their class; speed; measured load; temperature and atmospheric conditions; date; and location.
- Temperature Management—records the temperatures measured at a location; date and time.
- Cargo Management—records the loads measured by the sensors at a given location; date; and time.

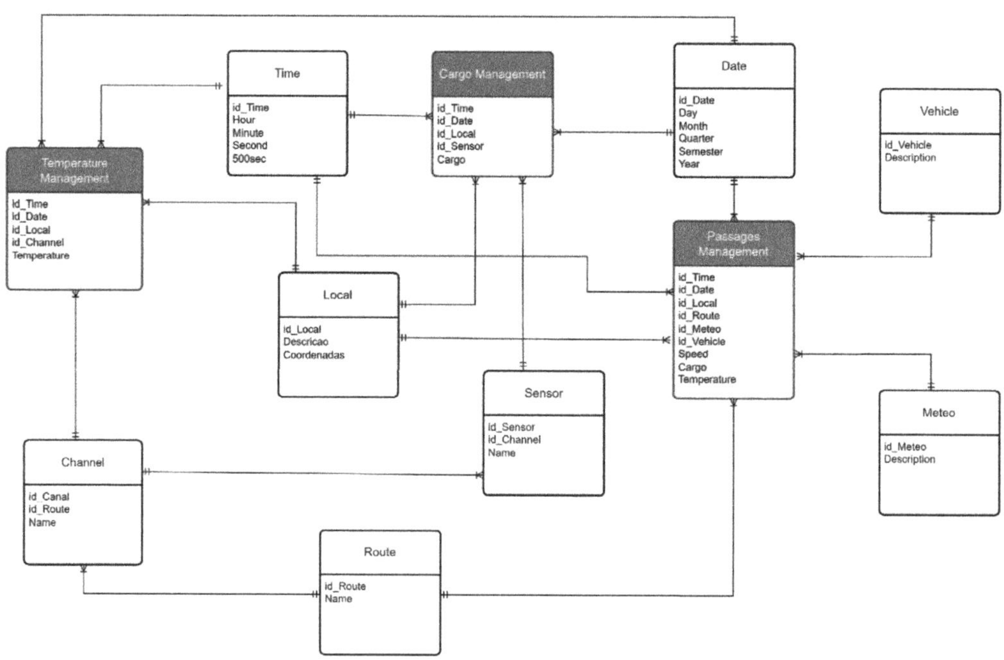

Figure 8. Dimensional Model.

The dimensional model allows you to answer analytical questions to better understand the evolution of the road condition. Here are some examples of such questions:

1. What are the loads recorded over a period of time at a given location?
2. Does the condition of the road—dry; wet; ice; or snow—affect the average speed recorded?
3. What is the frequency of passages recorded over the hours of a day in a month?
4. How has traffic developed in recent years, by type of vehicle and location?
5. Does the recorded temperature affect the value recorded on loads?
6. What is the average speed of heavy vehicles on working days in the current month at a particular location?
7. What type of vehicles travel at weekends to a particular location?
8. What is the maximum vehicle speed recorded? On what date? Where?

An example of a dashboard created using Tableau that can be used to view the passage of vehicles is shown in Figure 9a. It is possible to select the date and location to be displayed. This can range from a few hours of a day to several days, weeks, even months or years. This visualisation allows comparison with other values, for example, passages of vehicles from 15 June 2023 (blue color) can be compared with those from 13 June 2023 (orange color). In the visualisation, there is a graph of the number of vehicles per class on the left and a graph of the number of vehicles per road lane on the right. If you select one of the bars in the graph, for example, Class2 in the graph on the left (where is the hand symbol), a new view appears; see Figure 9b. In this case, it is a display of the loads and speeds registered by vehicles of that class over the time selected above.

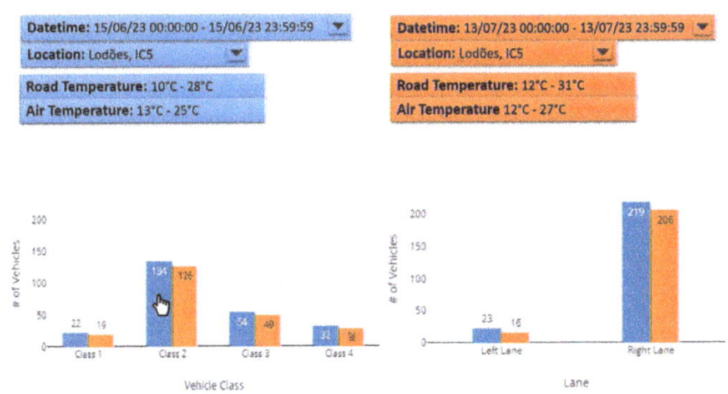

(a)

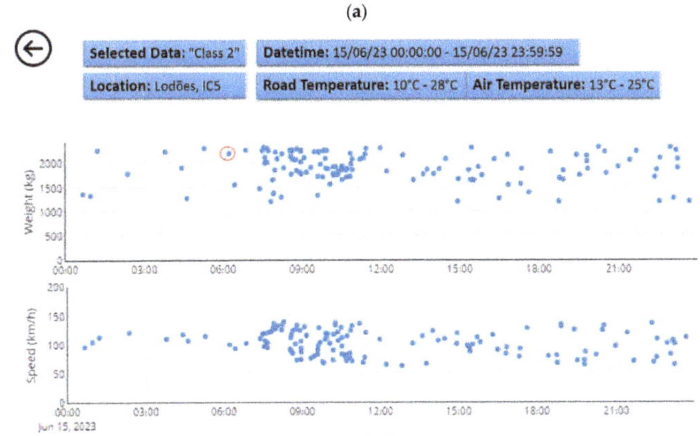

(b)

Figure 9. *Cont.*

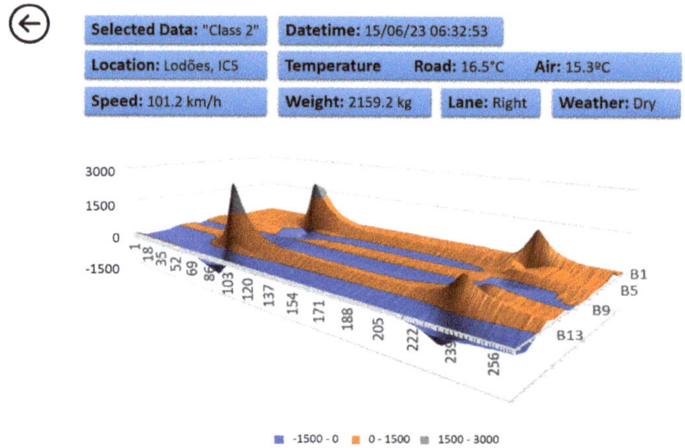

(c)

Figure 9. (**a**) General Dashboard (**b**) Vehicles passage details (**c**) One Passage details.

By selecting one of the points on the graph, in this case a vehicle (where is the red circle), you can see a 3D graph showing the vehicle passing in the lane through the channel with more sensors (B or E); see Figure 9c. You can also see the recorded temperature and the weather forecast for that location, day, and time. The 3D view allows you to see and analyse where the left and right wheels of the vehicle passed on the road. In this case, they passed close to sensors B3 and B13. It can be seen that when the wheel does not pass directly over a sensor, the readings are very different. The figure also shows the effect of loads on the road. The supporting structure of the road suffers permanent deformations, increasing the possibility of degradation. This set of visualisations allows you to answer the analytical questions 1, 3, 4, 6, and 7.

5. Results

The monitoring system developed is capable of successfully capturing, preprocessing, and storing data from optical strain and temperature sensors.

The real-time processing can reliably collect data continuously, with an uptime rate of nearly 100%. The only interruptions in data flow occur during maintenance or calibration processes, both of which do not occur frequently. Additionally, there is a brief downtime of approximately one second during the weekly sensor reading reset. The users can access the visualisations remotely, and the system runs smoothly with no performance issues.

The batch processing needs more pre-processing to collect the data and effectively filter out a substantial portion of noise, resulting in a data volume reduction of approximately 2% of the original raw data, i.e., the system can reduce 15 GB of daily data to less than 300 MB. These data are ingested into the HDFS so that they can be used by the distributed storage system. The next steps would involve completing the analysis performed with the Spark scripts to generate additional Delta tables (with three layers: Bronze, Silver, and Gold) and finalising the configuration of the HIVE database in the Gold Layer. Furthermore, establishing a connection to Tableau would enable more immersive visualisations and dashboards.

Although the system is online and capable of storing and pre-processing sensor data, its potential was not fully explored due to several obstacles encountered during the solution's development. These obstacles included several challenges in the selection, installation, and configuration of the tools to be integrated into the system, as well as time constraints for accumulating a substantial volume of sensor data required for building comprehensive historical analyses and predictive analyses.

However, following the BDSA allowed us to understand which layers and components were used (and used well) and also which components were not used, allowing us to prioritise the implementation of missing layers and components. In this particular case, the missing layers and components are:

- Big Data Analytics layer—it is important to implement all the missing components; in particular artificial intelligence; to predict road degradation.
- Data Publishers layer—to enable data sharing through a machine interface.
- Entities layer—to implement applications that can use the data and also allow external applications to access the data.
- Security, administration, and monitoring layer—when implementing the missing functionality; especially when enabling data sharing; it becomes important to ensure data security and protection; so this is a layer that needs to be managed and implemented whenever necessary.

6. Conclusions

This paper presents a proposal for a BDSA for road monitoring. Some specific layers and components have been integrated into two data flows: real-time and batch processing, from data collection to data analysis and visualisation to validate the system.

The case study presented showed how these technologies complement each other, with the overall objective of supporting the decision-making process. For this purpose, a section of road in Portugal was selected to install static Fibre Optical Sensors and all the necessary technological infrastructure to collect, process, store, and analyse the data collected, both in real-time and batch processing.

The case study implementation covered the layers of Data Source (sensors and web services for collecting weather data), Data Preparation (ETL for batch data and streaming ETL for real-time data), Data Storage (Real-Time storage, Data Lake, and (Big) Data Warehouse), and Big Data Analytics (data visualisation). It was possible to verify that the data stream was normal, that the selected technologies worked together, and that no integration or interoperability problems were identified.

In addition to the technical accomplishments outlined above, it is crucial to recognise the unique experimental environment in which the case study was undertaken. In addition to the technical achievements described above, it is crucial to acknowledge the unique experimental environment in which the case study was conducted. The journey from concept design to proof of concept was marked by several external factors that inevitably affected the pace and nature of this work, namely the reliance on third parties for sensor configuration, computer installation and configuration, permissions, etc., which posed unforeseen challenges and sometimes slowed progress.

In the future, it is expected that other components of the BDSA will be tested, as well as other technologies to be used. Some technologies will have to be chosen, and others can be replaced over time as more promising ones emerge. The challenge is to be able to store the volume of data generated to provide a history of at least five years.

Finally, in the BDSA Big Data Analytics layer, when we have a considerable data history, the AI component will be implemented and tested so that the solution will become an Adaptive Big Data System, combining predictive and optimisation algorithms to monitor and predict the health of roads to ensure their proper functioning.

Author Contributions: All authors have contributed to the conceptualisation, methodology, analysis, validation, and writing. Formal analysis, J.O.e.S. and F.R.; investigation, J.O.e.S., F.R., D.S., D.R., G.T. and J.R.; resources, D.R. and J.R.; data curation, J.O.e.S., F.R., D.S. and G.T.; writing—original draught preparation, D.S., G.T., D.R. and J.R.; supervision, J.O.e.S. and F.R. All authors have read and agreed to the published version of the manuscript.

Funding: This research was funded by PORTUGAL 2020 through the Operational Program for Competitiveness and Internationalization (POCI) and the European Regional Development Fund (ERDF) under the project "Rev@Construction—Digital Construction Revolution", with reference

POCI-01-0247-FEDER-046123, and by Fundação para a Ciência e a Tecnologia through the Ph.D. grant number 2022.14400.BD. This work was also partly financed by FCT/MCTES through national funds (PIDDAC) under the R&D Unit Institute for Sustainability and Innovation in Structural Engineering (ISISE), under reference UIDB/04029/2020, and under the Associate Laboratory Advanced Production and Intelligent Systems ARISE, under reference LA/P/0112/2020. This work was also partly supported by the FCT under the R&D Units Project Scope: UIDB/00319/2020.

Institutional Review Board Statement: Not applicable.

Informed Consent Statement: Not applicable.

Data Availability Statement: No new data were created or analyzed in this study. Data sharing is not applicable to this article.

Acknowledgments: The authors would like to acknowledge the staff from the authors' organisations involved in this work who contributed to achieving the objectives of this study. Acknowledgment to Freepik and Smashicons for the use of icons used in the figures, which were obtained from www.flaticon.com.

Conflicts of Interest: The authors declare no conflict of interest. The funders had no role in this study's design, in the collection, analysis, or interpretation of data, in the writing of this manuscript, or in the decision to publish this results.

References

1. Pramanik, S.; Bandyopadhyay, S.K. *Analysis of Big Data*; IGI Global: Hershey, PA, USA, 2023; pp. 97–115. [CrossRef]
2. Nicoletti, V.; Martini, R.; Carbonari, S.; Gara, F. Operational Modal Analysis as a Support for the Development of Digital Twin Models of Bridges. *Infrastructures* **2023**, *8*, 24. [CrossRef]
3. Shtayat, A.; Moridpour, S.; Best, B.; Shroff, A.; Raol, D. A review of monitoring systems of pavement condition in paved and unpaved roads. *J. Traffic Transp. Eng.* **2020**, *7*, 629–638. [CrossRef]
4. Amarasiri, S.; Gunaratne, M.; Sarkar, S. Use of Digital Image Modeling for Evaluation of Concrete Pavement Macrotexture and Wear. *J. Transp. Eng.* **2012**, *138*, 589–602. [CrossRef]
5. Khoudeir, M.; Brochard, J.; Legeay, V.; Do, M.T. Roughness Characterization through 3D Textured Image Analysis: Contribution to the Study of Road Wear Level. *Comput.-Aided Civ. Infrastruct. Eng.* **2004**, *19*, 93–104. [CrossRef]
6. Huidrom, L.; Das, L.K.; Sud, S.K. Method for automated assessment of potholes, cracks and patches from road surface video clips Selection and peer-review under responsibility of International Scientific Committee. *Procedia-Soc. Behav. Sci.* **2013**, *104*, 312–321. [CrossRef]
7. Astarita, V.; Caruso, M.V.; Danieli, G.; Festa, D.C.; Giofrè, V.P.; Iuele, T.; Vaiana, R. A Mobile Application for Road Surface Quality Control: UNIquALroad. *Procedia—Soc. Behav. Sci.* **2012**, *54*, 1135–1144. [CrossRef]
8. Douangphachanh, V.; Oneyama, H. A Study on the Use of Smartphones for Road Roughness Condition Estimation. *J. East. Asia Soc. Transp. Stud.* **2013**, *10*, 1551–1564. [CrossRef]
9. Eriksson, J.; Girod, L.; Hull, B.; Newton, R.; Madden, S.; Balakrishnan, H. The Pothole Patrol: Using a mobile sensor network for road surface monitoring. In *MobiSys'08, Proceedings of the 6th International Conference on Mobile Systems, Applications, and Services, Breckenridge, CO, USA, 17–20 June 2008*; Association for Computing Machinery: New York, NY, USA, 2008; pp. 29–39. [CrossRef]
10. González, A.; O'brien, E.J.; Li, Y.-Y.; Cashell, K. The use of vehicle acceleration measurements to estimate road roughness. *Veh. Syst. Dyn.* **2008**, *46*, 483–499. [CrossRef]
11. Menant, F.; Martin, J.-M.; Meignen, D.; Bétaille, D.; Ortiz, M. ScienceDirect Using probe vehicles for pavement monitoring: Experimental results from tests performed on a road network. *Transp. Res. Procedia* **2016**, *14*, 3013–3020. [CrossRef]
12. Mohan, P.; Padmanabhan, V.N.; Ramjee, R. Nericell: Rich monitoring of road and traffic conditions using mobile smartphones. In Proceedings of the 6th ACM Conference on Embedded Network Sensor Systems, Raleigh, NC, USA, 5–7 November 2008; pp. 323–336. [CrossRef]
13. Perttunen, M.; Mazhelis, O.; Cong, F.; Kauppila, M.; Leppänen, T.; Kantola, J.; Collin, J.; Pirttikangas, S.; Haverinen, J.; Ristaniemi, T.; et al. Distributed Road Surface Condition Monitoring Using Mobile Phones. In *Ubiquitous Intelligence and Computing, Proceedings of the UIC, Banff, AB, Canada, 2–4 September 2011*; Springer: Berlin/Heidelberg, Germany, 2011; pp. 64–78. [CrossRef]
14. Reza-Kashyzadeh, K.; Ostad-Ahmad-Ghorabi, M.J.; Arghavan, A. Study Effects of Vehicle Velocity on a Road Surface Roughness Simulation. *Appl. Mech. Mater.* **2013**, *372*, 650–656. [CrossRef]
15. Seraj, F.; van der Zwaag, B.; Dilo, A.; Luarasi, T.; Havinga, P. *Big Data Analytics in the Social and Ubiquitous Context*; Atzmueller, M., Chin, A., Janssen, F., Schweizer, I., Trattner, C., Eds.; Springer International Publishing: Cham, Switzerland, 2016. [CrossRef]
16. Souza, V.M.A. Asphalt pavement classification using smartphone accelerometer and Complexity Invariant Distance. *Eng. Appl. Artif. Intell.* **2018**, *74*, 198–211. [CrossRef]
17. Vittorio, A.; Rosolino, V.; Teresa, I.; Vittoria, C.M.; Vincenzo, P.G.; Francesco, D.M. Automated Sensing System for Monitoring of Road Surface Quality by Mobile Devices. *Procedia—Soc. Behav. Sci.* **2014**, *111*, 242–251. [CrossRef]
18. Du, Y.; Liu, X.; Zhong, S. Robust reliable H∞ control for neural networks with mixed time delays. *Chaos Solitons Fractals* **2016**, *91*, 1–8. [CrossRef]

19. Ye, Z.; Lu, Y.; Wang, L. Investigating the Pavement Vibration Response for Roadway Service Condition Evaluation. *Adv. Civ. Eng.* **2018**, *2018*, 2714657. [CrossRef]
20. Lajnef, N.; Chatti, K.; Chakrabartty, S.; Rhimi, M.; Sarkar, P.; University, M.S. Smart Pavement Monitoring System. Report; 2013. Available online: https://www.fhwa.dot.gov/publications/research/infrastructure/pavements/12072/12072.pdf (accessed on 8 March 2023).
21. Imai, M.; Igarashi, Y.; Shibata, M.; Miura, S. Experimental study on strain and deformation monitoring of asphalt structures using embedded fibre optic sensor. *J. Civ. Struct. Health Monit.* **2014**, *4*, 209–220. [CrossRef]
22. Zhang, Z.; Bao, X. Distributed optical fibre vibration sensor based on spectrum analysis of Polarization-OTDR system. *Opt. Express* **2008**, *16*, 10240–10247. [CrossRef]
23. Garcia, V. Improved Overlay Tester for Fatigue Cracking Resistance of Asphalt Mixtures (FHWA 0-6815-1). 2016. Available online: http://ctis.utep.edu (accessed on 9 March 2023).
24. Hu, J.; Wang, K.; Bektas, F. Monitoring of Joint Cracking Development in Concrete Pavement with Concrete Embedment Strain Gages. In *Design, Analysis, and Asphalt Material Characterization for Road and Airfield Pavements*; ASCE Publications: Reston, VA, USA, 2014; pp. 147–154. [CrossRef]
25. Kivi, A.K. Evaluation of Structural Dome Formwork Systems in Concrete Pavement Applications. Master's Thesis, University of Waterloo, Waterloo, ON, Canada, 2013. Available online: https://uwspace.uwaterloo.ca/handle/10012/7993 (accessed on 8 March 2023).
26. Liao, W.; Zhuang, Y.; Zeng, C.; Deng, W.; Huang, J.; Ma, H. Fiber optic sensors enabled monitoring of thermal curling of concrete pavement slab: Temperature, strain and inclination. *Measurement* **2020**, *165*, 108203. [CrossRef]
27. Duong, N.S.; Blanc, J.; Hornych, P.; Menant, F.; Lefeuvre, Y.; Bouveret, B. Monitoring of pavement deflections using geophones. *Int. J. Pavement Eng.* **2020**, *21*, 1103–1113. [CrossRef]
28. Benedetto, A.; Pensa, S. Indirect diagnosis of pavement structural damages using surface GPR reflection techniques. *J. Appl. Geophys.* **2007**, *62*, 107–123. [CrossRef]
29. Plati, C.; Loizos, A. Using ground-penetrating radar for assessing the structural needs of asphalt pavements. *Nondestruct. Test. Eval.* **2012**, *27*, 273–284. [CrossRef]
30. Bertolini, L.; D'Amico, F.; Napolitano, A.; Bianchini Ciampoli, L.; Gagliardi, V.; Romer Diezmos Manalo, J. A BIM-Based Approach for Pavement Monitoring Integrating Data from Non-Destructive Testing Methods (NDTs). *Infrastructure* **2023**, *8*, 81. [CrossRef]
31. Saarenketo, T. *Electrical Properties of Road Materials and Subgrade Soils and the Use of Ground Penetrating Radar in Traffic Infrastructure Surveys*; Oulu University Press: Oulu, Finland, 2006; p. 471.
32. Yang, S. Health Monitoring of Pavement Systems Using Smart Sensing Technologies. Master's Thesis, Iowa State University, Ames, IA, USA, 2014. Available online: https://core.ac.uk/download/pdf/38931372.pdf (accessed on 10 March 2023).
33. Ma, T.; Wang, N.; Han, C.; Wang, S.; Tong, Z. Asphalt Pavement Residual Life Assessment Based on SmartRock Sensors. *IEEE Trans. Instrum. Meas.* **2023**, *72*, 1–12. [CrossRef]
34. Hubbard, P.G.; Ou, R.; Xu, T.; Luo, L.; Nonaka, H.; Karrenbach, M.; Soga, K. Road deformation monitoring and event detection using asphalt-embedded distributed acoustic sensing (DAS). *Struct. Control. Health Monit.* **2022**, *29*, e3067. [CrossRef]
35. Barriera, M.; Pouget, S.; Lebental, B.; Van Rompu, J. In Situ Pavement Monitoring: A Review. *Infrastructures* **2020**, *5*, 18. [CrossRef]
36. Assumpcão, P.; Oliveira, C.; Ortiz, P.; Melo, W.; Carmo, L. A Secure Cloud-based Architecture for monitoring Cyber-Physical Critical Infrastructures. In Proceedings of the 6th Cyber Security in Networking Conference, CSNet 2022, Rio de Janeiro, Brazil, 24–26 October 2022. [CrossRef]
37. Peffers, K.; Tuunanen, T.; Rothenberger, M.A.; Chatterjee, S. A design science research methodology for information systems research. *J. Manag. Inf. Syst.* **2007**, *24*, 45–77. [CrossRef]
38. Oliveira, J. Big Data, Analytics e Inteligência Artificial, três Grandes Tendências para 2020. O Jornal Económico. 2020. Available online: https://jornaleconomico.pt/noticias/big-data-analytics-e-inteligencia-artificial-tres-grandes-tendencias-para-2020-538616/ (accessed on 6 March 2023).
39. Trends. Interest in Big Data over Time. 2016. Available online: https://trends.google.pt/trends/explore?date=all&q=big%20data (accessed on 6 March 2023).
40. IBM. Put AI to Work in the Real World. 2023. Available online: https://www.ibm.com/us-en (accessed on 6 March 2023).
41. Chen, H.; Chiang, R.H.; Storey, V.C. Business Intelligence and Analytics: From Big Data to Big Impact. *MIS Q.* **2012**, *36*, 1165. [CrossRef]
42. Someh, I.; Shanks, G.; Davern, M. Reconceptualizing synergy to explain the value of business analytics systems. *J. Inf. Technol.* **2019**, *34*, 371–391. [CrossRef]
43. Davenport, T.H. From analytics to artificial intelligence. *J. Bus. Anal.* **2018**, *1*, 73–80. [CrossRef]
44. Faúndez, M.O.; de la Fuente-Mella, H. Data Analysis and Domain Knowledge for Strategic Competencies Using Business Intelligence and Analytics. *Mathematics* **2022**, *11*, 34. [CrossRef]
45. Santos, M.Y.; Oliveira e Sá, J.; Costa, C.; Galvão, J.; Andrade, C.; Martinho, B.; Lima, F.V.; Costa, E. A big data analytics architecture for industry 4.0. *Adv. Intell. Syst. Comput.* **2017**, *570*, 175–184. [CrossRef]

46. Armbrust, M.; Ghodsi, A.; Xin, R.; Zaharia, M.; Berkeley, U. Lakehouse: A New Generation of Open Platforms that Unify Data Warehousing and Advanced Analytics. In Proceedings of the 11th Annual Conference on Innovative Data Systems Research (CIDR '21), Virtual Event, 11–15 January 2021.
47. Monteiro, B.; Lima, R.; Ramos, L.C.S.; Prado Da Fonseca, V.; Claudio, L.; Ramos, S.; Eustaquio, T.; De Oliveira, A.; Petriu, E.M. Heart Rate Detection Using a Multimodal Tactile Sensor and a Z-Score Based Peak Detection Algorithm Stable Design of Fuzzy Controllers View Project Heart Rate Detection Using a Multimodal Tactile Sensor and a Z-Score Based Peak Detection Algorithm. *CMBES Proc.* **2019**, *42*. Available online: https://proceedings.cmbes.ca/index.php/proceedings/article/view/850/843 (accessed on 8 March 2023).

infrastructures MDPI

Article

Installation and Use of a Pavement Monitoring System Based on Fibre Bragg Grating Optical Sensors

Francisco J. P. Rebelo [1], Joel R. M. Oliveira [1,*], Hugo M. R. D. Silva [1], Jorge Oliveira e Sá [2], Vânia Marecos [3] and João Afonso [4]

1 Department of Civil Engineering, Institute for Sustainability and Innovation in Structural Engineering, University of Minho, 4800-058 Guimaraes, Portugal; id9826@alunos.uminho.pt (F.J.P.R.); hugo@civil.uminho.pt (H.M.R.D.S.)
2 ALGORITMI Research Centre, Department of Information Systems, University of Minho, 4800-058 Guimaraes, Portugal; jos@dsi.uminho.pt
3 Laboratório Nacional de Engenharia Civil, 1700-066 Lisboa, Portugal; vmarecos@lnec.pt
4 Mota-Engil Group, 4300-454 Porto, Portugal; joao.afonso@mota-engil.pt
* Correspondence: joliveira@civil.uminho.pt; Tel.: +351-253-510200

Abstract: The evolution of technological tools, namely affordable sensors for data collection, and the growing concerns about maintaining roads in adequate conditions have promoted the development of continuous pavement monitoring systems. This paper presents the installation and use of an innovative pavement monitoring system, which was developed to measure the effects of vehicle loads and temperature on the performance of a pavement structure. The sensors used are based on fibre Bragg grating optical technology, collecting data about the strains imposed in the pavement and the temperature at which those measurements are made. The site selection for the system's installation and the essential installation details to ensure successful data collection are addressed. A calibration procedure was implemented by performing falling weight deflectometer tests and passing preweighed heavy vehicles over the sensors. In addition to validating the system installation, the results obtained in the calibration confirmed the importance of adequately choosing the distance between sensors. Differences of 50 mm in the position of the load may cause differences of about 20% to 25% in the resulting strains. These results confirmed the importance of increasing the sensor concentration in wheel paths. Furthermore, for loads between 25 kN and 65 kN, raising the temperature by 8 °C caused an increase of about 20% in the horizontal tensile strains measured in the pavement. In summary, it was possible to conclude that this innovative system is capable of capturing the effects of temperature and vehicle speed on the response of the pavement, which may be considered an advantage of this type of monitoring system when compared to those that are only used to determine the loads applied to the pavement or to characterise the type of vehicle.

Keywords: road pavements; monitoring; fibre-optic sensor; fibre Bragg grating (FBG); strains

1. Introduction

Roadways are essential infrastructures for the development of any country, representing a large part of the investment made in public assets. Due to the growing population and consequent goods consumption, transportation needs, especially on highways, are also increasing, demanding a higher investment in constructing and maintaining this essential infrastructure [1]. The preservation of road pavement state depends on several factors, including its design process, composition, and sound maintenance and rehabilitation policies. These policies are even more critical with the growing concern for sustainability.

During their lifecycle, many road pavements have to support loads significantly higher than those for which they were designed, which leads to faster degradation than initially expected [2]. Ai et al. [3] studied the influence of various factors such as axle configuration, axle load, speed, and temperature on different types of asphalt pavements with specific vehicles, showing that pavement strains increased with increasing temperature and axle load. Combining these effects leads to fatigue damage of the surface layers, decreasing the structure's lifetime.

Thus, improved road monitoring techniques are necessary to ensure traffic safety and quality throughout the pavement lifecycle [4]. Some of the main road pavement distresses can be assessed through visual inspection. However, the extensive length of any road network hinders this task, making it difficult to correctly and timely determine the pavement condition. Thus, structural health monitoring techniques can be used to save human resources. One advantage of these techniques is the possibility of obtaining a real-time understanding of road conditions by collecting data on the structural integrity of the entire infrastructure and not just of its upper layers [5].

Over recent decades, various monitoring methods have been applied to road pavements, potentially analysing parameters such as strain, stress, temperature, moisture, and deflection. In 1991, Sebaaly et al. [6] applied different types of sensors (pressure cell, strain gauge, moisture sensor, and transverse vehicle location sensor) to evaluate the effect of passing trucks. In Huff et al. [7], the response given by piezoelectric sensors was studied to obtain dynamic pavement deflection data. During one year of monitoring, Bayat et al. [8] also used strain gauges and temperature sensors on a test track to measure strains induced by both the temperature and the application of different vehicle loads. Also, in the work presented by Duong et al. [8], an asphalt pavement section subjected to heavy traffic (around 4500 trucks per day) was instrumented with strain gauges, geophones and temperature probes and monitored continuously for 18 months, concluding that temperature and the degree of bonding between the various pavement layers play a significant role in the response given.

More recently, optical sensors with FBG technology have been used instead of strain gauges in different structural monitoring systems, including pavements, because they present several advantages compared to other measuring systems. Liu et al. [9] tested the dynamic strain response of different asphalt pavement structures according to the base course. The sections were instrumented with FBG sensors, and the results generated exciting conclusions, including that the loads' position significantly influences the peak value of dynamic strain response. In the study presented by Kara De Maeijer et al. [2], focusing on heavy-duty pavements, a prototype monitoring system based on FBG sensors was installed on a CyPaTs test track and in the port of Antwerp. This system comprised a series of sensors that captured strain and temperature information between four asphalt layers to better understand how these layers respond under heavy loads. The sensors were applied prior to the laying of the asphalt layers. The results proved the potential of this solution to monitor pavement responses and demonstrated the best procedures for applying the sensors on the pavement.

Tan et al. [10] compared the results obtained with FBG sensors and strain gauges to improve the optical sensors' calibration method. By monitoring the dynamic response in three-point bending tests on small beams, the tensile strain of the beam bottom given by the strain gauge was compared with the theoretical results and the results given by the FBG sensors. According to specific and well-defined loading conditions, it was possible to relate the peak values between the two technologies and calibrate the equation obtained for the FBG sensors.

Over the past few years, temperature compensation in FBG sensors has been extensively studied using solutions based on distributed optical fibre sensors. The studies that have been conducted prove the importance of the temperature effect on FBG sensors, highlighting the need for this process to be performed with caution using different techniques and ways of combining optical sensors [11–14]. Wang et al. [15] performed laboratory tests on samples to propose a method to improve the temperature compensation, considering the interfacial interaction between the structure and the bonded FBG according to different temperature and loading conditions. Leal-Junior et al. [16] developed an FBG sensor based on a single polymer diaphragm for measuring pressure and temperature and compared the results obtained with a transient heat conduction model. This study concluded that the root mean square error was higher in the presence of variation in both parameters, the error being 3.88 °C and 5.13 kPa for temperature and pressure, respectively.

Han et al. [17] studied the deformation compatibility between the embedded strain sensor and asphalt layer, concluding that it is the key to ensuring the precise measurement of mechanical response. However, good deformation coordination may be difficult to maintain under different environments due to the viscoelasticity of asphalt mixture. They concluded that strain sensors with stiffness moduli similar to the asphalt layers are recommended in the dynamic response monitoring of pavement structures. Similar conclusions were obtained by Liu et al. [18] in a study where FBG sensors were compared to resistive sensors through laboratory and finite element modelling processes. They also observed that FBG sensors are more appropriate to measure horizontal strains when compared to resistive sensors.

The work discussed in the present manuscript aims to present an innovative pavement monitoring system based on FBG optical sensors implemented on a highway section in Portugal. This solution allows real-time monitoring of the mechanical conditions of the structure, measuring the effects of each traffic load on the pavement near the bottom of the asphalt layer, where cracking problems usually begin. This information was unavailable through other existing monitoring solutions and is essential to improve our knowledge regarding the causes of pavement distress. The subsequent data collection and analysis will help in scheduling future maintenance procedures. Over the last few months, several steps were carried out to prepare for the installation of the monitoring system. These steps included defining the system's architecture, selecting the type and quantities of sensors, the laboratory calibration of the sensors (by performing four-point bending and wheel tracking tests) [19], and validation of the installation procedures in a small trial section built at the National Laboratory of Civil Engineering.

2. Characteristics of Optical Sensors

2.1. Fibre Bragg Grating Sensors

In the second half of the twentieth century, the development of optical technology and its sensors revolutionised the telecommunications industry due to its enormous advantages over the dominant technologies. As time went by, and the level of knowledge of the technology increased, its application was extended to several areas, including the monitoring of structures [20].

The various characteristics of fibre-optic sensors make their use in pavement monitoring a very suitable solution. These include immunity to electromagnetic fields, high sensitivity, small size, and resistance to harsh environments when adequately protected [5,15]. Fibre Bragg grating sensors are one type of sensor based on fibre optics and are incredibly reliable [21]. The multiplexity of this technology is another feature differentiating FBG sensors because several Bragg wavelengths can be defined for numerous sensors placed in series on a single fibre-optic cable. In this situation, only the spectral band of each sensor must be respected so that the reflected signals do not cross. This feature allows a single optical fibre over tens of meters, instrumented with many sensors, which can even measure several parameters in addition to deformation (strains) and temperature, such as acceleration and pressure [16,22].

However, in addition to being expensive, the sensors and the optical fibre become fragile if not adequately protected. It is expected that these disadvantages will tend to disappear or become less relevant as technology evolves.

This type of sensor is created by exposing a portion of optical fibre a few millimetres long to a UV laser beam. The marking will change some of the optical characteristics of the fibre, with great importance for the refractive index. This change in the refractive index will cause only a portion of the incoming light to be reflected (based on the new refractive index), creating a characteristic wavelength for each sensor [23].

When the fibre-optic cable is stretched or compressed, the characteristic Bragg wavelength of the reflected light will change. This value is read by a piece of equipment called an interrogator, which recognizes the characteristic wavelength and the new characteristic Bragg wavelength. The interrogator then converts the wavelength variation into length units, and it is possible to interpret the variation in fibre length or strain.

This technology has the characteristic that only the Bragg wavelength defined within a narrow spectrum will be reflected and received by the interrogator, causing the other signals from other sensors with different wavelengths to propagate along the optical fibre with only residual variations [5].

For the sensor to function, it is essential that the marking remains intact, not disturbing the defined Bragg wavelength. Equations (1)–(4) [20,24,25] can be used to calculate the strains imposed by the external deformations and temperature variations when using FBG sensors. Equation (1) proves that the central wavelength of the reflected signal corresponds to the Bragg condition:

$$\lambda = 2 \times n_{eff} \times \Lambda, \tag{1}$$

where λ is the Bragg wavelength, n_{eff} is the effective refractive index, and Λ is the microstructure period. The parameters n_{eff} and Λ are susceptible to external disturbances such as temperature and deformation, and changes in these indicators will determine changes in the Bragg wavelength.

Equation (2) translates the strain variation due to an external deformation:

$$\frac{\Delta\lambda_1}{\lambda_1} = \frac{\Delta\lambda_\varepsilon}{\lambda_1} = (1 - P_e) \times \varepsilon, \tag{2}$$

where $\Delta\lambda_\varepsilon$ is the wavelength due to the deformation, ε is the longitudinal deformation, and P_e is the effective photo-elastic constant from the fibre's core.

Equation (3) translates the variation in wavelength due to temperature variations:

$$\frac{\Delta\lambda_2}{\lambda_2} = \frac{\Delta\lambda_T}{\lambda_2} = (\alpha + \xi) \times \Delta T, \tag{3}$$

where $\Delta\lambda_T$, ΔT, α, and ξ are, respectively, the wavelength variation due to temperature, the temperature variation, the coefficient of thermal expansion, and the thermo-optic coefficient related to the change in refractive index with temperature.

Equation (4) represents the junction of the deformation and temperature effects in an FBG sensor.

$$\frac{\Delta\lambda}{\lambda} = \frac{\Delta\lambda_\varepsilon}{\lambda_1} + \frac{\Delta\lambda_T}{\lambda_2} = (1 - P_e) \times \varepsilon + (\alpha + \xi) \times \Delta T. \tag{4}$$

These sensors cannot distinguish whether wavelength changes occur due to temperature or external load effects (deformations), so it is necessary to decouple both effects. Therefore, specific procedures have been created to compensate for the effect of temperature on strain measurements. Typically, an additional FBG temperature sensor is installed (only affected by temperature) in a series with FBG strain sensors, and its results are used as input for a temperature compensation algorithm.

Consequently, it is possible to particularise the deformation caused only by external loads in the FBG sensor by knowing the combined effect of load and temperature deformation in the sensor and compensating for the component corresponding only to the temperature deformation. However, calibration is essential to ensure the quality of the results obtained by these sensors.

2.2. Fibre Bragg Grating Coating and Protection

The literature review demonstrated that one of the main difficulties in using FBG sensors is ensuring the correct stress transmission between the asphalt material and the sensors. Nevertheless, many solutions to overcome this problem have been developed in the last few years.

For example, Kara De Maeijer et al. [26] tested two ways to apply FBG sensors on asphalt layers. The first consisted of implementing sensors on prefabricated asphalt samples in the base layer, and the second consisted of installing the same sensors directly on the surface of the previously mentioned asphalt layer. The survival rate of the sensors was 100%, indicating that these methodologies can be applied in monitoring heavy-duty asphalt pavements.

Xiang and Wang [27] took a different approach concerning the FBG sensor coating by developing an asphalt mastic containing an FBG sensor inside to monitor distributed strains in beams. The theoretical analysis of strain transmission was used to improve the encapsulation design and reduce measurement error. The prototype was also tested at standard temperature and traffic loads, indicating that the proposed solution obtained satisfactory laboratory and in situ results.

Zhou et al. [24] developed an FBG sensor packaged in fibre-reinforced polymer for 3D structural strain monitoring, with optimised dimensions according to the road structure. Comparisons of the actual results with the simulations proved that the developed FBG sensor provided effective and reliable information about the strain distribution on the pavement.

The approach used in the current work consisted of initially performing laboratory tests on beams and slabs with FBG sensors with different encapsulations and resins from which the sensors' readings were calibrated. The laboratory study also defined the solution for installation on the actual pavement. Therefore, a fibreglass rod was instrumented with a series of FBG sensors and used as a support for the optic cable, and the set was then inserted in a groove cut in the pavement that was subsequently filled with an appropriate resin (QuiniResin Fix) [28] selected during the laboratory work. This approach for the post-installation of FBG sensors in road pavements, transversally to the traffic direction and embedded in a fibreglass rod (for protection), can be considered an innovative solution since no other similar application has been found in the literature, namely, in a recent review on this subject [29].

3. Methods for Developing the Pavement Monitoring System

3.1. Data Acquisition System (Interrogator)

The optical interrogator is a piece of equipment essential for operating the monitoring system with FBG sensors and can perform static (e.g., temperature) or dynamic (e.g., traffic load) measurements. This optoelectronic device works as a measurement unit/data acquisition system, reading the signals reflected by the FBG sensors.

A single device can receive data from dozens of FBG sensors due to the multiplexing characteristic of the technology and the various reading channels of the device. This feature allows the same device to collect data on different parameters, even at different acquisition rates. A BraggMETER interrogator (Figure 1) was used in the present work, with a maximum acquisition rate of 1000 readings per second. Additionally, this type of FBG technology presents a very high resolution/repeatability (<1.5 pm) and stability/reproducibility (5 pm) [23].

The Catman data acquisition program is the native software used by this equipment. This tool is used to visualise and analyse the information collected by the interrogator from the various sensors. The analysis can be performed in real time or post-processing. The adaptability of the software to the needs of each situation is a critical factor for its use, as it is possible to customise it according to the objectives of each use through the creation of different graphs, tables, and other forms of visualisation. Nevertheless, other data processing and analytical tools can also be used for the same purpose [30].

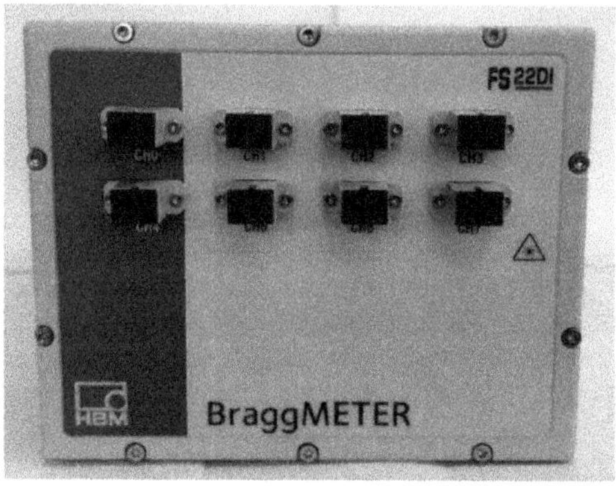

Figure 1. Optical interrogator used in the present work.

3.2. Pavement Monitoring System Architecture

After defining the sensor technology to be used and identifying some of the main challenges that its installation would involve, the monitoring system's architecture was defined, covering features such as:

- The type of information to be collected (which determines the type of sensors to be used);
- The location of the sensors;
- The use of protection, coatings, or resins;
- The application procedures;
- The interrogator connection to the communication network infrastructure.

Given the strength and flexibility of fibreglass rods, this material was used to protect the optical sensors' physical integrity. Thus, the rod (with a circular cross-section) was machined to create a notch where the fibre-optic cable containing the sensors would be inserted and then glued with a resin suitable for this purpose, as illustrated in Figure 2. A diameter of approximately 4 mm was chosen for the fibreglass rod. The fibre-optic cable had a diameter of approximately 125 μm.

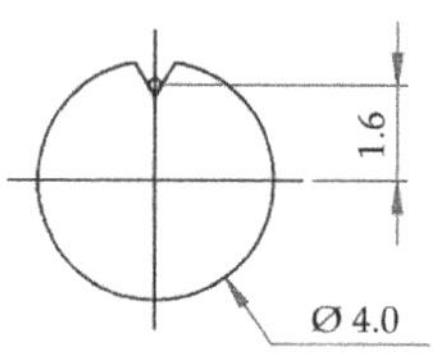

Figure 2. Fibreglass rod and fibre-optic cable cross-section.

The number of sensors each fibreglass rod would have, the respective spacing between the sensors, and the number of rods to be installed in the pavement section were defined according to the specific characteristics of the pavement of this project's selected site. Thus, two types of rods were defined according to the number and spacing of sensors. The first type comprised fifteen strain sensors and one temperature sensor, and the second type included seven strain sensors and one temperature sensor. In order to monitor the performance of two traffic lanes and the measurement of the traffic speed, each lane was instrumented with two fibreglass rods (i.e., one of each type).

In order to increase the valuable information obtained from the monitoring system and optimise its cost, each traffic lane was instrumented with one of the rods with more sensors and another with fewer sensors. Each lane should have at least two rods to determine the speed of the passing vehicles. Furthermore, one of these rods should comprise many sensors to cover the lateral distance more efficiently and allow a better understanding of the transversal strain distribution resulting from the pavement response when subjected to vehicle loads.

The following criteria were taken into account to choose the spacing between the sensors:

- Each traffic lane is about 3.35 m wide;
- The width of a heavy vehicle is approximately 2.55 m;
- The average width of one heavy vehicle tyre is about 30 cm.

In order to obtain a better characterisation of the loads and consequent strains registered in the pavement, through the analysis of the transversal strain basins, a higher density of FBG sensors was considered for the positions corresponding to the common pavement wheel path.

Figure 3 schematically shows the final architecture of each type of instrumented rod used in the pavement monitoring system.

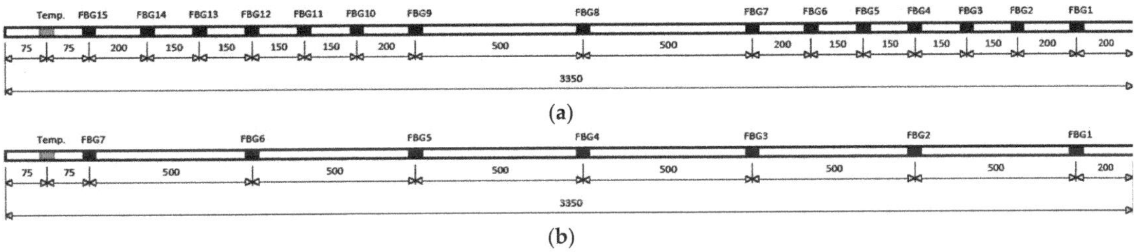

Figure 3. Architecture of the monitoring system with the position of each FBG sensor (in mm): (**a**) rod B with fifteen strain FBG sensors and one temperature FBG sensor; (**b**) rod A with seven strain FBG sensors and one temperature FBG sensor.

As shown later, these fibreglass rods instrumented with FBG sensors were installed in a groove cut in the pavement. The width of the groove and the resin used to fill it were defined in previous phases of the project, both in the laboratory and in the intermediate physical model installed at the National Laboratory of Civil Engineering (LNEC) facilities. Thus, the groove in the pavement section would have a width of 1 cm, and the resin chosen to fill it was the QuiniResin Fix, which is a high-performance polymer adhesive used in various highway applications (filling of grooves, sealing of cracks, bonding of several elements) [28].

3.3. Site Selection for Pavement Monitoring System Installation

The site selected to implement the pavement monitoring system (Figure 4) is located in the country's northeast on the IC5 highway, which belongs to the Portuguese national road network.

This site would have to meet several criteria given the stated objectives and characteristics of the monitoring system, among which the following can be highlighted:

- Easy access and connection to an electrical power supply;
- Access to an underground infrastructure to route the fibre-optic cables;
- Presence of a roadside technical cabinet nearby to install the interrogator;
- Access to a communication network in the technical cabinet to transfer the data collected in the interrogator to an external database.

An additional advantage of the selected site is being relatively close to an assistance and maintenance road administration centre, allowing easy and quick access to the equipment should any operation be necessary.

The location where the monitoring system was installed has a three-lane carriageway, two lanes in the uphill gradient direction and one in the downhill gradient direction. The monitoring system was installed in the two uphill lanes to compare the strains measured in a fast (left) and a slow (right) traffic lane, as shown in Figure 5.

Figure 4. Location selected for the installation of the monitoring system.

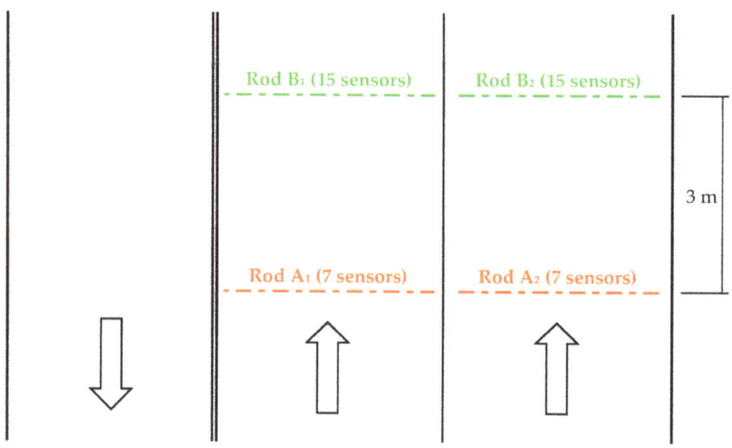

Figure 5. Schematic representation of the monitoring system location with the position of the instrumented fibreglass rods.

Considering that heavy vehicles generally circulate in the slow lane, the monitoring system would be used to compare the effects of different loads and speeds on pavement performance.

3.4. Pavement Monitoring System Installation

The research carried out in the laboratory at an early stage of this work was essential to identify the potential problems that could be faced in this application. Thus, on the one hand, the design of the monitoring system should ensure the mechanical strength necessary to withstand the vehicle loads and, on the other hand, the system should be able to assure an efficient transmission of strains/stresses between the pavement, the filling resin, and the instrumented rod. Therefore, the monitoring system was installed according to the recommendations from the laboratory study and the small trial performed at the national civil engineering laboratory. The details of the installation are given below.

First, two grooves were made in the pavement with a depth of approximately 14 cm and 3 m apart (Figure 6), where four instrumented rods of the monitoring system were later installed. Rods A_1 and A_2, with seven strain sensors, were installed in the first groove, while rods B_1 and B_2, with fifteen strain sensors, were installed in the second groove, as shown in Figure 5.

(**a**) (**b**) (**c**)

Figure 6. Initial phases of the pavement monitoring system installation: (**a**) markings for the precise positioning of the system; (**b**) the process of cutting the grooves; (**c**) visual inspection of one of the grooves.

The groove depth was defined to install the sensors near the bottom of the asphalt layers where the maximum tensile strains causing fatigue cracking occur. The distance between the two grooves is essential to assess the traffic speed by knowing the elapsed time between the consecutive strain peaks measured in rod A and B sensors due to traffic. The selection of a smaller distance would have reduced the reliability of speed calculation, while higher distances could have caused some data loss due to vehicles changing lanes (e.g., during overtaking manoeuvres).

As mentioned previously, the width of the grooves was already set at 1 cm, as well as the type of resin that would be used to fill them after installing the instrumented rods. The cutting of the pavement grooves was performed without water to ensure the resin perfectly adhered to the groove and the instrumented rod. Furthermore, after the dry cutting, the groove was cleaned using an air blast to apply the resin on a dry surface free of dust and dirt, ensuring the best performance of the resin.

After cleaning the grooves, the four instrumented rods and their fibre-optic cables were installed. Considering that the grooves were cut across the entire width of both traffic lanes, two distinct rods were positioned in each groove (one on the fast traffic lane and the other on the slow traffic lane). Thus, it was necessary to overlap the fibre-optic cables of the rods on the fast (left-hand side) lane over the rods on the slow (right-hand side) lane and join the four cables on the right-hand side of the road.

In order to ensure that the fibreglass rod would not move during the resin application, several plastic fixing pieces were developed by the system supplier. These pieces were built to fit the groove width and were fixed to the rod (Figure 7) to prevent it from rotating or moving once laid, thus ensuring the sensors would be installed in the expected position.

As can be observed in Figure 8, the four cables were routed through auxiliary grooves made in the shoulder towards the corrugated pipe that connects to the technical cabinet where the interrogator was located.

After the instrumented fibreglass rods and cables had been positioned correctly in the grooves, the resin was prepared by mixing two components: a polymer compound and a hardener. Filling the grooves with resin (Figure 9a) was simple since this material has good workability and can be handled at ambient temperature. Special care was taken when overlapping fibreglass rods A_2 and B_2 with the cables from fibreglass rods A_1 and B_1, placing enough resin among them to avoid future interference in the strains to be measured on the slow (right-hand side) traffic lane.

Figure 7. Positioning of the fixing pieces in the instrumented rods.

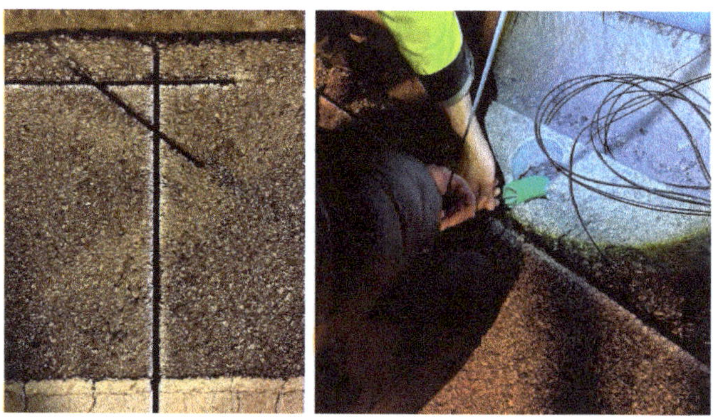

(**a**) (**b**)

Figure 8. Installation details: (**a**) groove made on the road shoulder to protect the cables; (**b**) introduction of the cables on the corrugated pipe that connects to the technical cabinet.

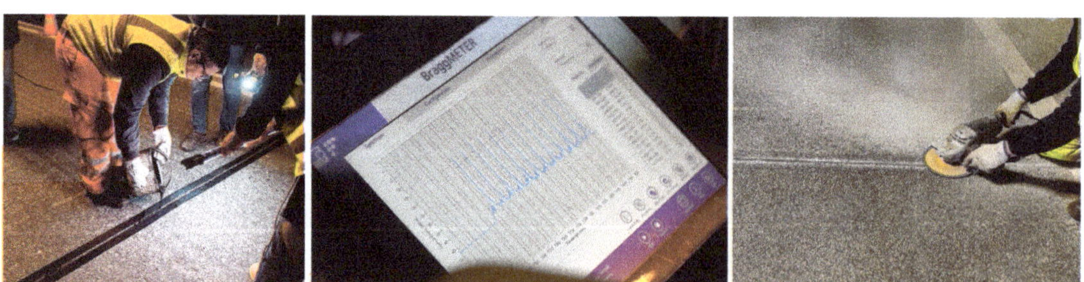

(**a**) (**b**) (**c**)

Figure 9. Final phases of pavement monitoring system installation: (**a**) filling of the grooves with resin; (**b**) FBG sensor signal strength confirmation; (**c**) rectification of the resin used in the grooves.

The relatively low ambient temperature influenced the resin curing time, extending the time available for mixing, handling, and applying the resin. Therefore, filling the grooves with resin did not present any particular problems.

After filling the grooves with resin, the four cables were routed into the technical cabinet through a corrugated pipe and connected to the optical interrogator. The technical cabinet is linked to the assistance and maintenance road administration centre through an Ethernet connection, assuring the data transmission into a server for collection and subsequent analysis. Once the connection of the optical cables was finished, the first tests were carried out by the FBG sensors' supplier (HBK company) with a portable interrogator (Figure 9b), which confirmed the excellent signal strength of the forty-four strain sensors and four temperature sensors.

The last step of the installation was the resin rectification (Figure 9c) to level the pavement surface where the two grooves were made before reopening both lanes to regular traffic. Naturally, the resin only reached the necessary hardening point for traffic reopening after a few hours due to the low ambient temperature, increasing the duration of the entire installation process. The monitoring system installation took about nine hours to complete.

3.5. *Pavement Monitoring System Calibration*

The data collection process began after the monitoring system based on FBG sensors was installed. After a few initial adjustments, the system continuously picked up the pavement's strain data. Due to the high sampling rate (i.e., equal to or higher than 500 samples per second) required to record all the relevant information from dynamic loads applied on the pavement, a vast amount of data was generated that needed to be filtered, treated, and analysed.

Regarding the analysis, knowing that pavements are structures whose behaviour is influenced by various factors, a proper calibration procedure should be carried out to validate the information provided by the monitoring system and understand the performance of the pavement. This calibration procedure aims to analyse the behaviour of the pavement as a whole and, in particular, the most influential factors, such as the magnitude and position of the surface applied loads and the pavement temperature. Therefore, calibration tests were performed with a falling weight deflectometer (FWD) and based on heavy vehicles of known weights passing over the monitoring system.

3.5.1. Calibration with Falling Weight Deflectometer Tests

The first calibration tests were performed using the falling weight deflectometer (FWD), as shown in Figure 10. The main objective of these tests was to analyse the behaviour of the pavement at different temperatures and for different load levels.

(a)

(b)

Figure 10. Falling weight deflectometer: (**a**) view of the entire equipment; (**b**) carrying out a test.

Five test campaigns were carried out throughout the day at different temperatures in this first type of calibration. Each test campaign included six positions per rod, twelve in total, where the equipment plate was laid (Figure 11). Four increasingly higher loads were applied in each position, according to the weight falling height.

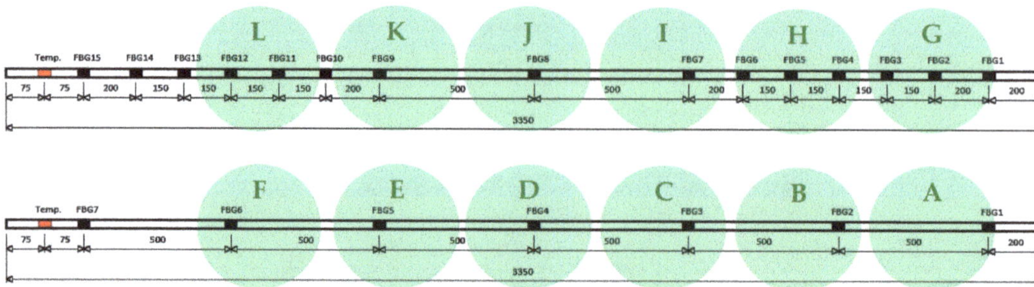

Figure 11. Scheme of each load application position in the FWD calibration tests (in mm), with letters A to L representing the load application points during the tests.

The tests were performed on both instrumented rods (A2 and B2) installed on the slow (right-hand side) lane to allow the traffic to circulate in the other lane. With this procedure, it was possible to understand if the higher number of sensors in rod B2 could provide more reliable results.

3.5.2. Calibration for Heavy Vehicles Passing with Known Loads

In addition to the tests performed with the FWD, the calibration process also aimed to study the effect of heavy vehicle loads on the pavement's behaviour. This task used two trucks with known weights (Figure 12).

(a) (b)

Figure 12. Heavy vehicles used in the calibration tests: (**a**) Heavy Vehicle 1; (**b**) Heavy Vehicle 2.

Thus, knowing the weather conditions observed during the calibration test, it is possible to establish a relationship between the strains measured on the pavement and the weight of each vehicle. Vehicles 1 and 2 had a total weight of 29.66 tons and 15.66 tons, respectively.

3.6. Analysis of the Results Obtained during the Monitoring System Calibration

The interpretation of the results obtained during the system calibration depended on the synchronisation of data obtained by different means (e.g., FWD results and strain measurements in the sensors) and the subsequent data analysis to be performed. The four instrumented rods were divided into four channels to facilitate data analysis.

The Catman software (catmanEasy version 5.6.1.12) was used to visualise the real-time strains in the FBG sensors and collect the on-site data, subsequently analysed using Matlab (version R2023a Update 3) and Microsoft Excel (for Microsoft 365 MSO version 2211 Build 16. 0. 15831. 20098) software.

The falling weight deflectometer test results and the corresponding FBG strains registered in the monitoring system were analysed to determine the amplitude of the peak strains corresponding to each load application. Temperature also plays a critical role in this analysis. The distance of each sensor to the FWD plate centre was estimated by knowing the position of the equipment in each load application, which was then used to obtain the transverse strain basins and understand the pavement behaviour.

The calibration tests performed with heavy vehicles were used to assess parameters such as the vehicle's speed and obtain strain basins in both transverse and longitudinal directions due to the dynamic effect of the moving loads.

The pavement response under different types of axle loads is another variable that will be addressed in the results section by analysing the strain peaks caused by the passage of single- and double-wheeled axles.

4. Results and Discussion

4.1. Falling Weight Deflectometer Test Results

4.1.1. Sensor Sensitivity

The position of each sensor in the pavement was determined during the installation, considering the distance between the pavement marking and the beginning of the instrumented rod and the internal distance among the sensors specified in Figure 3. Thus, based on the position at which the FWD loading plate was located (Figure 11), it was possible to estimate the distance between the centre of the plate and the FBG sensors. Consequently, Figure 13 compares the maximum strain values measured when applying the 25, 35, 45, and 65 kN loads at horizontal distances of 1 cm and 6 cm from a specific sensor (measured from the centre of the FWD plate).

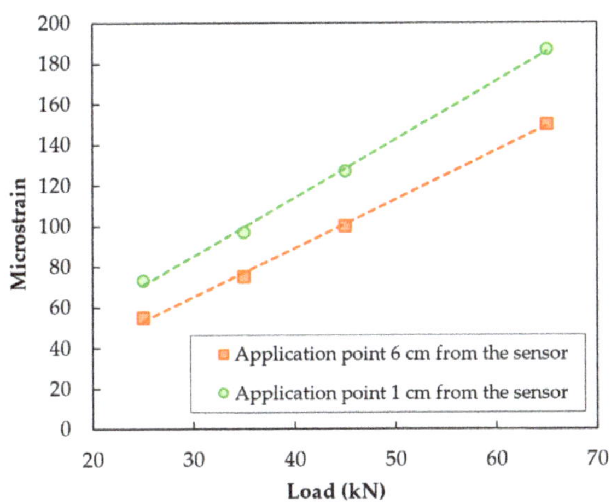

Figure 13. Influence of the distance from the load to the sensor on the measured strains.

Increasing the distance between the load and the sensor by 5 cm causes a reduction in the maximum strain of between 20% and 25%. The results show that the pavement response, measured by the sensors' horizontal strains, is very sensitive to the distance from which the load is applied. Thus, the number of sensors included in the instrumented rods (and the consequent internal distance) will be critical in detecting the accurate peak strain values resulting from each vehicle load application on the pavement.

Furthermore, it is possible to observe that a linear relationship exists between the strain and the load, which is discussed later in this manuscript.

4.1.2. Temperature Influence

Several FWD tests were performed during a significant part of the day to evaluate the influence of ambient and pavement temperature on the strain level measured by the FBG sensors embedded in the pavement under different weather conditions. This information is essential for asphalt pavements since they exhibit viscoelastic behaviour and are susceptible to temperature and loading frequency.

Figure 14 presents the maximum strain values measured by the monitoring system in sensor 5 of rod A_2 for FWD tests performed in position E (Figure 11) to exemplify the strain variation with temperature. The horizontal distance between sensor 5 of rod A_2 and the FWD load application point E is approximately 2 cm. The lowest pavement temperature registered by the monitoring system during the FWD calibration tests was 25 °C, measured at 11:00 a.m., and the highest temperature registered during those tests was 33 °C at 4:30 p.m. Thus, Figure 14 compares the maximum strain values measured when applying the 25, 35, 45, and 65 kN loads at two testing temperatures (i.e., 25 °C and 33 °C) measured by the FBG temperature sensor of the monitoring system.

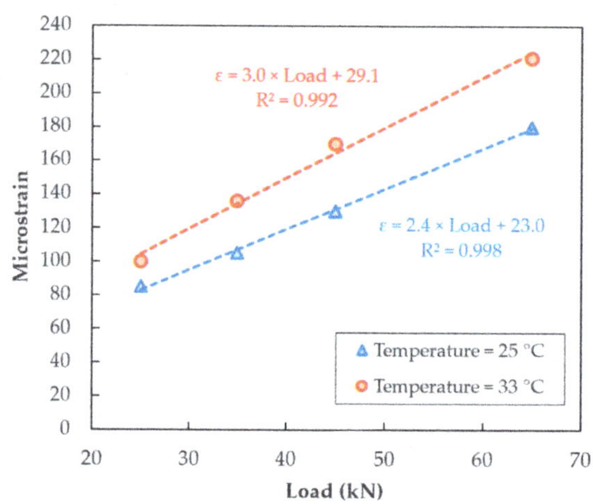

Figure 14. Influence of pavement testing temperature on the measured strains.

Pavements can accumulate large amounts of thermal energy from solar radiation, which may cause significant daily and annual temperature variations that influence pavement performance. This study measured a pavement temperature variation of 8 °C at a depth of about 13.5 cm from 11:00 a.m. to 4:30 p.m.

The load influence on the maximum strain values measured by the monitoring system was fitted through linear equations at both test temperatures. On average, for loads between 25 kN and 65 kN, the mentioned temperature rise of 8 °C caused an increase of about 20.3% in the horizontal tensile strains measured in the pavement. Therefore, FBG sensors are sensitive to pavement performance changes at different temperatures.

4.1.3. Relationship between Loads and Strains

All the loads (25, 35, 45, and 65 kN) applied in the twelve positions shown in Figure 11 were analysed for all the FWD test repetitions carried out throughout the day to study the evolution of the pavement strains with increasing loads.

Figure 15 shows the peak strain values measured by the monitoring system for six loading positions when the highest pavement temperature was recorded. Figure 15a concerns the loads applied on positions D, E, and F of rod A_2, while Figure 15b concerns the loads applied on positions G, H, and I of rod B_2.

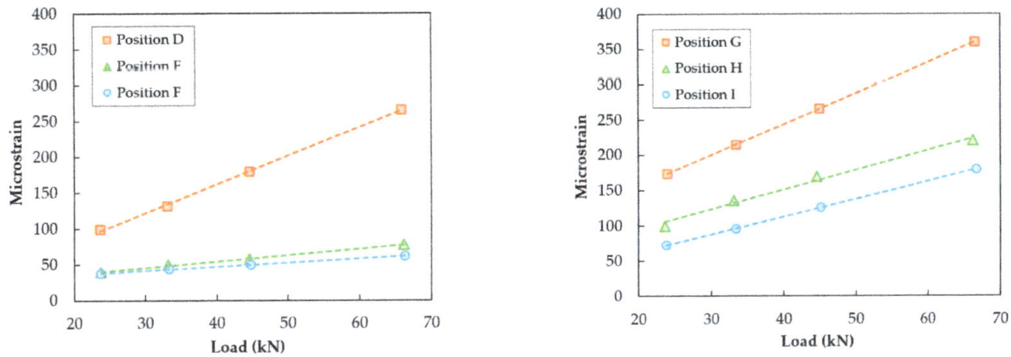

(**a**) (**b**)

Figure 15. Strain results obtained in the fifth FWD test campaign: (**a**) loads applied on positions D, E, and F; (**b**) loads applied on positions G, H, and I.

As can be concluded from the results shown in Figure 15, there is an evident linear variation in the strains with the load applied in all situations. However, the distance from the centre of the loading plate to the closest sensors varies, influencing the evolution of the strain values obtained for the different loads. The load applied in position G is very close to the vertical alignment of sensor 2 on rod B_2, while the distance of the closest sensor to the centres of the load in positions H (sensor 5) and I (sensor 7) is 3 cm and 4 cm, respectively. Many factors influence the strains recorded in pavements due to their complex behaviour, and sensors placed in diverse pavement locations may present different strain values under similar loading conditions. Nevertheless, a decrease in the strain value measured in the sensor is generally associated with a higher distance to the centre of the load, as can be seen in the three mentioned positions and as previously discussed.

Although it is possible to see a linear relationship between load and strain, positions E and F showed a marginal increase between the various loads applied. This result can be explained by the horizontal distance from the load application point to the closest sensor of rod A_2. In position D, the load centre is vertically aligned with sensor 4; position E represents a distance of 15 cm between the centre of the plate and sensor 5; finally, position F represents a distance of 21 cm between the centre of the plate and sensor 6. Despite the slight variation in the strains measured for the last two positions, they always increase with the load applied on the pavement surface, which is satisfactory. The results highlight the importance of the distance between the sensor and the position where the load is applied if the load value is to be estimated.

4.1.4. Transverse Strain Basins

The loading effect of traffic on road pavement, also simulated by the FWD tests, causes changes in its stress and strain state response noted up to a certain distance from the load application position. Therefore, the multiple sensors of the monitoring system were used in this work to evaluate the transverse strain basins resulting from loads applied by the falling weight deflectometer in specific positions.

Initially, Figure 16a represents the strain variation over time measured in sensor 2 of fibreglass rod A_2 for a 65 kN load applied with the FWD in position B (14 cm apart). Figure 16b shows a similar result registered in sensor 5 of fibreglass rod B_2 for the same load applied in position H (2 cm apart).

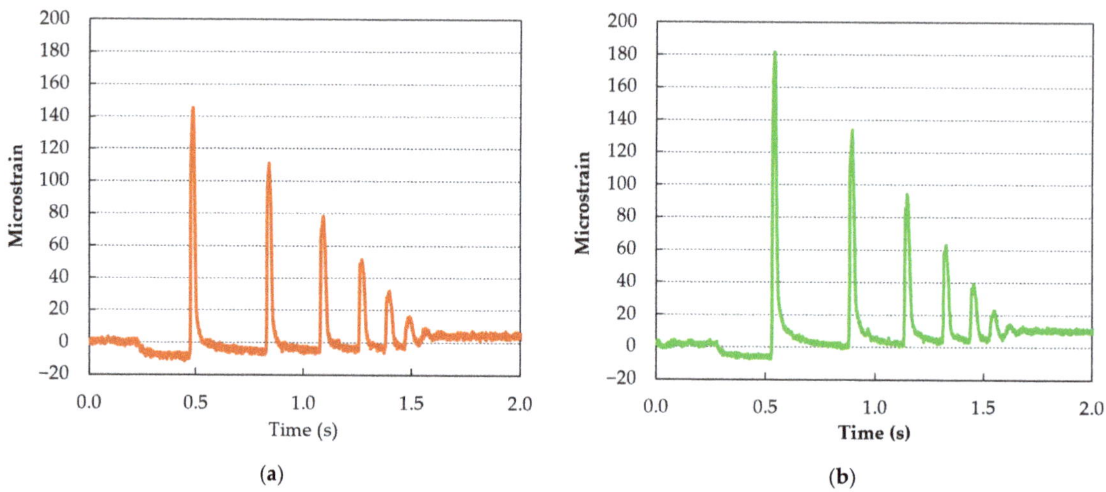

Figure 16. Strain variation over time measured by the monitoring system for a 65 kN load applied by the FWD: (**a**) sensor 2 of fibreglass rod A_2; (**b**) sensor 5 of fibreglass rod B_2.

The dynamic effect of the falling weight load on the pavement response is evident in these results because FBG sensors could register the strains caused by the several rebounds of the weight on the pavement. The maximum strain recorded in sensor 2 of rod A_2 was 145 $\mu\varepsilon$, while the maximum strain measured in sensor 5 of rod B_2 was 182 $\mu\varepsilon$. The smaller distance between the loading position H and sensor 5 justifies the increased strain values measured in that sensor. Thus, on average, the fibreglass rods with fifteen sensors (B_1 and B_2) are expected to measure higher peak strain values than those with seven sensors (A_1 or A_2) because the distance from a random traffic load position to the nearest sensor will be statistically lower when using more sensors in the monitoring system.

Subsequently, Figure 17 presents the maximum strain values measured in all the FBG sensors of each rod (A_2 and B_2) for the 65 kN load applied in positions B and H, respectively. The points represent the strain values registered in each sensor, and the line connecting these points can be described as the transverse strain basin.

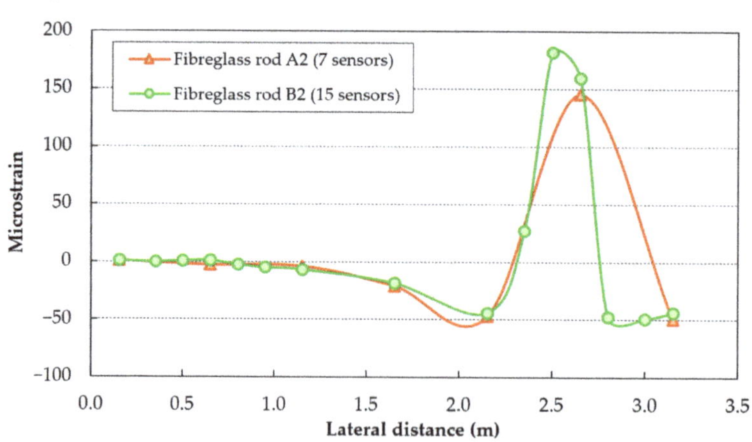

Figure 17. Comparison between the transverse strain basins measured by fibreglass rods A_2 and B_2 (with seven and fifteen sensors, respectively) for a 65 kN load applied by the FWD.

In fibreglass rod A_2 (in brown), with seven sensors, only sensor 2 (the second value from the right-hand side) recorded tensile (positive) strains because it was nearest to the load application point. The adjacent sensors (1 and 3) registered compression strains because they were 36 cm and 64 cm away from the load application point. These compression strains are expected to occur adjacent to the tyre loads in the bottom part of the asphalt layers, as described by Sudarsanan and Kim [31].

Regarding fibreglass rod B_2 (in green), with more sensors, three sensors (4, 5, and 6) registered tensile (positive) strains. The difference that an increase in the number of sensors causes in the results, namely in the strain basin shape, is significant, improving the understanding of the actual pavement response to load applications. In this case, sensor 4 was 13 cm away, sensor 5 was 2 cm away, and sensor 6 was 17 cm away from the centre of the loading plate. The shorter distance between sensors 4, 5, and 6 of rod B_2 also allowed a more precise definition of the strain peak than rod A_2 (sensor 2), measuring a strain value 24% higher than rod A_2.

The higher number of sensors included in rods B_1 and B_2 will be of great value in the comprehensive characterisation of the pavement performance over time. The more detailed strain basins also allow a better understanding of the compressive (negative) strains generated in the regions surrounding the loaded area of the pavement, which is not adequately characterised by the rods with fewer sensors due to the higher distances among the sensors. However, those rods with fewer sensors are essential to evaluate the traffic speed, as explained later in this manuscript, which affects the pavement response.

4.2. Results from the Dynamic Loading Effect of Heavy Vehicles

4.2.1. Strains Caused by Heavy Vehicles

Figure 18 shows the strains caused by the passage of Heavy Vehicle 1, registered in sensor 6 of rod B_1. This vehicle has a total mass of 29.66 tons, distributed over three axles (a single-wheeled front axle and two double-wheeled rear axles).

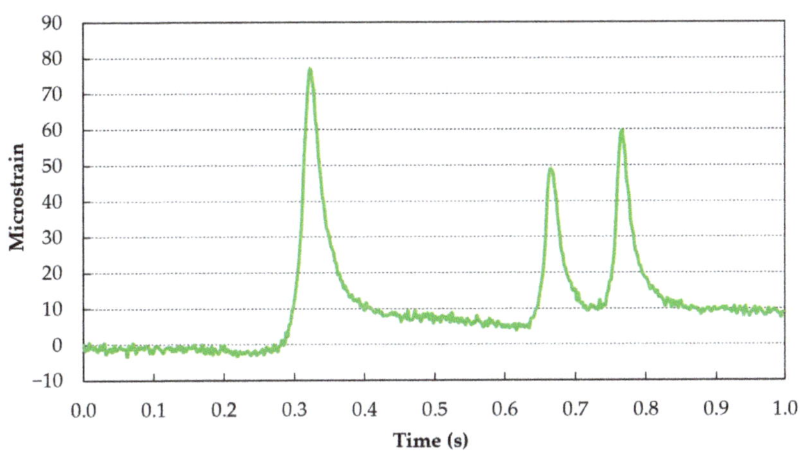

Figure 18. Strain record of the passage of Vehicle 1 over sensor 6 of rod B_1.

As can be seen from the graph, the front axle caused more significant strain than the rear axles, which may be related to the higher load distribution enabled by the rear axles that are close to each other and both of which possess a double-wheeled configuration, thus assuring a larger contact area between the tyres and the pavement and a lower stress applied to the surface. Among the rear axles, the last axle also caused peak strains higher than the first one, which is explained by the viscoelastic behaviour of the asphalt pavement, where the time between loads of both axles is not enough to allow the pavement to fully recover from the deformation (strain) imposed by the first axle before the load of the second axle is applied.

The vehicle speed calculation (Figure 19) considers the distance between the two instrumented rods (3 m) and the time it takes for one axle to be recorded by rod A_1 (first peak in the graph) and rod B_1 (second peak), which in this case is 0.204 s. Thus, as demonstrated in Equation (5), the speed of this truck (Vehicle 1) was 52.9 km/h.

$$s = \frac{\Delta d}{\Delta t} = \frac{3\text{ m}}{0.204\text{ s}} = 14.7\text{ m/s} = 52.9\text{ km/h} \quad (5)$$

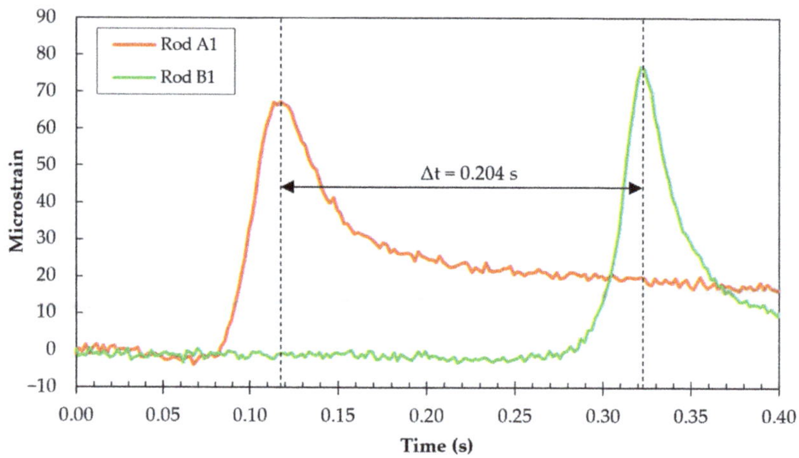

Figure 19. Overlapping of the strain signals registered by two sensors from rods A_1 and B_1 during the passage of Vehicle 1.

Figure 20 shows the strains caused by Heavy Vehicle 2 in one of its passages over the monitoring system and registered by sensor 3 of rod B_2. These graphs can be used to estimate the distance between the vehicle's axles after computing the vehicle speed.

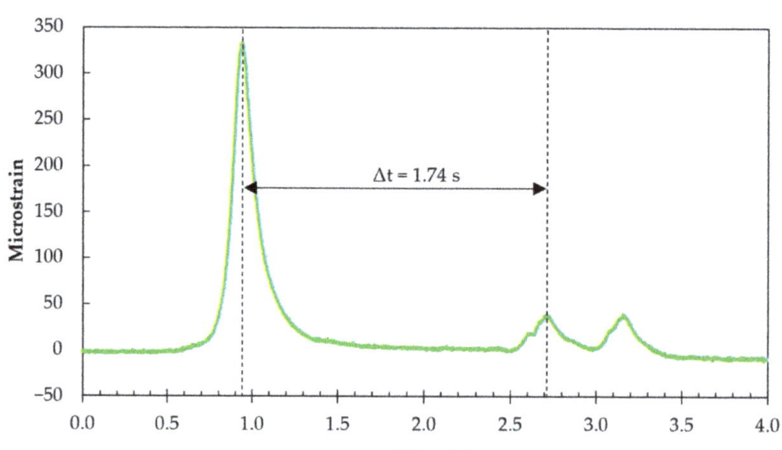

Figure 20. Strain record of the passage of Vehicle 2 over sensor 3 of rod B_2.

Using the same method presented previously in Figure 19 for Vehicle 1, the average speed of Vehicle 2 was calculated to be approximately 9.8 km/h. At that speed, it took about 1.74 s for the first and second axles to cross the same instrumented rod. Therefore, these values make it possible to determine that the distance between the first and second axles of Vehicle 2 is 4.74 metres.

This vehicle did not carry any significant load, which may explain that, unlike Vehicle 1, there was no increase in the strains from the first to the second rear axle. Furthermore, the vehicle's low speed (circulating in a protected area) allows the pavement to recover from the deformation caused by the passage of the previous axle. In addition, as in Vehicle 2, the rear axles are double, leading to better weight distribution and consequently, the peak strains measured when those axles pass over the system are significantly lower than those observed for the front axle.

4.2.2. Influence of the Type of Axle on the Strain Lateral Distribution

Both heavy vehicles have three axles (one front and two rear axles), and in both situations, the two rear axles are double-wheeled, and the front axle is simple. Thus, this analysis was only performed for Vehicle 2. Although the actual weight of each axle was unknown, the total weight of this vehicle was 15.66 tons.

Figure 21 presents the maximum strains recorded by the monitoring system when the front single-wheeled axle and first rear double-wheeled axle of Vehicle 2, respectively, passed over the fifteen sensors of fibreglass rod B_2.

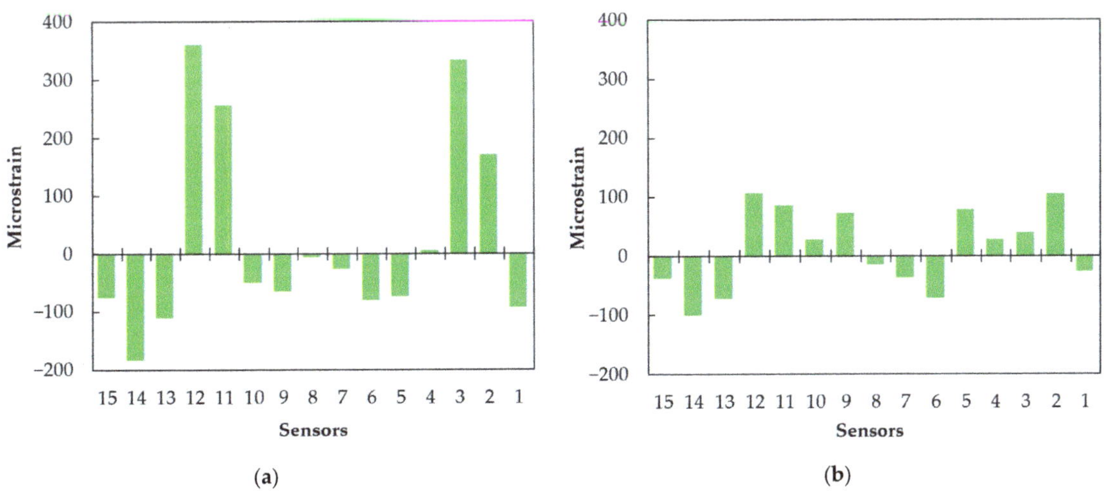

Figure 21. Transverse variation in maximum strains recorded on rod B_2 due to the dynamic loading effect of Vehicle 2: (**a**) single-wheeled front axle; (**b**) double-wheeled rear axle.

As observed in these graphs, it is possible to identify the sensors most loaded by both wheels and determine the lateral distribution of the vehicle loads. In this case, sensors 2 and 3 of Figure 21a relate to the left wheel, and sensors 11 and 12 to the right wheel of the front axle. This single-wheeled axle produced positive strains in an area with a radius of approximately 15 cm (spacing between the two sensors).

The first rear axle of this heavy-duty vehicle is shown in Figure 21b. By comparison with Figure 21a, it is possible to understand the different pavement behaviour under these double-wheeled axle loads. This axle type causes four sensors to register positive strain values for each set of wheels, resulting in an approximately 60 cm diameter area of tensile strain under the wheels. The increased load distribution area justifies the significantly lower strain values registered for the rear axle.

4.2.3. Transverse and Longitudinal Deformation Basins

An overview of the horizontal strains measured in the transverse direction caused by the three axles of Vehicles 1 and 2 is shown in Figures 22 and 23.

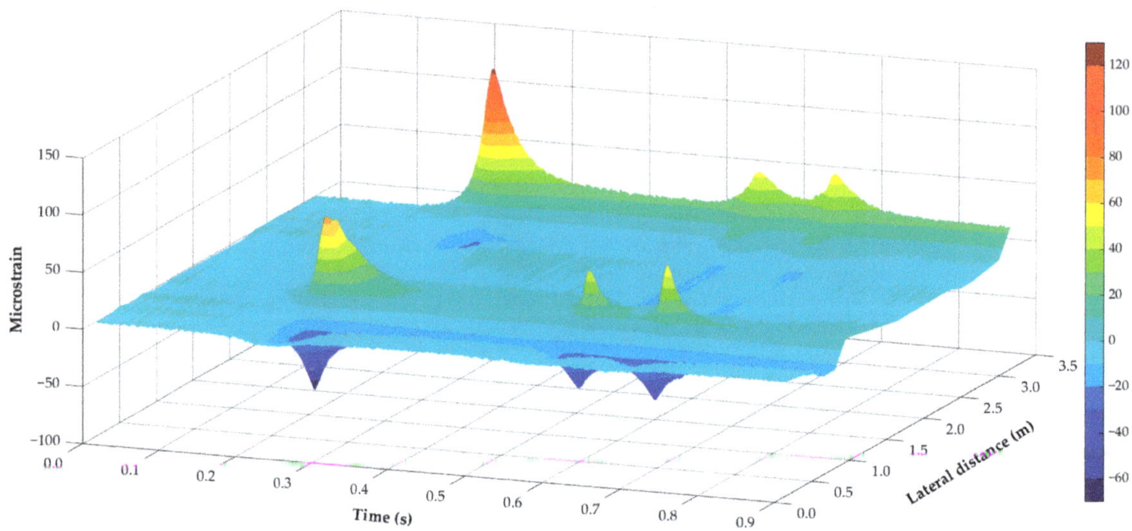

Figure 22. Strain distribution over time in all sensors of rod B_1 during Vehicle 1 monitoring.

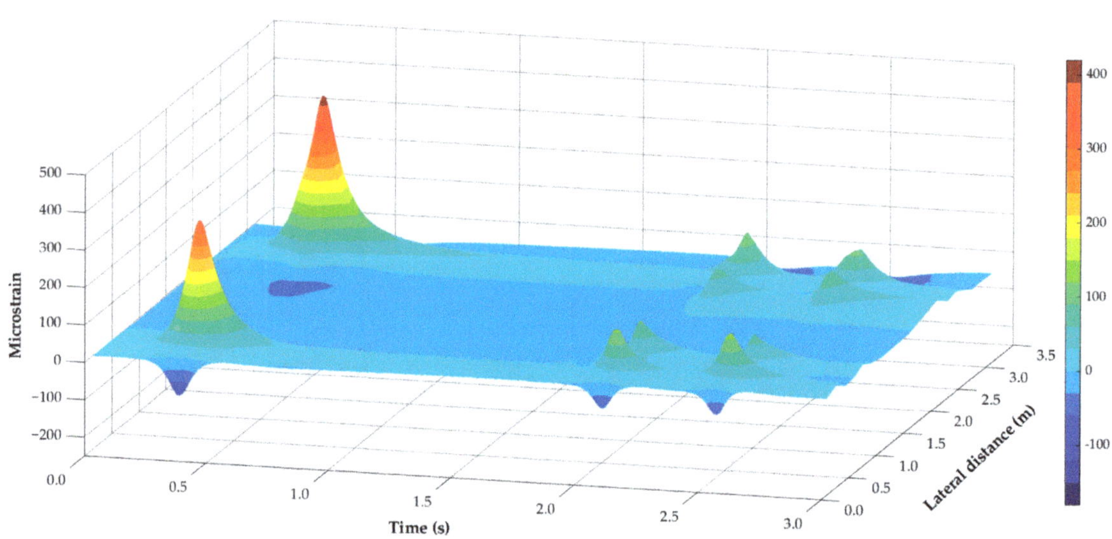

Figure 23. Strain distribution over time in all sensors of rod B_2 during Vehicle 2 monitoring.

The 3D graph allows a global visualisation of the information collected in all FBG sensors of each rod (B_1 and B_2), providing a more comprehensive perspective of the strains imposed in that pavement section over the period during which the vehicle passage was recorded. This type of analysis focuses more on evaluating the loading effect of the passing vehicle on the pavement response according to the observed weather conditions.

In this graph, it is possible to observe the compressive (negative) strains generated in the areas adjacent to those loaded by the vehicle wheels. On the other hand, comparing the deformation (strain level) before and after the front axle loads are applied, it is clear that the pavement deformation imposed by these loads has a delayed recovery, typical of asphalt materials, due to their viscoelastic nature.

Figure 23 shows the difference in the results obtained between axles with single and double wheels, with a considerably higher loaded area in the rear axles of the heavy vehicle. Despite circulating over the monitoring system with a much lower load than Vehicle 1, Vehicle 2 caused significantly higher strains in all axles. As previously demonstrated, its speed was much lower, influencing the results obtained by the monitoring system. Slower loading frequencies result in a lower stiffness modulus of the asphalt layers, which implies the development of higher strain levels within the pavement structure. Furthermore, Vehicle 2 circulated in the slow traffic lane, whose pavement may have a lower bearing capacity than the fast traffic lane (where the passage of Vehicle 1 was registered), as usually occurs due to the increased damage caused in the pavement by heavy and low-speed loads typically passing in the slow traffic lane.

This analysis with 3D graphs may be a valuable tool in studying pavement performance evolution over time. Furthermore, it can be used to analyse the effects of different axle and vehicle configurations on pavement behaviour, including the effect of pavement temperature and vehicle speed, and help improve the pavement performance models used to schedule pavement maintenance operations.

5. Conclusions

The process of installing a continuous pavement monitoring system based on FBG optical sensors presented here, along with a series of calibration tests performed in the following weeks, allowed some conclusions regarding its operation to be drawn:

- The type of sensors used in this work is very accurate; slight differences in the position of the load (in the order of 50 mm) may cause significant differences (20% to 25%) in the strains obtained for the same load, justifying the shorter distances (150 mm) between the sensors used in two of the instrumented rods, namely near the wheel tracks;
- One of this technology's critical issues is the temperature calibration of the sensors, as they are susceptible to temperature variations. However, a calibration factor can be applied to each sensor to correct the readings using the software (Catman) provided by the supplier of the sensors. Moreover, it was observed that a temperature rise of 8 °C increased the measured tensile strains by about 20%;
- The FWD tests performed with different loads for calibration of the monitoring system showed that a linear relationship could be established between the applied load and the strains obtained, which will be used in the future to analyse the data gathered from this monitoring system, to estimate the loads applied to the pavement surface;
- The effects of the type and number of axles of each vehicle on the response of the pavement at each load application were analysed in this work, using 3D representations of the strains over time; the fibreglass rods instrumented with 15 strain sensors were essential for the accurate representation of this information, yielding reliable knowledge of the pavement behaviour;
- The rods instrumented with 15 strain sensors provide a more comprehensive analysis of the transverse variation in the strains in a pavement section, which can be associated with the temperature data measured by the specific sensors installed for that purpose to assess the evolution of the pavement performance over its lifecycle, generating valuable information to develop pavement performance models.

The main limitation of this study is the lack of universality among the calibration results due to restrictions in the time available to perform the tests, as these experiments demanded the road's closure, which could only be possible during a short period. Thus, only four FWD loads could be applied to the pavement in each testing location, and the temperature variations only represent a small spectrum of the whole range values observed in this location in a day or a year.

Briefly, this monitoring sensor technology will be used as a monitoring technology to assess pavement performance by measuring variations in the pavement response to load application during the different seasons and throughout the pavement's life. Nevertheless, at this stage of software development, using the collected traffic data to estimate the average load per vehicle is still not viable with this system.

Author Contributions: Conceptualisation, F.J.P.R., J.R.M.O., H.M.R.D.S., J.O.e.S., V.M. and J.A.; methodology, F.J.P.R., J.R.M.O., H.M.R.D.S., J.O.e.S., V.M. and J.A.; validation, J.R.M.O. and H.M.R.D.S.; formal analysis, V.M., J.R.M.O. and H.M.R.D.S.; investigation, F.J.P.R., J.R.M.O., H.M.R.D.S., J.O.e.S., V.M. and J.A.; writing—original draft preparation, F.J.P.R., J.R.M.O., H.M.R.D.S. and J.O.e.S.; writing—review and editing, F.J.P.R., H.M.R.D.S. and J.R.M.O.; supervision, J.R.M.O., H.M.R.D.S. and J.O.e.S. All authors have read and agreed to the published version of the manuscript.

Funding: This research was funded by PORTUGAL 2020 through the Operational Program for Competitiveness and Internationalization (POCI) and the European Regional Development Fund (ERDF) under the project "Rev@Construction—Digital Construction Revolution", with reference POCI-01-0247-FEDER-046123, and by Fundação para a Ciência e a Tecnologia through the PhD grant number 2022.14400.BD. This work was also partly financed by FCT/MCTES through national funds (PIDDAC) under the R&D Unit Institute for Sustainability and Innovation in Structural Engineering (ISISE), reference UIDB/04029/2020, the Associate Laboratory Advanced Production and Intelligent Systems ARISE, reference LA/P/0112/2020, and the R&D Unit ALGORITMI, reference UIDB/00319/2020.

Institutional Review Board Statement: Not applicable.

Informed Consent Statement: Not applicable.

Data Availability Statement: Data sharing does not apply to this article.

Acknowledgments: The authors would like to acknowledge the staff from the authors' organisations involved in this work, who contributed to achieving the objectives of this study.

Conflicts of Interest: The authors declare no conflict of interest. The funders had no role in the study's design; in the collection, analyses, or interpretation of data; in the writing of the manuscript; or in the decision to publish the results.

References

1. Chen, J.; Dan, H.; Ding, Y.; Gao, Y.; Guo, M.; Guo, S.; Han, B.; Hong, B.; Hou, Y.; Hu, C.; et al. New innovations in pavement materials and engineering: A review on pavement engineering research 2021. *J. Traffic Transp. Eng. (Engl. Ed.)* **2021**, *8*, 815–999. [CrossRef]
2. Kara De Maeijer, P.; Voet, E.; Windels, J.; Van den Bergh, W.; Vuye, C.; Braspenninckx, J. Fiber Bragg grating monitoring system for heavy-duty pavements. In Proceedings of the 7th Euroasphalt & Eurobitume Congress, Virtual, 16–18 June 2021.
3. Ai, C.; Rahman, A.; Xiao, C.; Yang, E.; Qiu, Y. Analysis of measured strain response of asphalt pavements and relevant prediction models. *Int. J. Pavement Eng.* **2017**, *18*, 1089–1097. [CrossRef]
4. Hou, Y.; Li, Q.; Zhang, C.; Lu, G.; Ye, Z.; Chen, Y.; Wang, L.; Cao, D. The State-of-the-Art Review on Applications of Intrusive Sensing, Image Processing Techniques, and Machine Learning Methods in Pavement Monitoring and Analysis. *Engineering* **2021**, *7*, 845–856. [CrossRef]
5. Braunfelds, J.; Senkans, U.; Skels, P.; Janeliukstis, R.; Salgals, T.; Redka, D.; Lyashuk, I.; Porins, J.; Spolitis, S.; Haritonovs, V.; et al. FBG-Based Sensing for Structural Health Monitoring of Road Infrastructure. *J. Sens.* **2021**, *2021*, 8850368. [CrossRef]
6. Sebaaly, P.; Tabatabaee, N.; Kulakowski, B.; Scullion, T. *Instrumentation for Flexible Pavements—Field Performance of Selected Sensors, Volume 1: Final Report*; Pennsylvania Transportation Institut: University Park, PA, USA, 1992.
7. Huff, R.; Berthelot, C.; Daku, B. Continuous primary dynamic pavement response system using piezoelectric axle sensors. *Can. J. Civ. Eng.* **2011**, *32*, 260–269. [CrossRef]
8. Duong, N.S.; Blanc, J.; Hornych, P.; Bouveret, B.; Carroget, J.; Le Feuvre, Y. Continuous strain monitoring of an instrumented pavement section. *Int. J. Pavement Eng.* **2019**, *20*, 1435–1450. [CrossRef]
9. Liu, H.; Ge, W.; Pan, Q.; Hu, R.; Lv, S.; Huang, T. Characteristics and analysis of dynamic strain response on typical asphalt pavement using Fiber Bragg Grating sensing technology. *Constr. Build. Mater.* **2021**, *310*, 125242. [CrossRef]
10. Tan, Y.Q.; Wang, H.-P.; Sun, Z.-J.; Li, Y.-W.; Shi, X. Calibration method of FBG sensor based on asphalt pavement indoor small size test. In Proceedings of the 2011 International Conference on Transportation, Mechanical, and Electrical Engineering (TMEE), Changchun, China, 16–18 December 2011; pp. 1390–1394.

11. Frank, M.H.; Jason, K.R.; Peter, D.F. A strain-isolated fibre Bragg grating sensor for temperature compensation of fibre Bragg grating strain sensors. *Meas. Sci. Technol.* **1998**, *9*, 1163. [CrossRef]
12. Song, M.; Lee, S.B.; Choi, S.S.; Lee, B. Simultaneous Measurement of Temperature and Strain Using Two Fiber Bragg Gratings Embedded in a Glass Tube. *Opt. Fiber Technol.* **1997**, *3*, 194–196. [CrossRef]
13. Patrick, H.J.; Williams, G.M.; Kersey, A.D.; Pedrazzani, J.R.; Vengsarkar, A.M. Hybrid fiber Bragg grating/long period fiber grating sensor for strain/temperature discrimination. *IEEE Photonics Technol. Lett.* **1996**, *8*, 1223–1225. [CrossRef]
14. Jaehoon, J.; Hui, N.; Namkyoo, P.; Byoungho, L. Simultaneous measurement of strain and temperature using a single fiber Bragg grating written in an erbium-ytterbium-doped fiber. In *Proceedings of Technical Digest. Summaries of Papers Presented at the Conference on Lasers and Electro-Optics. Postconference Edition. CLEO '99. Conference on Lasers and Electro-Optics (IEEE Cat. No.99CH37013), 28 May 1999*; IEEE: Piscataway, PA, USA, 1999; p. 386.
15. Wang, H.-P.; Dai, J.-G.; Wang, X.-Z. Improved temperature compensation of fiber Bragg grating-based sensors applied to structures under different loading conditions. *Opt. Fiber Technol.* **2021**, *63*, 102506. [CrossRef]
16. Leal-Junior, A.G.; Díaz, C.A.R.; Frizera, A.; Marques, C.; Ribeiro, M.R.N.; Pontes, M.J. Simultaneous measurement of pressure and temperature with a single FBG embedded in a polymer diaphragm. *Opt. Laser Technol.* **2019**, *112*, 77–84. [CrossRef]
17. Han, D.; Liu, G.; Xi, Y.; Zhao, Y. Theoretical analysis on the measurement accuracy of embedded strain sensor in asphalt pavement dynamic response monitoring based on FEM. *Struct. Control Health Monit.* **2022**, *29*, e3140. [CrossRef]
18. Liu, Z.; Gu, X.; Wu, C.; Ren, H.; Zhou, Z.; Tang, S. Studies on the validity of strain sensors for pavement monitoring: A case study for a fiber Bragg grating sensor and resistive sensor. *Constr. Build. Mater.* **2022**, *321*, 126085. [CrossRef]
19. Rebelo, F.; Dabiri, A.; Silva, H.; Oliveira, J. Laboratory Investigation of Sensors Reliability to Allow Their Incorporation in a Real-Time Road Pavement Monitoring System. In Proceedings of the 3rd ISIC International Conference on Trends on Construction in the Post-Digital Era, Guimarães, Portugal, 26–29 September 2022; pp. 490–501.
20. Barrias, A.; Casas, J.R.; Villalba, S. A Review of Distributed Optical Fiber Sensors for Civil Engineering Applications. *Sensors* **2016**, *16*, 748. [CrossRef]
21. Majumder, M.; Gangopadhyay, T.K.; Chakraborty, A.K.; Dasgupta, K.; Bhattacharya, D.K. Fibre Bragg gratings in structural health monitoring—Present status and applications. *Sens. Actuators A Phys.* **2008**, *147*, 150–164. [CrossRef]
22. Lei, Y.; Hu, X.; Wang, H.; You, Z.; Zhou, Y.; Yang, X. Effects of vehicle speeds on the hydrodynamic pressure of pavement surface: Measurement with a designed device. *Measurement* **2017**, *98*, 1–9. [CrossRef]
23. Hottinger Bruel & Kjaer. NewLight Optical Fiber Sensors. Available online: https://www.hbm.com/en/4599/new-light-optical-fiber-sensors/?product_type_no=newLight (accessed on 23 June 2023).
24. Zhou, Z.; Liu, W.; Huang, Y.; Wang, H.; Jianping, H.; Huang, M.; Jinping, O. Optical fiber Bragg grating sensor assembly for 3D strain monitoring and its case study in highway pavement. *Mech. Syst. Signal Process.* **2012**, *28*, 36–49. [CrossRef]
25. Chen, J.; Liu, B.; Zhang, H. Review of fiber Bragg grating sensor technology. *Front. Optoelectron. China* **2011**, *4*, 204–212. [CrossRef]
26. Kara De Maeijer, P.; Van den Bergh, W.; Vuye, C. Fiber Bragg Grating Sensors in Three Asphalt Pavement Layers. *Infrastructures* **2018**, *3*, 16. [CrossRef]
27. Xiang, P.; Wang, H. Optical fibre-based sensors for distributed strain monitoring of asphalt pavements. *Int. J. Pavement Eng.* **2018**, *19*, 842–850. [CrossRef]
28. Quinimar. QuiniResin Fix. Available online: https://quinimar.pt/pt/quiniresin.htm (accessed on 10 June 2023).
29. Wang, J.; Han, Y.; Cao, Z.; Xu, X.; Zhang, J.; Xiao, F. Applications of optical fiber sensor in pavement Engineering: A review. *Constr. Build. Mater.* **2023**, *400*, 132713. [CrossRef]
30. Hottinger Bruel & Kjaer. Catman Data Acquisition Software: Connect. Measure. Visualise. Analyse. Available online: https://www.hbm.com/en/2290/catman-data-acquisition-software/?product_type_no=DAQ%20Software (accessed on 15 June 2023).
31. Sudarsanan, N.; Kim, Y.R. A critical review of the fatigue life prediction of asphalt mixtures and pavements. *J. Traffic Transp. Eng.* **2022**, *9*, 808–835. [CrossRef]

infrastructures

MDPI

Article

Estimation of Pavement Condition Based on Data from Connected and Autonomous Vehicles

David Llopis-Castelló [1,*], Francisco Javier Camacho-Torregrosa [1], Fabio Romeral-Pérez [2] and Pedro Tomás-Martínez [3]

[1] Highway Engineering Research Group, Universitat Politècnica de València, Camino de Vera s/n, 46022 Valencia, Spain; fracator@upv.es
[2] Xouba Ingeniería S.L., C/de Cristóbal Bordiú 33 Entreplanta A, 28003 Madrid, Spain; fabio@xouba.io
[3] Ministry of Transport and Sustainable Mobility, Paseo de la Castellana 67, 28046 Madrid, Spain; ptomas@transportes.gob.es
* Correspondence: dallocas@upv.es

Abstract: Proper road network maintenance is essential for ensuring safety, reducing transportation costs, and improving fuel efficiency. Traditional pavement condition assessments rely on specialized equipment, limiting the frequency and scope of inspections due to technical and financial constraints. In response, crowdsourcing data from connected and autonomous vehicles (CAVs) offers an innovative alternative. CAVs, equipped with sensors and accelerometers by Original Equipment Manufacturers (OEMs), continuously gather real-time data on road conditions. This study evaluates the feasibility of using CAV data to assess pavement condition through the International Roughness Index (IRI). By comparing CAV-derived data with traditional pavement auscultation results, various thresholds were established to quantitatively and qualitatively define pavement conditions. The results indicate a moderate positive correlation between the two datasets, particularly in segments with good-to-satisfactory surface conditions (IRI 1 to 2.5 dm/km). Although the IRI values from CAVs tended to be slightly lower than those from auscultation surveys, this difference can be attributed to driving behavior. Nonetheless, our analysis shows that CAV data can be used to reliably identify pavement conditions, offering a scalable, non-destructive, and continuous monitoring solution. This approach could enhance the efficiency and effectiveness of traditional road inspection campaigns.

Keywords: pavement; road maintenance; International Roughness Index; connected and autonomous vehicles

1. Introduction

The proper maintenance of road networks is crucial for preserving and enhancing citizens' quality of life [1,2]. Otherwise, the costs associated with the transportation of goods and people would increase due to poor road conditions, which lead to higher fuel consumption and, consequently, increased greenhouse gas emissions. Additionally, poor pavement conditions pose significant dangers to road users, cause greater tire wear, and can damage vehicles [3,4].

Therefore, it is essential for highway authorities to develop a pavement management system to analyze the lifecycle of road infrastructure and create optimal pavement conditions. This requires evaluating pavement condition and developing predictive models to understand how pavement deterioration will evolve [5].

In this context, the Spanish Highway Administration, like many national highway agencies, has implemented a pavement management system designed to effectively and efficiently manage road maintenance. This system facilitates the creation of inventories, the maintenance of a database for surveys and inspections, and the assessment of pavement condition using various indices. There are, however, areas for improvement regarding the availability of information on pavement condition, as auscultation systems require significant investment. Additionally, while visual inspections are necessary before making decisions, they may introduce variability in assessments. Exploring the use of evolutionary models to predict pavement condition, estimate pavement life, and determine the optimal timing for interventions could further enhance management efficiency and effectiveness.

As a result, according to the Spanish Road Association, Spanish roads are in a "poor" state of conservation, nearing the "very poor" threshold [6]. One out of thirteen kilometers of the Spanish road and highway network shows significant deterioration in over 50% of the pavement surface, featuring potholes, rutting, and longitudinal and transverse cracks. This lack of maintenance is severely impacting Spanish road infrastructure, with a 36% loss in asset value between 2001 and 2017 for national roads and a 38% loss for regional and local roads. This progressive deterioration leads to uncomfortable driving conditions, road safety issues, inter-territorial and European competitiveness losses, exponential increases in pavement repair costs, higher vehicle maintenance costs, and increased pollutant emissions.

Current methods for evaluating pavement condition involve conducting inspections with specialized equipment to assess pavement condition and driving comfort [7–10]. Due to economic and technical constraints, administrations cannot cover the entire road network on an annual basis. This typically results in detailed data collection only for high-volume roads, with less attention given to lower-priority roads.

Despite advancements in image processing and specialized vehicle instrumentation [11–16], a significant gap exists in terms of scalability and continuous monitoring across all road types. The current methods, including those based on image processing, are limited by their dependency on equipment setup, cost, and infrequent data collection cycles. Additionally, the reliance on specific vehicles or routes further restricts the ability to gather consistent, real-time pavement data.

An alternative to these instrumented vehicles is crowdsourcing data from connected and autonomous vehicles (CAVs). Original Equipment Manufacturers (OEMs) integrate sensors, accelerometers, and mobile network connections in vehicles, providing a data source on current road conditions. Integrating and utilizing vehicle data have enabled the assessment of road markings, traffic signs, and crash mitigation through surrogate safety measures [17–21]. In pavement condition evaluation, vehicles use a system that leverages individual wheel speed through rotational sensors combined with transmission information to provide data on ride quality or comfort [22]. This information can be used to estimate the International Roughness Index (IRI) using a fleet of crowdsourced vehicles.

Unlike auscultation methods, whose results depend significantly on the path of the specialized equipment at the time of measurement and have very low data collection frequency, data from CAVs constitute a more reliable type of real-time road condition information, with data from hundreds or thousands of vehicles at each road point.

Thus, this study fills the gap by leveraging crowdsourced data from CAVs to provide continuous, scalable, and cost-effective pavement monitoring. The novelty of this approach lies in its ability to utilize data from a vast fleet of vehicles, offering higher data frequency and broader network coverage than traditional methods.

2. Materials and Methods

This study aims to analyze the relationship between International Roughness Index (IRI) data obtained from pavement assessments made via specialized equipment and IRI values derived from data collected by Connected and Autonomous Vehicles (CAVs) to explore the feasibility of using the latter for assessing the roughness of rural roads.

To achieve this objective, this study will first provide a description of available roughness data, identifying the specific road where these data were collected. Next, a descriptive and graphical analysis of IRI values obtained from both pavement assessments and CAVs will be conducted to assess their initial correlation.

Following the descriptive analysis, a statistical analysis using paired sample tests will be performed by matching IRI data from both sources at each kilometer and hectometer point along the road. The null hypothesis for this statistical test assumes there is no difference in means between the two datasets, subject to prior evaluation of data normality.

Finally, this study proposes establishing various thresholds to quantitatively and qualitatively determine pavement conditions based on data collected from connected and autonomous vehicles.

2.1. Road Segment

The road section used for the development of this study is a segment of road N-310, from station 144 + 990 (San Clemente) to station 198 + 710 (Villanueva de la Jara), in the province of Cuenca (Spain).

This road segment is a two-lane rural road with a total cross-section width of 10 m composed of asphalt pavement. The width of each lane and shoulder is 3.5 m and 1.5 m, respectively. Specifically, along this segment, there are a total of six traffic-counting stations (see Table 1). The Average Annual Daily Traffic (AADT) along this road segment in 2021 ranged from approximately 1000 to 3000 vehicles per day.

Table 1. Annual Average Daily Traffic (AADT) on the road segment in question in 2021 [23].

Start Station	Final Station	Length (m)	AADT (veh/day)
144 + 990	154 + 140	9150	1993
154 + 140	158 + 960	4820	1819
158 + 960	174 + 158	15,198	1379
174 + 158	180 + 950	6792	3224
180 + 950	191 + 280	10,330	2680
191 + 280	198 + 710	7430	1026

2.2. Auscultation Data

The pavement condition data were provided by the Highway Department of Castilla–La Mancha, a division of the Highway Administration of the Spanish Ministry of Transport and Sustainable Mobility.

The field data were collected on 15 June 2023. Specifically, the following variables were obtained: (i) section identifier (IdSection), (ii) highway (IdRoad), (iii) initial station (PKIHito and PKIDist), (iv) final station (PKFHito and PKFDist), (v) right wheel track IRI (IRI_der), (vi) left wheel track IRI (IRI_izq), and (vii) average IRI (IRI_med).

In addition to these data, another set of georeferencing data were provided, with the following information: (i) section identifier (IdSection), (ii) highway (IdRoad), (iii) initial station (PKIHito and PKIDist), (iv) final station (PKFHito and PKFDist), and (v) UTM coordinates (UTMx and UTMy). From these data, IRI values could be assigned to the highway and the corresponding measurement unit, as the data related to the station points might not be sufficiently accurate. In short, IRI data are available every hectometer.

2.3. Data from Connected and Autonomous Vehicles

Data regarding pavement condition from connected and autonomous vehicles were downloaded on the same day as the pavement assessment, specifically on 15 June 2023. These data were provided by NIRA Dynamics, which manages a vehicle fleet consisting of close to two million cars around the world.

These vehicles are equipped with software that collects real-time data from the existing onboard sensors. Thus, the vehicles serve as a continuous sensor of infrastructure conditions whenever they travel on it with the minimum sample size required to ensure that the measurement is not biased by a single vehicle. For this study, data from the connected vehicles were collected not only on the specified day but also aggregated over a 30-day period to provide a more comprehensive overview. This product is called the long-term value of road roughness.

Original Equipment Manufacturers (OEMs) integrate enhanced sensors, accelerometers, and mobile connections into vehicles to provide a rich data source on current road conditions. Wheel speed is leveraged through rotation sensors in combination with drivetrain information to assess pavement quality across vehicle fleets. These measurements are processed through NIRA Dynamics' sensor fusion algorithms, which uses the calibration data to translate raw sensor outputs into roughness indicators, including the International Roughness Index (IRI). This method ensures that the IRI values are consistent with standard pavement assessment methodologies.

Data provided by the supplier were aggregated into segments of approximately 20 m based on mapping from navigation and mapping companies. Once a segment was obtained, it was georeferenced and assigned to the road under study. Specifically, the data structure available for each road segment includes (i) segment identifier (IdSection), (ii) International Roughness Index (IRI) value in dm/hm (IRI), (iii) UTM coordinates of points within the segment (geometry), (iv) UTM coordinates of the starting point of the road segment (GeometryInitialPoint), and (v) UTM coordinates of the endpoint of the road segment (GeometryFinalPoint). As with the auscultation data, a value of IRI per hectometer was available.

3. Results

3.1. Descriptive Analysis

Table 2 presents a statistical summary of the IRI data obtained, expressed in dm/hm. As can be seen, the mean IRI value obtained from the data recorded by the CAVs (IRI_cavs) closely resembles, in overall terms, the mean IRI from auscultations (IRI_med), calculated as the average of the maximum and minimum values. However, the positional parameters—minimum value, maximum value, and percentiles—indicate that the minimum IRI values (IRI_min) dataset is the most similar to the distribution of the IRI_cavs data.

Table 2. Statistical summary for IRI.

	IRI_cavs	IRI_auscultation		
		IRI_max	IRI_min	IRI_med
Average	1.70024	2.04155	1.56469	1.80170
Standard deviation	0.55467	0.64284	0.44336	0.51644
Min.	0.86038	0.86000	0.74000	0.83000
25th percentile	1.32116	1.57750	1.27000	1.43000
50th percentile	1.56672	1.95000	1.49000	1.74250
75th percentile	1.94227	2.39250	1.76000	2.07500
Max.	4.19950	5.46000	4.29000	4.47500

To further explore the aforementioned points, density distributions of all the IRI datasets are represented (Figure 1). Firstly, it is noteworthy that all the distributions exhibit positive skewness, meaning the data cluster at lower IRI values, with a tail towards higher values. This suggests that the mean values of the datasets are higher than the median or the 50th percentile (see Table 2). Additionally, it can be observed that the density curve most similar to that described by the IRI_cavs data is that of the minimum values obtained from auscultation (IRI_min). However, in the tail of the distribution, the IRI dataset from CAVs (IRI_cavs) more closely resembles the distribution of the mean auscultation data (IRI_med).

Nevertheless, similarity between distributions does not necessarily imply a higher correlation between the datasets, especially considering that the data are actually paired by the location of the observations.

Figure 2 includes the correlation matrix between the different IRI datasets analyzed, as well as the corresponding scatter plots comparing the IRI values from CAVs with those from the other datasets. As expected, the correlations between the auscultation datasets are strong (>0.8). Regarding the correlation between IRI_cavs and IRI from auscultation, it is noteworthy that they are very similar across all the datasets considered. However, the correlation between IRI_cavs and the datasets of minimum and mean IRI is very similar (>0.5), despite identifying earlier that the density distribution most similar to the IRI_cavs data was that of the minimum auscultation values.

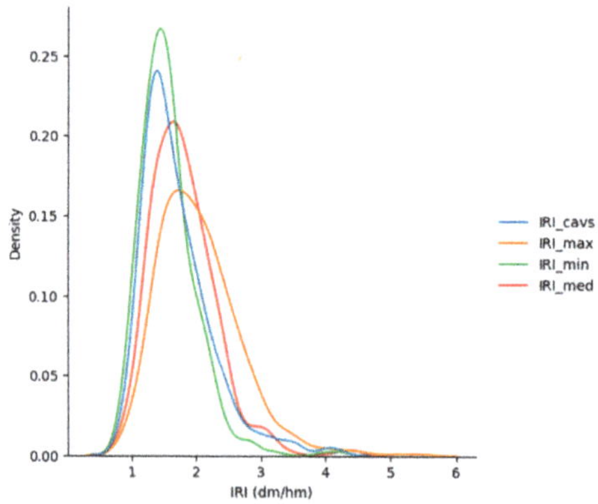

Figure 1. Density distribution of IRI datasets.

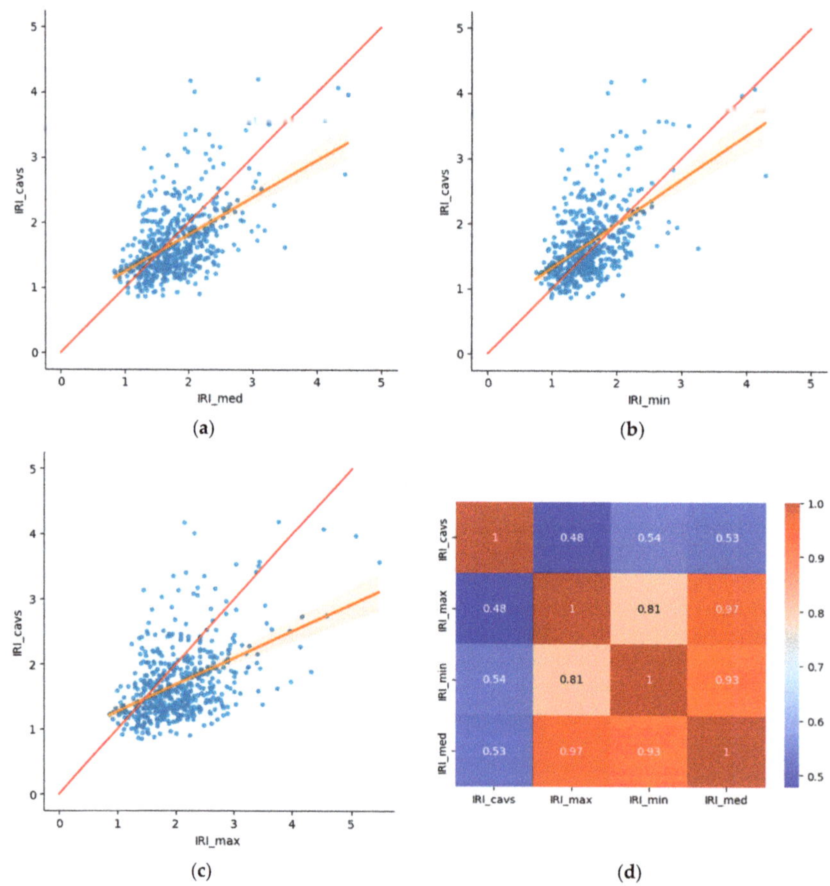

Figure 2. Correlation analysis: (**a**) IRI_cavs and IRI_med, (**b**) IRI_cavs and IRI_med, (**c**) IRI_cavs and IRI_med, and (**d**) correlation matrix.

Additionally, a scatter plot was created, which includes the point density in the represented region (see Figure 3). Specifically, only the relationship between IRI_cavs and IRI_min is represented, which is indicative of the other cases. As observed, most observations are situated between values of IRI of 1 dm/hm and 2.5 dm/hm, and these values appear to concentrate very close to the line representing IRI_cavs = IRI_min. Specifically, the values obtained from vehicles seem slightly lower than those obtained from auscultation, particularly as the IRI value increases (see Figure 2). This phenomenon could be explained by the behavior of road users, who tend to avoid the most deteriorated parts of the road. Given that these parts usually correspond to lane-centered driving and that auscultation methods try to trace this trajectory, it was expected that the values of the roughness data obtained from CAVs would be slightly lower than those of the data obtained via auscultation, mainly as users' discomfort increases.

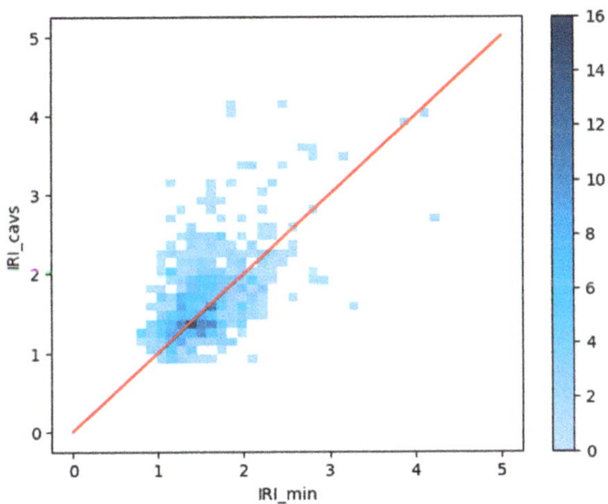

Figure 3. Point histogram.

3.2. Statistical Analysis

This section aims to determine whether IRI data from Connected and Autonomous Vehicles (CAVs) can be considered equivalent to those obtained through auscultation methods. For this purpose, the chosen statistical test is the paired samples comparison. The primary aim of this type of test is to compare IRI values obtained from auscultation and those derived from CAVs for each hectometer point.

Specifically, to determine the statistical test to be applied, it is necessary first to evaluate the normality of the data. As mentioned during the descriptive analysis, all the datasets exhibit strong positive skewness. Therefore, according to the results of the Shapiro–Wilk test, it cannot be confirmed with 95% confidence that the datasets follow a normal distribution, as the p-values are less than 0.05 (Table 3).

Table 3. Assessment of data normality: Shapiro–Wilk test.

	IRI_cavs	IRI_med	IRI_min	IRI_max
W	0.876208	0.917700	0.893479	0.920585
p-value	4.959857×10^{-20}	2.608296×10^{-16}	1.298214×10^{-18}	5.271826×10^{-16}

This implies that the paired data test using the usual t-test cannot be performed. Thus, the following alternative tests were proposed: (i) the Wilcoxon signed-rank test and (ii) the Kruskal–Wallis test. Both these tests are non-parametric and therefore require fewer assumptions compared to the t-test for dependent samples. Specifically, the Wilcoxon test checks whether the mean values of two dependent groups differ significantly from each other. On the other hand, the Kruskal–Wallis test determines whether the medians of two or more groups are different.

Tables 4 and 5 present the results of the Wilcoxon signed-rank test and the Kruskal–Wallis test, respectively, conducted for each pair of datasets. As a result, at a 95% confidence level, neither of the null hypotheses of the considered tests can be confirmed. Therefore, we concluded that the IRI data from CAVs are not equivalent to the data obtained through auscultation.

Table 4. Wilcoxon test.

	IRI_cavs and IRI_med	**IRI_cavs and IRI_min**	**IRI_cavs and IRI_max**
W	49,426	50,286	28,106
p-value	1.705195×10^{-8}	6.959174×10^{-8}	5.549335×10^{-32}

Table 5. Kruskal-Wallis test.

	IRI_cavs and IRI_med	**IRI_cavs and IRI_min**	**IRI_cavs and IRI_max**
W	20,068,770	12,903,457	99,849,646
p-value	7.470667×10^{-6}	0.000328	1.644166×10^{-23}

Despite not being equivalent, a relationship between both datasets could be established to, for example, estimate auscultation data using the IRI values from CAVs.

3.3. Proposal of Thresholds for Pavement Evaluation

Considering that it is not possible to assert, with a 95% confidence level, that the IRI values obtained from connected and autonomous vehicles (CAVs) are similar to those obtained from auscultation equipment, the relationship between IRI_cavs and IRI_min was analyzed in greater depth since there exists a moderate correlation between both datasets.

The qualitative evaluation of pavement condition established by the Spanish Highway Administration based on the IRI value for conventional road network routes served as the basis for this assessment (see Table 6).

Table 6. Qualitative levels of pavement condition according to the Spanish Highway Administration.

Pavement Condition	**IRI (dm/hm)**
Very good	$IRI \leq 1.5$
Good	$1.5 < IRI \leq 2.0$
Satisfactory	$2.0 < IRI \leq 2.5$
Fair	$2.5 < IRI \leq 3.0$
Poor	$3.0 < IRI \leq 4.0$
Very poor	>4.0

Using these qualitative pavement condition levels and the IRI_min value, the IRI data from CAVs were classified into different subsets, with each subset's outcome represented in a box–whisker plot (see Figure 4). As pavement condition worsens, the mean and median values of the IRI_cavs subset increase, demonstrating a clear relationship between the IRI value obtained from CAV data and pavement condition. It is important to note that most available IRI_cavs values range between 1.0 and 2.5 dm/hm; hence, for pavement conditions rated as fair, poor, or very poor, variability is quite high. A larger volume of data is needed to study these pavement condition levels in greater detail.

Additionally, the confusion matrix generated when classifying pavement condition based on the variables IRI_cavs and IRI_min was estimated. To achieve this, the number of thresholds previously shown in Table 6 was reduced due to the low representativeness of levels associated with high IRI values, resulting in the thresholds presented in Table 7.

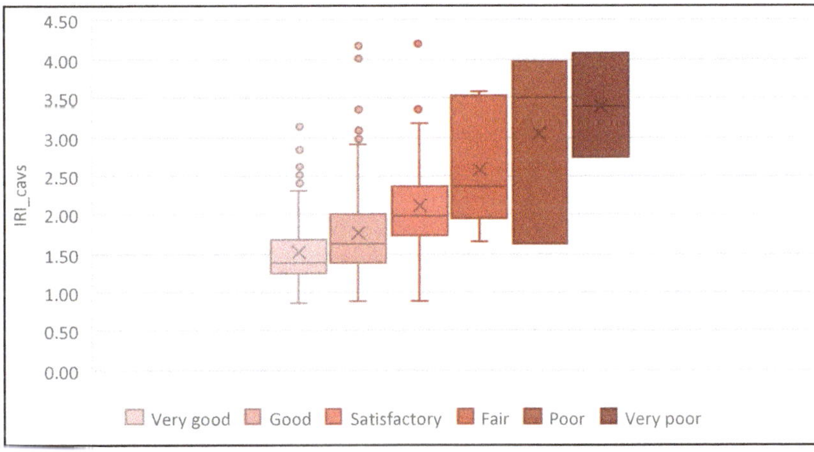

Figure 4. Box–whisker diagrams for IRI_cavs according to pavement condition level.

Table 7. Qualitative levels of pavement condition (adaptation from Spanish Highway Administration).

Pavement Condition	IRI (dm/hm)
Good	IRI ≤ 2.0
Fair	2.0 < IRI ≤ 3.0
Poor	IRI > 3.0

Table 8 shows the confusion matrix considering the IRI values obtained from auscultation to be true. Each row of the matrix sums to 100%, indicating for a given IRI_min threshold the proportion of data classified into the proposed thresholds according to IRI_cavs. Notably, more than 80% of the data classified as corresponding to good pavement condition according to IRI_min were similarly classified according to IRI_cavs. However, 50% of the data evaluated as corresponding to fair pavement condition according to IRI_min were classified as corresponding to good condition according to IRI_cavs, potentially leading to inadequate pavement management. These results were anticipated based on Figure 3 and the statistical analysis, suggesting that the thresholds used to determine pavement condition from auscultation-derived IRI values may differ from those for CAV-derived IRI values.

Table 8. Confusion matrix according to the qualitative levels defined by the Spanish Highway Administration.

		IRI_cavs		
		Good	**Fair**	**Poor**
IRI_min	Good	82.18%	16.26%	1.56%
	Fair	47.14%	38.57%	14.29%
	Poor	20.00%	20.00%	60.00%

Thus, new thresholds were calibrated to determine pavement conditions based on the IRI values from vehicles. The objective function was designed to maximize the values on the diagonal of the confusion matrix, with fixed IRI_min thresholds according to Table 7 and varying IRI_cavs thresholds. As a result, the thresholds presented in Table 9 were obtained. The updated confusion matrix is shown in Table 10, indicating that approximately

70% of the data were appropriately classified. However, we recommend recalibrating these thresholds and validating the results with a larger dataset that includes a greater range of pavement condition levels, particularly with data on pavements in poor condition.

Table 9. Qualitative pavement condition levels according to IRI_cavs.

Pavement Condition	IRI (dm/hm)
Good	IRI $\leq$ 1.75
Fair	1.75 < IRI $\leq$ 3.5
Poor	IRI > 3.5

Table 10. Confusion matrix considering the thresholds defined in Tables 7 and 9 for IRI_min and IRI_cavs, respectively.

			IRI_cavs		
			Good	Fair	Poor
			IRI $\leq$ 1.75	1.75 < IRI $\leq$ 3.5	IRI > 3.5
IRI_min	Good	IRI $\leq$ 2.0	70.82%	28.73%	0.45%
	Fair	2.0 < IRI $\leq$ 3.0	24.29%	70.00%	5.71%
	Poor	IRI > 3.0	20.00%	20.00%	60.00%

4. Discussion

This study delves into the potential of using data from connected and autonomous vehicles (CAVs) to assess pavement condition, specifically through the International Roughness Index (IRI). The results presented here contribute to understanding how CAV data can complement or even substitute for traditional manual survey methods in pavement management systems.

4.1. Comparison of IRI Data Sources

The moderate positive correlation observed between the IRI values derived from CAVs and those from manual surveys underscores the utility of CAV data in assessing pavement conditions. However, the discrepancies between the datasets, particularly noticeable at higher IRI values, suggest the need for cautious interpretation. Variations can arise from differences in measurement precision, vehicle-sampling biases, and environmental factors affecting data collection. For instance, while CAVs provide extensive coverage across road networks, variations in vehicle speeds and path selections may influence the accuracy and representativeness of the data, especially in segments with more severe pavement degradation.

These practical considerations indicate that while CAVs offer a valuable and scalable alternative for road condition monitoring, their results should be carefully interpreted, particularly in the case of severely deteriorated roads where traditional methods might still provide a more precise evaluation.

4.2. Data Distribution and Pavement Condition

The density distributions of the IRI datasets, which skew towards lower values, predominantly align with segments categorized as "Very Good" to "Satisfactory" according to established thresholds. This distribution pattern reflects the suitability of CAV data for identifying well-maintained pavement sections but highlights the need for increased data sampling in poorer conditions. This expansion would enhance the robustness of correlation analyses across a broader spectrum of pavement states, thereby improving the reliability of condition assessments.

From a road maintenance perspective, these results suggest that CAV-derived data could be integrated into existing pavement management strategies to provide continuous updates on well-maintained roads, with additional efforts required for roads in poorer condition.

4.3. Implications for Pavement Management

Integrating CAV data into pavement management systems offers several operational advantages. By providing real-time and continuous data streams, CAVs enable timely updates on pavement conditions without the logistical and cost constraints associated with periodic manual surveys. This capability enhances the scalability and efficiency of maintenance planning and decision-making processes, potentially reducing operational costs and minimizing disruptions to road users. Furthermore, the comprehensive coverage provided by CAVs facilitates a more equitable distribution of monitoring efforts across entire road networks, ensuring that critical maintenance needs are identified and addressed promptly.

These practical implications underscore the potential for road authorities to adopt CAV-based systems as part of a proactive maintenance strategy, allowing for more efficient resource allocation and timely intervention before road conditions worsen.

4.4. Practical Considerations and Recommendations

The establishment of distinct IRI thresholds for CAV-derived data has proven to be effective in categorizing pavement conditions into qualitative states. However, this study highlights the importance of refining these thresholds through extensive validation exercises involving diverse pavement conditions and geographic contexts. Such validations are crucial for ensuring the accuracy and applicability of CAV data in supporting informed decision-making by road authorities and stakeholders. Moreover, advancements in sensor technology and data analytics present opportunities to enhance the precision and reliability of CAV-based pavement assessments, warranting continued research and innovation in this field.

While CAV data offer several advantages, practical limitations remain, particularly regarding real-time applications. Variability in data due to vehicle behavior—such as changes in driving patterns to avoid road defects—and environmental conditions like weather can introduce noise into data, making calibration essential to minimize these influences.

4.5. Future Research Directions

Future research efforts should focus on addressing several key areas to further advance the integration of CAV data into pavement management:

- Enhanced data sampling—increasing the diversity and volume of CAV data collected, particularly in segments exhibiting greater pavement deterioration, to improve the robustness of correlation analyses.
- Validation and calibration—conducting extensive validation studies across varied environmental and traffic conditions to refine and validate proposed IRI thresholds derived from CAV data.
- Sensor technology advancements—exploring advancements in sensor technologies and data-processing algorithms to enhance the accuracy, reliability, and real-time capabilities of CAV-based pavement assessments.
- Lifecycle analysis—developing predictive models that leverage CAV data to forecast pavement deterioration and optimize maintenance strategies over the lifecycle of road infrastructure.

In conclusion, while CAV data have substantial potential for revolutionizing pavement management practices, further research efforts are essential to address relevant technical, methodological, and operational challenges. By advancing these fronts, the transportation sector can leverage CAV technologies to achieve more sustainable, cost-effective, and resilient pavement management solutions in the future.

5. Conclusions

This study presents an analysis of the relationship between International Roughness Index (IRI) values obtained through auscultation methods and those gathered by Connected and Autonomous Vehicles (CAVs). The most noteworthy conclusions are as follows:

- The density distributions of the IRI datasets exhibit positive skewness, with observations predominantly clustering at lower IRI values. Thus, expanding the dataset to include higher IRI values would refine the correlation analysis between the two datasets, particularly under poor pavement conditions.
- The results of the correlation analysis indicate there is a moderate positive correlation between IRI values recorded by connected and autonomous vehicles and those obtained through manual surveys
- The majority of observations fall within an IRI range of 1 dm/km to 2.5 dm/km, suggesting road segments with a surface roughness level ranging from "Very Good" to "Satisfactory" based on thresholds established by the Spanish Highway Administration.
- IRI values derived from vehicles are slightly lower than those obtained through manual surveys as IRI values increase. This could be attributed to driver behavior, where drivers tend to avoid more deteriorated paths, resulting in non-centered lane travel.
- Despite not being directly comparable to auscultation data, the IRI values from connected and autonomous vehicles can be used to establish distinct IRI thresholds to qualitatively assess pavement condition.
- The confusion matrix obtained (see Table 10) indicates that the defined IRI thresholds from connected and autonomous vehicles effectively identify pavement condition.
- Considering that IRI values from connected and autonomous vehicles aggregate data from hundreds or thousands of vehicles, they are deemed highly reliable for initial pavement condition assessments. Thus, employing IRI data from these vehicles presents a cost-effective and non-destructive alternative to traditional methods, potentially enhancing and optimizing conventional field data collection campaigns.

Author Contributions: Conceptualization, D.L.-C. and P.T.-M.; methodology, D.L.-C. and F.J.C.-T.; formal analysis, D.L.-C. and F.J.C.-T.; data curation, F.R.-P.; writing—original draft preparation, D.L.-C.; writing—review and editing, F.J.C.-T., P.T.-M. and F.R.-P. All authors have read and agreed to the published version of the manuscript.

Funding: This research received no external funding.

Data Availability Statement: The data presented in this study are available on request from the corresponding author due to privacy reasons.

Acknowledgments: The authors would like to thank the Highway Department of Castilla La Mancha, a division of the Highway Administration of the Spanish Ministry of Transport and Sustainable Mobility, for providing the auscultation data.

Conflicts of Interest: The authors declare no competing interests. The author Fabio Romeral-Pérez was employed by the company Xouba Ingeniería S.L., C/de Cristóbal Bordiú 33 Entreplanta A. There is no conflict of interest between any of the authors and the company Xouba Ingeniería S.L., C/de Cristóbal Bordiú 33 Entreplanta A.

References

1. Bull, A.; CEPAL, N. *Traffic Congestion: The Problem and How to Deal with It*; ECLAC: Santiago, Chile, 2003.
2. Hajj, E.Y.; Loria, L.; Sebaaly, P.E. Performance evaluation of asphalt pavement preservation activities. *Transp. Res. Rec.* **2010**, *2150*, 36–46. [CrossRef]
3. Santero, N.J.; Horvath, A. Global warming potential of pavements. *Environ. Res. Lett.* **2009**, *4*, 034011. [CrossRef]
4. Lee, J.; Abdel-Aty, M.; Nyame-Baafi, E. Investigating the effects of pavement roughness on freeway safety using data from five states. *Transp. Res. Rec.* **2020**, *2674*, 127–134. [CrossRef]
5. Pérez-Acebo, H.; Bejan, S.; Gonzalo-Orden, H. Transition probability matrices for flexible pavement deterioration models with half-year cycle time. *Int. J. Civ. Eng.* **2018**, *16*, 1045–1056. [CrossRef]
6. Asociación Española de la Carretera. *Análisis de la Relación Entre el Estado de Conservación del Pavimento, el Consumo de Combustible y las Emisiones de los Vehículos*; Asociación Española de la Carretera: Madrid, Spain, 2020.
7. Pierce, L.M.; McGovern, G.; Zimmerman, K.A. *Practical Guide for Quality Management of Pavement Condition Data Collection*; Federal Highway Administration: Washington, DC, USA, 2013.

8. Attoh-Okine, N.; Adarkwa, O. *Pavement Condition Surveys–Overview of Current Practices*; Delaware Center for Transportation, University of Delaware: Newark, DE, USA, 2013.
9. Seraj, F.; Van Der Zwaag, B.J.; Dilo, A.; Luarasi, T.; Havinga, P. RoADS: A road pavement monitoring system for anomaly detection using smart phones. In *International Workshop on Modeling Social Media*; Springer International Publishing: Cham, Switzerland, 2014; pp. 128–146.
10. Kamranfar, P.; Lattanzi, D.; Shehu, A.; Stoffels, S. Pavement Distress Recognition via Wavelet-Based Clustering of Smartphone Accelerometer Data. *J. Comput. Civ. Eng.* **2022**, *36*, 04022007. [CrossRef]
11. Mahmoudzadeh, A.; Golroo, A.; Jahanshahi, M.R.; Firoozi Yeganeh, S. Estimating pavement roughness by fusing color and depth data obtained from an inexpensive RGB-D sensor. *Sensors* **2019**, *19*, 1655. [CrossRef] [PubMed]
12. Baek, J.W.; Chung, K. Pothole classification model using edge detection in road image. *Appl. Sci.* **2020**, *10*, 6662. [CrossRef]
13. Fan, R.; Wang, H.; Wang, Y.; Liu, M.; Pitas, I. Graph attention layer evolves semantic segmentation for road pothole detection: A benchmark and algorithms. *IEEE Trans. Image Process.* **2021**, *30*, 8144–8154. [CrossRef] [PubMed]
14. Tedeschi, A.; Benedetto, F. A real-time automatic pavement crack and pothole recognition system for mobile Android-based devices. *Adv. Eng. Inform.* **2017**, *32*, 11–25. [CrossRef]
15. Llopis-Castelló, D.; Paredes, R.; Parreño-Lara, M.; García-Segura, T.; Pellicer, E. Automatic classification and quantification of basic distresses on urban flexible pavement through convolutional neural networks. *J. Transp. Eng. Part B Pavements* **2021**, *147*, 04021063. [CrossRef]
16. Ravi, R.; Bullock, D.; Habib, A. Pavement distress and debris detection using a mobile mapping system with 2D profiler lidar. *Transp. Res. Rec.* **2021**, *2675*, 428–438. [CrossRef]
17. Xie, K.; Yang, D.; Ozbay, K.; Yang, H. Use of real-world connected vehicle data in identifying high-risk locations based on a new surrogate safety measure. *Accid. Anal. Prev.* **2019**, *125*, 311–319. [CrossRef] [PubMed]
18. Mahlberg, J.A.; Sakhare, R.S.; Li, H.; Mathew, J.K.; Bullock, D.M.; Surnilla, G.C. Prioritizing roadway pavement marking maintenance using lane keep assist sensor data. *Sensors* **2021**, *21*, 6014. [CrossRef] [PubMed]
19. Mahlberg, J.A.; Li, H.; Cheng, Y.T.; Habib, A.; Bullock, D.M. Measuring roadway lane widths using connected vehicle sensor data. *Sensors* **2022**, *22*, 7187. [CrossRef] [PubMed]
20. Saldivar-Carranza, E.; Li, H.; Mathew, J.; Hunter, M.; Sturdevant, J.; Bullock, D.M. Deriving operational traffic signal performance measures from vehicle trajectory data. *Transp. Res. Rec.* **2021**, *2675*, 1250–1264. [CrossRef]
21. Hunter, M.; Saldivar-Carranza, E.; Desai, J.; Mathew, J.K.; Li, H.; Bullock, D.M. A proactive approach to evaluating intersection safety using hard-braking data. *J. Big Data Anal. Transp.* **2021**, *3*, 81–94. [CrossRef]
22. Magnusson, P.; Svantesson, T. Road Condition Monitoring. U.S. Patent 10,953,887 B2, 23 March 2021.
23. Ministerio de Transportes y Movilidad Sostenible. Mapa de Tráfico. 2021. Available online: https://www.mitma.es/carreteras/trafico-velocidades-y-accidentes-mapa-estimacion-y-evolucion/mapas-de-trafico/2021 (accessed on 1 September 2024).

infrastructures

MDPI

Article

Research on Asphalt Pavement Surface Distress Detection Technology Coupling Deep Learning and Object Detection Algorithms

Hong Zhang [1,*], Yuanshuai Dong [1,2,3], Yun Hou [2,3], Xiangjun Cheng [2,3], Peiwen Xie [3] and Keming Di [4]

[1] Department of Civil Engineering, Chongqing Jiaotong University, Chongqing 400074, China; dys_bj@mails.cqjtu.edu.cn
[2] China Highway Engineering Consulting Corporation, Beijing 100089, China; 13042365036@mails.cqjtu.edu.cn (Y.H.); 202321194@mail.sdu.edu.cn (X.C.)
[3] China Highway Engineering Consulting Corporation DATA Co., Ltd., Beijing 100089, China; cheng@mails.cqjtu.edu.cn
[4] Shanxi Provincial Highway Bureau Linfen Branch, Linfen 041000, China; jianan@mails.cqjtu.edu.cn
* Correspondence: zhanghong1997@mails.cqjtu.edu.cn; Tel.: +86-18-2815-70337

Abstract: To address the challenges posed by the vast scale of highway maintenance in China and the high costs associated with traditional inspection vehicles. This study focuses on a routine maintenance project for national and provincial roads in Shanxi Province, with an emphasis on the selection and design of hardware for lightweight, portable pavement inspection devices. A monocular camera was used to capture pavement surface images, resulting in a dataset of 85,511 training samples. Additionally, the YOLOv5 object detection algorithm, combined with convolutional deep learning techniques, was employed to classify and identify pavement surface distresses in the collected images. Through multiple iterations of model tuning and validation, the proposed detection system achieved a false negative rate of 1.13%, a recall rate of 97.35%, and a precision rate of 98.30%. Its high accuracy provides a technical reference for the development and design of portable pavement distress detection devices.

Keywords: pavement engineering; asphalt pavement maintenance; pavement surface inspection; deep learning; object detection

1. Introduction

By the end of 2024, the total highway mileage in China had reached 5.4368 million kilometers, with 99.9% of this network being maintained. Over the past several years, the scope of road maintenance has consistently expanded, demonstrating a steady upward trend. However, large-scale specialized inspection equipment faces significant limitations in terms of detection costs, data acquisition, and processing, creating a major bottleneck in road maintenance management. Additionally, China's extensive rural road network remains largely underserved by maintenance interventions due to inspection constraints. To address these challenges, this project aims to develop a compact, portable, and cost-effective device for the rapid collection, detection, and assessment of road conditions. The system will also facilitate timely data processing and analysis, ensuring a continuous flow of data to support informed decision-making in road maintenance management.

In the 1970s, the French road management authorities developed the GERPHO system. In the 1980s, a Japanese research team designed the Komatsu system, which was based on analog video technology [1]. By the 1990s, American researchers had introduced the Pavement Condition Evaluation Service (PCES) system, utilizing a line-scan digital camera for pavement distress detection [2,3]. Compared to its predecessors, the PCES system enabled simultaneous image acquisition and processing. However, its functionality was limited, as it was unable to distinguish between different types of pavement distresses [4]. During the same period, rapid advancements in Charge-Coupled Device (CCD) image sensor technology led a Canadian company to develop the Automatic Road Analyzer (ARAN) system [5]. Despite its innovations, the ARAN system faced high hardware costs and lacked synchronization between pavement data collection and distress identification [6,7]. Since the early 21st century, 3D laser scanning technology has rapidly evolved. Researchers have integrated this technology into pavement distress detection systems, such as the DHDV system in the United States, which achieved true automation in detecting pavement distresses. However, its high hardware requirements and maintenance costs have hindered widespread adoption [8]. More recently, studies have utilized 3D point cloud data and YOLOv5 detection models to identify various types of pavement distresses, including longitudinal cracks, transverse cracks, alligator cracks, and potholes. Notable contributions in this field include the work of Ravi Radhika and Ayman Habib from the United States [9], as well as Sami Abdullah from Australia and Sakib Saadman from Bangladesh [10–13]. Early pavement distress detection methods predominantly relied on traditional texture feature extraction techniques, such as Local Binary Patterns (LBP) and Gabor filters, which often yielded suboptimal performance in practical applications. In contrast, recent advancements in deep learning technologies have enabled the direct learning of feature representations from data, significantly improving the accuracy of pavement distress identification. By applying deep learning-based image processing algorithms to pavement images, researchers can accurately detect and localize distresses. The detected distress regions can then be segmented to extract geometric features such as area, length, and width. Statistical analysis of all detected distresses allows for the calculation of the distress rate (DR) for the evaluated road section, ultimately leading to the derivation of the Pavement Condition Index (PCI). Several other scholars have also made significant contributions to pavement distress detection [13–18].

In comparison to foreign countries, China's development of automatic pavement inspection technology began relatively late. However, since the early 21st century, advancements in hardware capabilities and significant progress in image processing technologies have enabled substantial achievements by domestic research institutes and universities. Several Chinese scholars have conducted extensive research on automatic pavement inspection systems, making significant contributions to image processing techniques for pavement images. Their work has greatly advanced the application of digital image processing technology in road inspection [19–21]. In China, pavement inspection technologies are primarily classified into four categories: deep learning-based methods, 3D laser-based techniques, vibration signal-based detection, and pavement texture analysis [21–28].

With the advancement of the 2025 initiative, the digitalization, informatization, and automation of road maintenance have been rapidly progressing. In recent years, a variety of lightweight inspection devices have emerged on the market. Companies such as Baidu, Qianxun, and Shanghai Tonglu Cloud, as well as universities including Tongji University and Southeast University, have actively engaged in research and development. The hardware used in these devices is largely similar, with most relying on monocular cameras, stereo cameras, or industrial cameras to capture road surface data. The collected images are subsequently processed for various recognition tasks to generate pavement inspection data. This study aims to develop a pavement distress detection system by selecting suitable hardware components and utilizing a monocular camera. An enhanced YOLOv5 algorithm is adopted as the target detection model, incorporating convolutional deep learning techniques to classify pavement distress images captured by the monocular camera. By constructing a distress dataset and conducting debugging, validation, and comparative experiments, the proposed model undergoes training and technical validation. Ultimately, the system facilitates the extraction of key physical characteristics of pavement distresses, including length, width, and area, thereby achieving accurate pavement distress detection.

2. Hardware Selection and Algorithm Principles

2.1. Hardware Selection

Based on the demand assessment, the hardware platform must meet the following requirements: (1) The CPU should have a minimum of four cores. (2) The encoding and decoding capabilities must support the processing of two channels of 4K 30 fps video streams. (3) The system should be equipped with at least 8 GB of RAM. (4) A minimum of two USB 3.0 ports should be available. (5) The platform must support hard disk read/write operations and include an HDMI interface for display connectivity.

2.1.1. Hardware Platform

A comparative analysis was performed on several embedded platforms from NVIDIA, including the TX2 and Jetson Xavier, as well as processing platforms such as Raspberry Pi, BeagleBone, and Huawei, as shown in Figure 1. Initially, considering both processor performance and physical size, the TX2 was deemed too large, while platforms such as the Raspberry Pi lacked sufficient computational power and stability. As a result, the Jetson AGX Xavier platform was selected for this study. However, due to its high cost, future iterations of the system may consider adopting the Xavier NX series to optimize cost control.

Figure 1. Nvidia Xavier platform.

Considering the installation environment, the camera will be mounted on a data collection vehicle. The vertically installed lens is typically positioned 1 to 3 m above the ground, while the side-mounted lens is generally placed 1 to 10 m from the guardrail. Given these conditions, a short-focus lens with a focal length of 8 mm is selected for the pavement camera.

Currently, positioning systems include GPS, BeiDou, Galileo, and GLONASS. The full deployment of the BeiDou satellite network in China provides robust support for positioning-related equipment and significantly improves location accuracy. In crack detection, the algorithm must accurately determine the location of detected pavement distresses, enabling detailed analysis of distress distribution across entire road segments. These data are crucial for effective road maintenance planning and decision-making. To ensure high-precision positioning, the system will employ a multi-system, multi-frequency positioning board that supports BeiDou navigation to further enhance accuracy.

2.1.2. Camera Selection

Industrial cameras are available with various interface types, including USB, Ethernet (GigE), CameraLink, IEEE 1394, and CoaXPress. When conducting outdoor data collection, industrial cameras offer superior resistance to environmental variations, dust, and vibrations, ensuring greater reliability in complex conditions. To achieve a 1 mm crack resolution over a 4 m wide pavement section, a minimum of 4000 pixels is required. Since pavement inspection typically focuses on a single lane per capture, a 2D area-scan camera is preferred over a line-scan camera.

Industrial camera sensors are typically classified into two types: CMOS and CCD, with CMOS sensors becoming increasingly prevalent. Additionally, industrial cameras use two exposure methods: (1) Global Shutter, which captures the entire image simultaneously; and (2) Rolling Shutter, which exposes different rows at different times, potentially causing distortion when capturing moving objects. Given the vehicle's motion speed, the image capture response time must be as fast as possible, and the exposure time should be minimized while ensuring image clarity. Therefore, a global shutter CMOS camera is selected to guarantee distortion-free imaging

FLIR offers both global shutter and rolling shutter cameras, such as the Flea3 USB3 series. However, these cameras have a resolution range limited to 1K to 2K pixels. In contrast, the ORYX 10GigE can capture 4K resolution, 12-bit images at over 60 FPS, but it uses a GigE interface.

In contrast, the Huarui A7A20MU30 from Zhejiang Dahua Technology achieves 4K resolution with a USB 3.0 interface, global shutter, and C-mount compatibility. It supports the USB3 Vision protocol and the GenICam standard, and features a pixel size of 3.45 μm × 3.45 μm. A USB-interface camera was selected for its plug-and-play functionality, which facilitates rapid prototype development and allows for easy future upgrades with similar or higher-performance cameras. Consequently, the Dahua 4K industrial camera was chosen for this study (as shown in Figure 2).

Figure 2. 4K industrial camera.

2.1.3. Positioning System

Currently, positioning systems primarily include GPS, GNSS (Global Navigation Satellite System), Galileo, and GLONASS. With the full deployment of the BeiDou satellite network in China, it provides substantial support for positioning-related equipment and significantly improves location accuracy. For crack detection, the algorithm requires precise geolocation of identified distresses, enabling detailed analysis of distress distribution across entire road segments and offering essential data for road maintenance decision-making. To enhance the reliability and accuracy of detection data, this study selects a high-precision, multi-system, multi-frequency positioning board that supports dual-mode BeiDou + GPS positioning. This module features low power consumption and a compact size (as shown in Figure 3), making it easy to integrate into vehicle-mounted and other automated inspection systems.

Figure 3. Gradation curves.

2.1.4. Movable Central Control System

The data acquisition platform functions as the central control and operation hub for the automated inspection system (as shown in Figure 4). It integrates a positioning data processing module, power module, display unit, and an embedded core processing platform, while incorporating a high-capacity 1 TB solid-state drive (SSD) to support large-scale 4K image data storage.

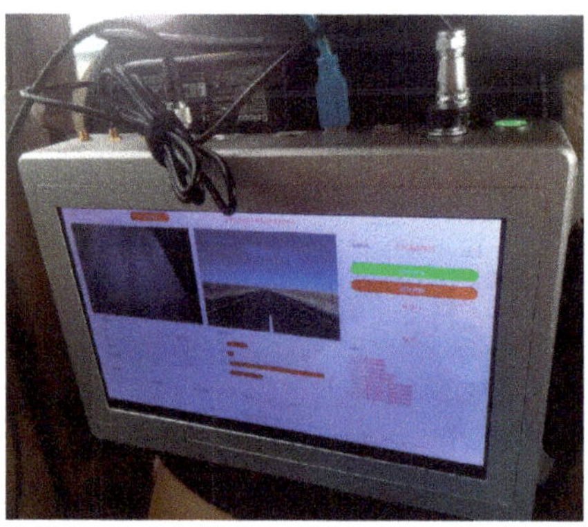

Figure 4. Movable central control system.

The device housing has been modified to include connection ports, ensuring efficient interconnectivity among all components. The display unit supports touchscreen operation and is equipped with customized software that includes functionalities such as algorithm parameter configuration, algorithm encapsulation and execution, result visualization, image list display, and processing progress tracking. Additionally, the system supports large-panel display operation for pavement distress detection, thereby significantly enhancing workflow efficiency for maintenance personnel.

2.1.5. Car Triangular Bracket

To facilitate the integration and utilization of various modular components within the pavement distress detection system—including the solid-state storage drive—and to ensure seamless interface connectivity, signal transmission, and external data output, a custom-designed mechanical enclosure was developed to house all system components in a unified structure.

A dedicated hard drive support and fixation bracket was designed, featuring a tripod with suction cups for stability. The industrial camera and high-resolution lens are screw-mounted onto the bracket, which is securely positioned in a triangular configuration at the rear of the vehicle (as shown in Figure 5). A balancing mechanism is incorporated between the camera and the mounting structure, enabling the camera to capture pavement images at a fixed angle relative to the vertical axis, ensuring optimal image acquisition.

Figure 5. Vehicle-mounted triangular bracket.

2.2. Algorithm Principle and Selection

The computational principles employed in this study primarily include deep learning theory and object detection algorithms.

2.2.1. Data Collection

The core principle of deep learning is to extract hierarchical feature representations through layered networks, which enables the learning and abstraction of complex features. Convolutional Neural Networks (CNNs) are widely used in image classification and object detection tasks. Their architecture is primarily composed of three components: the input layer, hidden layers, and output layer.

In this study, a CNN-based model is employed for training, which consists of two main stages. The first stage is the forward propagation phase, during which data flows from lower-level to higher-level representations. The second stage is backpropagation, in which, if the predicted results deviate from the expected outcomes, the error is propagated backward from higher to lower layers, facilitating model optimization and improved accuracy.

2.2.2. Object Detection Algorithms

Object detection algorithms can be broadly classified into two types. The first type consists of Region Proposal-based methods, such as R-CNN, Fast R-CNN, and Faster R-CNN, which follow a two-stage approach. These methods first generate region proposals using heuristic techniques, such as Selective Search or a CNN-based Region Proposal Network (RPN). Classification and regression are then performed on these proposals. The second type includes one-stage methods, such as YOLO and SSD, which directly predict object categories and locations using a single CNN network. A typical object detection model consists of a feature extraction backbone and a classification-detection network. The design of the backbone network is crucial in determining how features are fused across different network layers. With advancements in deep learning, backbone networks are continuously optimized to enhance feature extraction.

The YOLOv5 network, a representative one-stage detection model, comprises four main components: the input layer, backbone network (Backbone), feature fusion network (Neck), and output layer (Prediction). Compared to other models, YOLOv5 offers superior efficiency and speed, making it particularly well suited for high-frequency pavement distress detection. The architecture of YOLOv5 is shown in Figure 6. This study will perform a comparative analysis to select the optimal YOLOv5 model and integrate deep learning techniques to enhance the accuracy and robustness of pavement distress detection.

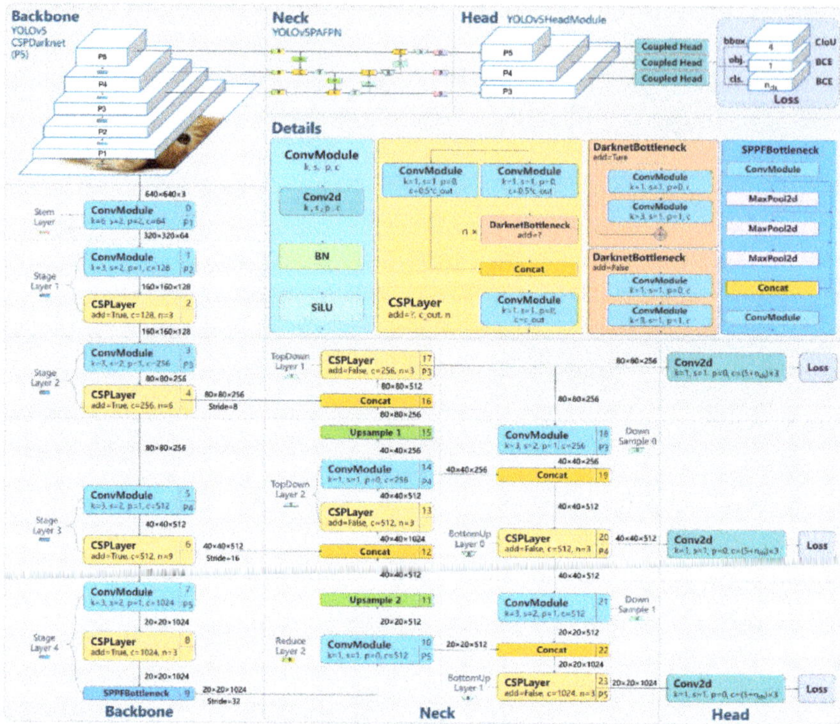

Figure 6. The network structure of YOLOv5.

YOLOv5 offers several backbone network architectures with varying depths, including YOLOv5s, YOLOv5m, YOLOv5l, and YOLOv5x. These four structures are governed by two key parameters: model depth and the channel width between layers. The number of model parameters increases progressively from YOLOv5s (the smallest) to YOLOv5x (the largest). Specific parameter settings and channel configurations are provided in the Table 1.

Table 1. Parameters and parameters of backbone networks at each level.

Backbone Network	Model Depth	The Width of the Interstory Passage	Parameter Size/KB
YOLOv5s	0.33	0.5	14,468
YOLOv5m	0.67	0.75	42,367
YOLOv5l	1.0	1.0	93,086
YOLOv5x	1.33	1.25	173,370

Using the same Cross Stage Partial (CSP) structure for comparison, the YOLOv5s network includes one residual block, while YOLOv5m utilizes two residual blocks, YOLOv5l employs three residual blocks, and YOLOv5x incorporates four residual blocks at the same locations. This design effectively controls the network's depth. Additionally, the network width in YOLOv5 is regulated by adjusting the number of convolutional kernels at different stages. Taking the Focus structure as an example, YOLOv5s uses 32 convolutional kernels, YOLOv5m increases this to 48, YOLOv5l to 64, and YOLOv5x to 80.

3. Research on the Construction of Disease Datasets and Scheme Design

3.1. Construction of Road Disease Detection Dataset

In this study, the pavement distress detection dataset was constructed with a strict proportional division into a training set, validation set, and test set. By collecting real-time traffic photos on the road after the equipment is assembled, the dataset can categorize pavement distresses into five types: transverse cracks, longitudinal cracks, alligator cracks, potholes, and patched areas. A total of 85,511 pavement images were collected from actual road surfaces, and 14,641 valid samples were selected to build the dataset. The training and test sets were split into a 9:1 ratio, resulting in 13,388 images for training and 1253 images for testing. Additionally, during training, a validation set was extracted from the training set at a 9:1 ratio to update the model weights, resulting in 1338 validation images. Annotated examples of pavement distresses in the dataset are illustrated in Figure 7.

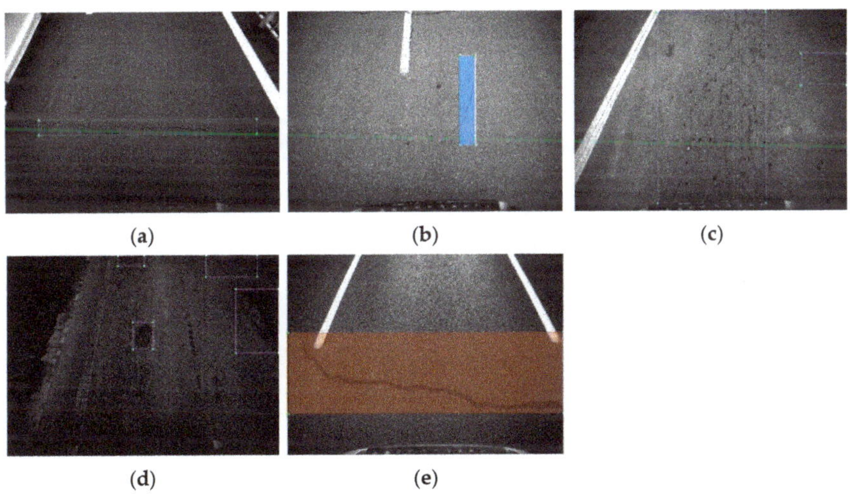

Figure 7. Legend of the road disease dataset: (**a**) Transverse crack disease (**b**) Longitudinal crack disease (**c**) Two reticular fissure diseases. (**d**) Four pit diseases (**e**) Repair disease.

3.2. Model Training Design

3.2.1. Platform Environment Deployment

In this study, the YOLOv5 object detection algorithm was trained using a self-constructed pavement distress dataset and applied to identify distress types on actual road surfaces. The platform and development environment consist primarily of a Linux-based system and the Python programming language.

3.2.2. Dataset Preparation

The YOLOv5 network does not directly read dataset images and annotations, as this may result in insufficient memory when processing a large number of images. To address this issue, the dataset and annotation files are first stored in a designated directory during training. The image filenames, labels, and bounding box coordinates are then extracted into a single text file, which YOLOv5 reads in batches. This approach prevents memory overflow and enhances data loading efficiency.

During the preprocessing stage, training images are stored in the JPEGImages directory, while annotation files are placed in the Annotations directory. The voc_label.py script is employed to automatically divide the dataset into training and validation sets, generating a labels directory. Within this directory, all annotation data are recorded in text files, including image filenames, label names, and bounding box coordinates (upper-left and lower-right points).

Before training, the dataset parameters must be modified in the test.yaml file. These parameters include the file paths for the training and test datasets, the number of classes, and the class names, which are provided as a string array. The structure of each dataset is illustrated in Figures 8–11 the following figure.

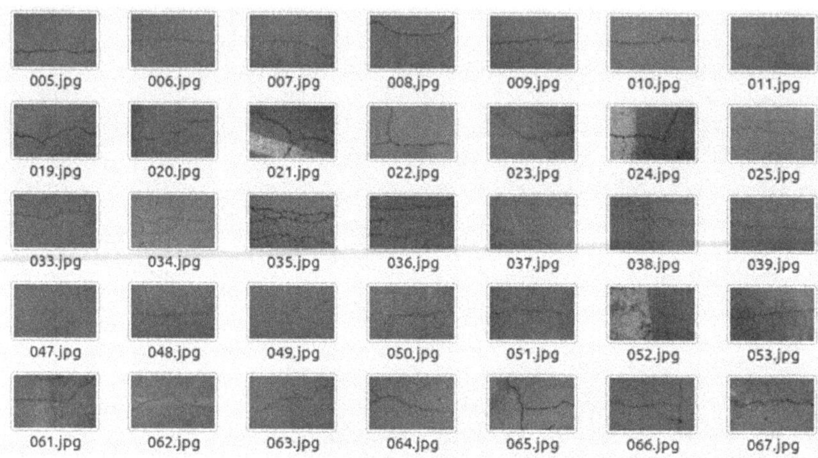

Figure 8. The CrackForest dataset.

Figure 9. The 4K dataset.

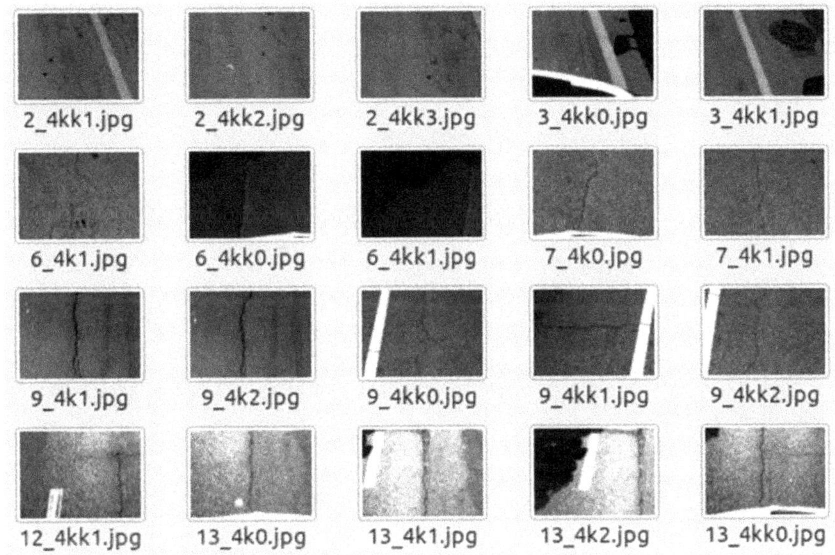

Figure 10. The 4K fixed-size split dataset.

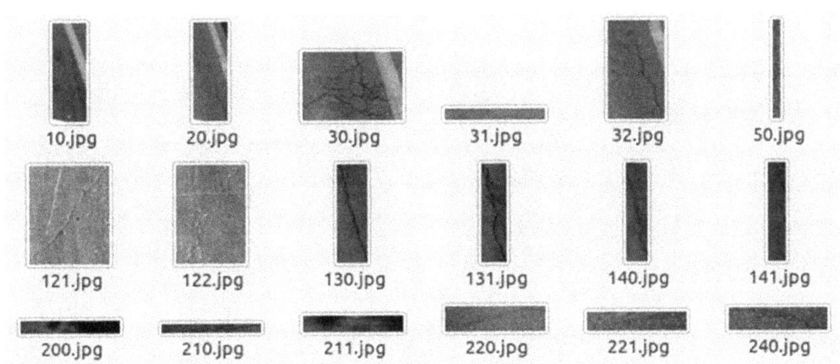

Figure 11. The 4K arbitrarily cropped dataset.

3.2.3. Training Parameter Adjustments

Before initiating training, the YOLOv5 algorithm provides configurable network parameters, including the number of training iterations, the batch size for image loading per training step, the configuration file for the backbone network, and the input image size.

3.3. Methods for Analyzing Test Results

In this study, recall, precision, and the false-negative rate are used as evaluation metrics for pavement distress detection. True Positive (TP) refers to a correctly predicted positive instance, True Negative (TN) to a correctly predicted negative instance, False Positive (FP) to a negative instance incorrectly predicted as positive, and False Negative (FN) to a positive instance incorrectly predicted as negative.

(1) Recall detection rate

Recall is the ratio of correctly identified instances to the total number of instances that should be identified in the dataset.

$$Recall = TP/(TP + FN) \tag{1}$$

(2) Precision detection rate

Precision is the ratio of correctly identified instances to the total identified instances.

$$Precision = TP/(TP + FP) \tag{2}$$

(3) False-negative rate

The false-negative rate is the ratio of incorrectly unidentified instances to the total number of actual samples.

$$FNR = FN/(TP + FN) \tag{3}$$

(4) Test results

Figure 12 shows the detection results of YOLOv5 networks under different backbone networks.

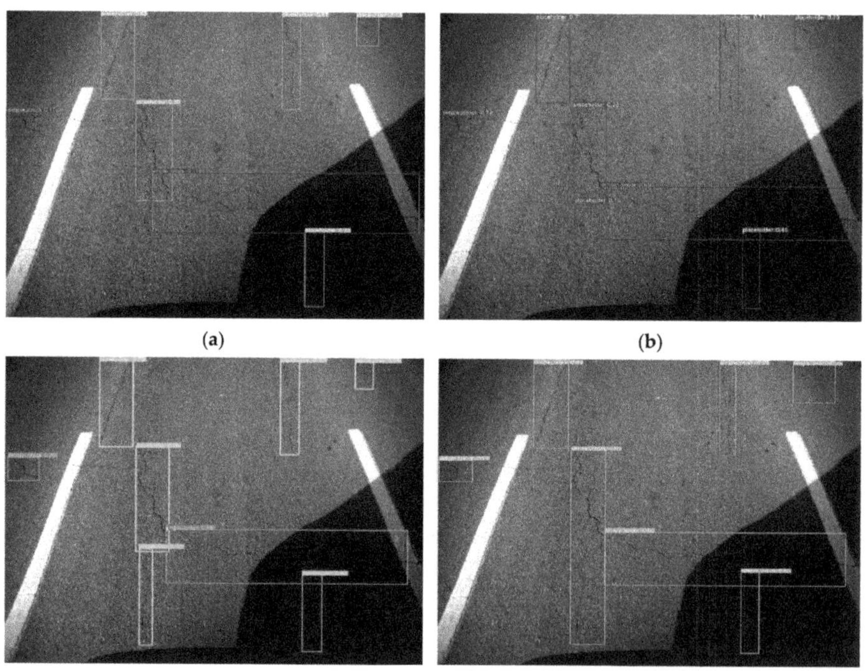

(a) (b) (c) (d)

Figure 12. Detection results of different models of YOLOv5: (**a**) YOLOv5s, (**b**) YOLOv5m, (**c**) YOLOv5l, (**d**) YOLOv5x.

4. Analysis of Asphalt Pavement Surface Damage Detection Technology

4.1. Demonstration and Analysis of the Experimental Results of Object Detection Algorithm

The initial validation experiment in this study aimed to assess the feasibility of the algorithm. This experiment compared the detection performance of the instance segmentation algorithm Mask R-CNN and the object detection algorithm YOLOv5 using the same set of pavement images. Mask R-CNN performs instance segmentation by first locating objects with bounding boxes and then conducting pixel-wise classification within each bounding box to segment the target objects. In contrast, YOLOv5, as an object detection algorithm, only locates objects using bounding boxes without performing pixel-level segmentation.

As shown in Figure 13, the results of the validation experiment indicate that Mask R-CNN exhibits significant deviation in the bounding box localization stage, leading to inaccuracies in the segmentation phase, where it struggles to fully segment the cracks. In contrast, YOLOv5 effectively locates pavement cracks and correctly identifies them individually, demonstrating its suitability for pavement distress detection.

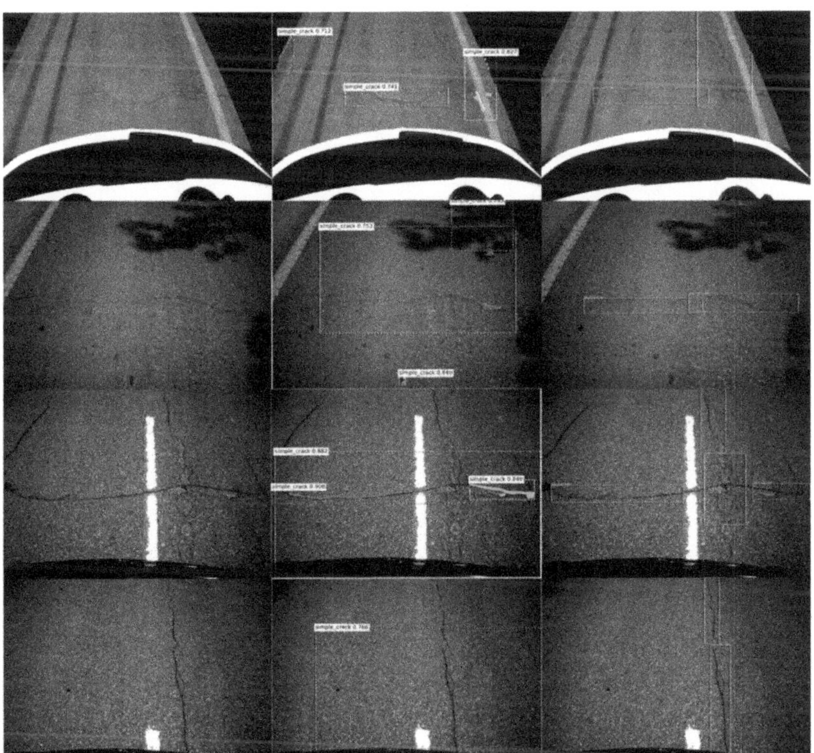

(a) (b) (c)

Figure 13. Comparison of Mask-RCNN and YOLOv5: (**a**) Original images, (**b**) Mask-RCNN, (**c**) YOLOv5.

Through four rounds of experimental validation, this study found that the algorithm's generalization ability is directly related to the size of the dataset. Based on these findings, the final validation phase focused on significantly expanding the pavement distress dataset while ensuring consistent categorization.

During the validation process, pavement distresses were strictly classified into five categories: transverse cracks, longitudinal cracks, alligator cracks, potholes, and patched areas. The dataset was divided into training and validation sets in a 9:1 ratio, resulting in a total of 13,388 image samples for training.

As shown in Figure 14, the results from the four validation experiments demonstrate clear and accurate distress localization and classification. These findings confirm that the YOLOv5 algorithm achieves a high level of model fitting and generalization when applied to a comprehensive pavement distress dataset. Statistical analysis reveals an overall false-negative rate of 1.13%, a recall rate of 97.35%, and a precision rate of 98.30%.

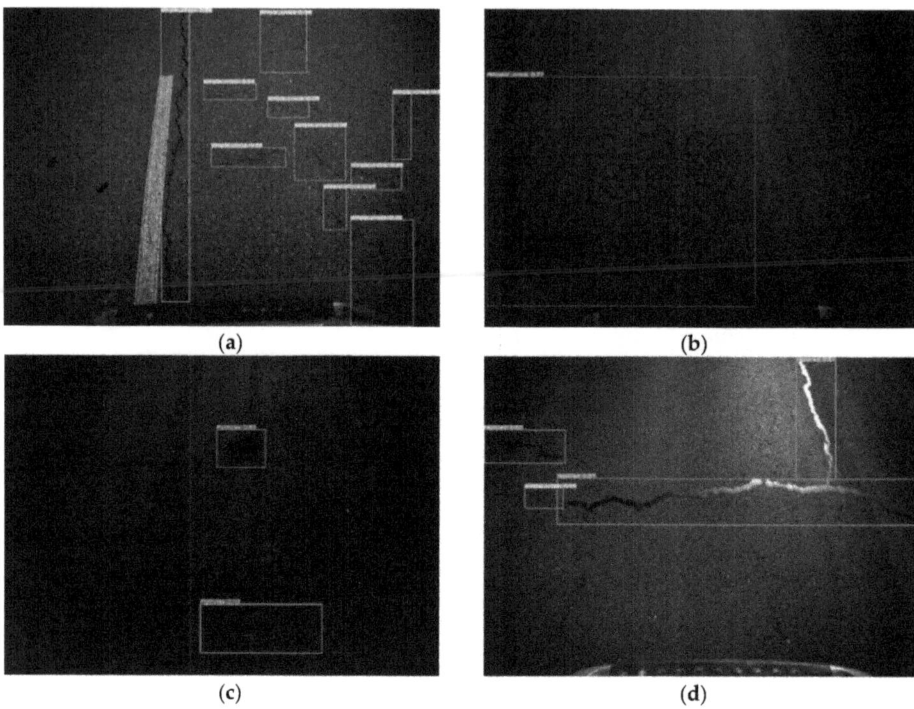

Figure 14. The results of the fourth verification: (**a**) Crack disease, (**b**) Reticular fissure diseases, (**c**) Pit diseases, (**d**) Repair disease.

The training results of the YOLOv5 object detection model developed in this study are shown in Figure 15. It can be observed that, during the 200-step iterative process, the GIOU loss and other loss functions exhibit a gradual decrease, while accuracy and recall steadily increase, eventually converging to stable values.

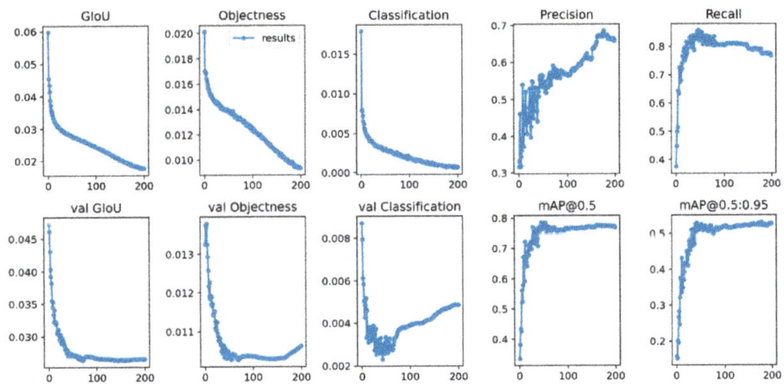

Figure 15. Model training results.

4.2. Analysis of Camera Calibration Algorithms

The calibration images captured for the experiment are shown in Figure 16. Among the 28 images used in this study, the position and angle of the calibration board vary across images, effectively covering the entire field of view of the camera. This diverse positioning strategy significantly reduces calibration errors.

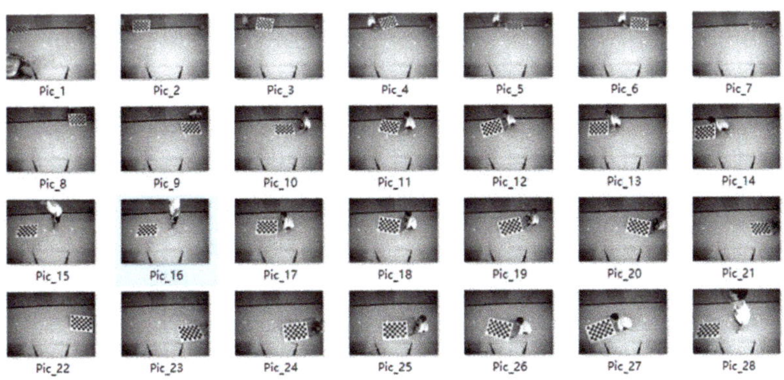

Figure 16. Images used for camera calibration.

First, a corner detection algorithm was applied to identify the subpixel locations of all corner points in each image. The yellow rectangular box represents the origin corner point, while the green markers indicate the subpixel coordinates of each detected corner. Using the complete set of corner point data from all images, the camera's intrinsic and extrinsic parameters, as well as distortion coefficients, were optimized using Zhang's calibration method, the least squares method, and the Levenberg–Marquardt (LM) algorithm. The 3D spatial positions of the monocular camera and all calibration boards are illustrated in Figure 17.

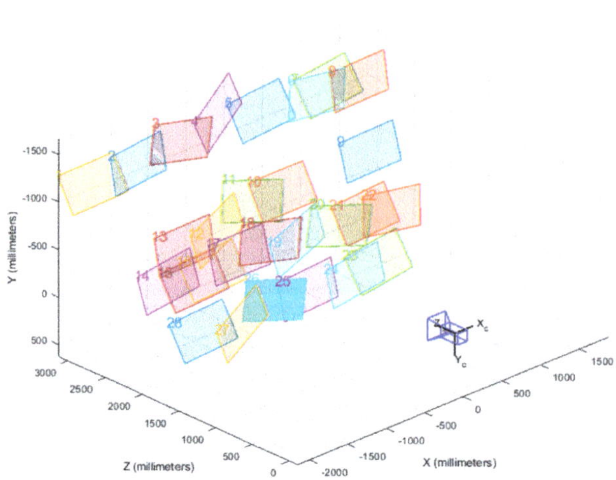

Figure 17. Three−dimensional spatial position of the monocular camera and the calibration plate.

4.3. Measurement Index Table

After obtaining the accurate intrinsic and extrinsic camera parameters, along with the distortion coefficients, an index table can be established. Since this process involves only matrix operations, no additional errors are introduced, and the index table's error remains consistent with the calibration error of 0.52 pixels.

Given that the camera has a 4K resolution with 3000 $\times$ 4096 pixels, the index table contains data for approximately 12 million points, fully representing the physical spatial information of the captured image, as illustrated in Figure 18. Each entry in the index table stores four key pieces of information for each pixel: its corresponding physical length, width, diagonal length, and area, as shown in Figure 19. Ultimately, the index table comprises 48 million precise parameters, which serve as a reference for subsequent pavement distress geometric measurement algorithms.

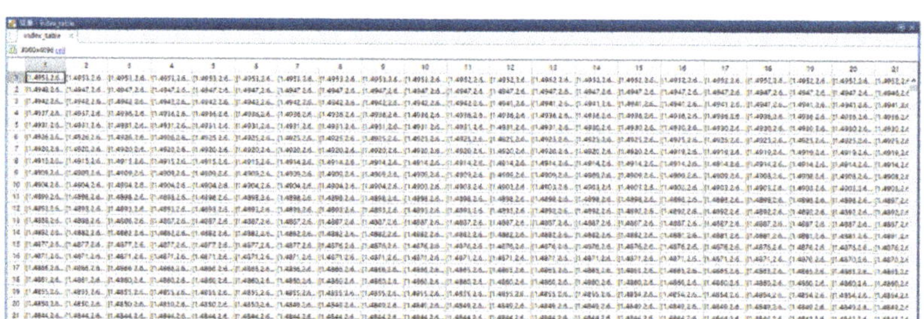

Figure 18. Index table.

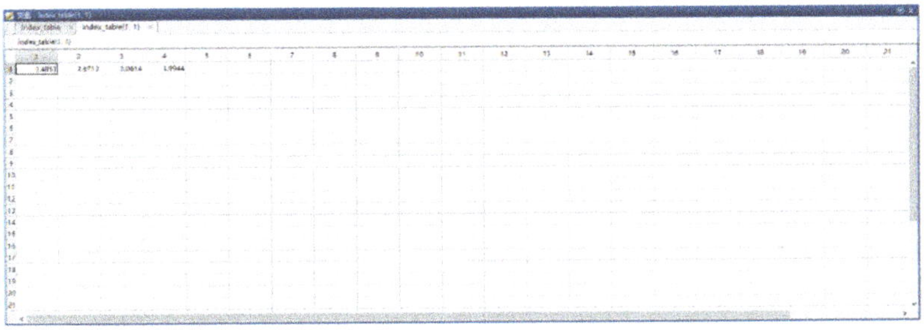

Figure 19. Four parameters for each pixel.

4.4. Analysis of Broken Geometry Information Measurement Algorithm

Due to the irregular geometric shape of pavement distresses, this study first calculates the length and then derives the average width. However, the accuracy of the distress skeletonization algorithm directly affects the width measurement error. To obtain the actual physical length of the distress, the binarized mask image must be processed into a skeletonized image, which consists of a single-pixel-wide representation of the distress region. The skeletonized image must be continuous and free of discontinuities or noise artifacts to ensure the most accurate length estimation.

The skeletonization algorithm used in this study is a lookup table-based method that iteratively applies logical operations to the outermost white pixels of the distress region. The algorithm determines whether a pixel belongs to the skeleton based on its relationship with neighboring pixels. If the pixel is part of the skeleton, it is retained; otherwise, it is removed. This process ultimately reduces the distress mask to a single-pixel-wide skeletonized representation.

This study categorizes pavement distresses into four major types: simple cracks, alligator cracks, potholes, and patched areas. Among these, potholes and patched areas are excluded from the length and width calculations. The study further classifies simple cracks and alligator cracks into six subtypes, as illustrated in Figure 20. Simple cracks are subdivided into transverse cracks, longitudinal cracks, diagonal cracks, and herringbone cracks, while alligator cracks are divided into simple alligator cracks and complex alligator cracks.

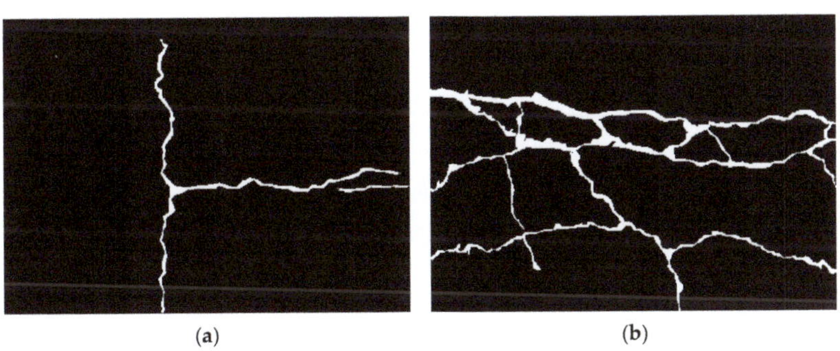

Figure 20. Four parameters for each pixel: (**a**) Simple cracks, (**b**) Reticulated cracks.

Figure 21 presents the skeletonized images extracted from the six types of pavement distress masks using the lookup table method. As shown in the figure, the width varies across different regions of the distress masks; however, the lookup table method accurately identifies the primary direction of distress propagation. Furthermore, the extracted skeleton images remain continuous, without disruptions or artifacts, ensuring that length calculations are not affected by noise or distortions. With a well-defined skeleton image, the index table can be used to retrieve the physical length corresponding to each pixel position, and these values are summed to obtain the actual physical length and average width of the distress region.

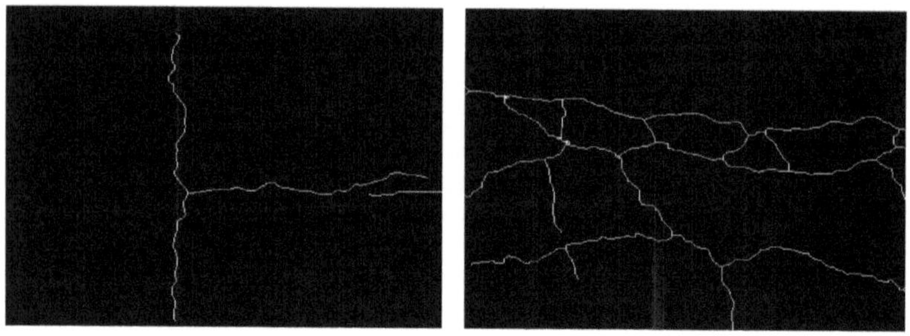

(a) (b)

Figure 21. Four parameters for each pixel: (**a**) Simple cracks, (**b**) Reticulated cracks.

In summary, the distress geometry measurement algorithm enables the precise extraction of physical spatial parameters based solely on the calibration-generated index table and the semantic segmentation-derived distress mask, providing an effective method for measuring pavement distress geometries in real-world conditions.

5. Conclusions

(1) Through the hardware selection and design process, a lightweight and portable pavement distress detection device was assembled, offering an efficient and practical solution for on-site pavement inspection.

(2) By integrating the YOLOv5 object detection algorithm with convolutional deep learning techniques, a model was trained using 85,511 pavement sample images. The final statistical results show an overall false-negative rate of 1.13%, a recall rate of 97.35%, and a precision rate of 98.30%, demonstrating the model's high accuracy and reliability.

(3) Algorithm validation and analysis confirmed that the distress geometry measurement algorithm can accurately extract physical spatial parameters using only the calibration-generated index table and the semantic segmentation-derived distress mask. The study concludes that the developed pavement distress detection device has significant potential for practical engineering applications.

6. Prospect

(1) This study utilizes a monocular camera, which effectively identifies two-dimensional pavement distresses, such as transverse and longitudinal cracks, alligator cracking, block cracking, and patched areas. However, it currently lacks the capability to accurately detect three-dimensional distresses, such as potholes and subsidence. In the future, a stereo camera system could be implemented, incorporating existing equipment algorithms and advanced technical approaches to enable comprehensive pavement distress detection.

(2) This study focuses solely on pavement distress detection for asphalt surfaces. Given the distinct differences between cement and asphalt pavement distresses, future research could explore established distress recognition methodologies to develop an automated detection system for cement pavement distresses.

Author Contributions: Formal analysis, Y.D.; Methodology, Y.D.; Resources, Y.D. and Y.H.; Supervision, Y.D., Y.H., X.C., P.X. and K.D.; Validation, Y.D.; Writing—review & editing, H.Z. and Y.D. All authors have read and agreed to the published version of the manuscript.

Funding: This research received no external funding.

Data Availability Statement: The data used to support the findings of this study are included within the article.

Acknowledgments: Thanks for China Highway Engineering Consulting Corporation support.

Conflicts of Interest: Authors Yuanshuai Dong, Yun Hou and Xiangjun Cheng were employed by the company China Highway Engineering Consulting Corporation. Authors Yuanshuai Dong, Yun Hou, Xiangjun Cheng and Peiwen Xie were employed by the company China Highway Engineering Consulting Corporation DATA Co., Ltd. The remaining authors declare that the research was conducted in the absence of any commercial or financial relationships that could be construed as a potential conflict of interest.

References

1. Sermanet, P.; Eigen, D.; Zhang, X.; Mathieu, M.; Fergus, R.; LeCun, Y. Overfeat: Integrated recognition, localization and detection using convolutional networks. In Proceedings of the 2nd International Conference on Learning Representations, ICLR 2014, Banff, AB, Canada, 14–16 April 2014.
2. Girshick, R.; Donahue, J.; Darrell, T.; Malik, J. Rich feature hierarchies for accurate object detection and semantic segmentation. In Proceedings of the 27th IEEE Conference on Computer Vision and Pattern Recognition, CVPR 2014, Columbus, OH, USA, 23–28 June 2014; IEEE Computer Society: Washington, DC, USA, 2014.
3. Uijlings, J.R.; Van De Sande, K.E.; Gevers, T.; Smeulders, A.W. Selective Search for Object Recognition. *Int. J. Comput. Vis.* **2013**, *104*, 154–171. [CrossRef]
4. He, K.; Zhang, X.; Ren, S.; Sun, J. Spatial Pyramid Pooling in Deep Convolutional Networks for Visual Recognition. *IEEE Trans. Pattern Anal. Mach. Intell.* **2015**, *37*, 1904–1916. [CrossRef] [PubMed]
5. Girshick, R. Fast R-CNN. In Proceedings of the 15th IEEE International Conference on Computer Vision, ICCV 2015, Santiago, Chile, 11–18 December 2015.
6. Ren, S.; He, K.; Girshick, R.; Sun, J. Faster R-CNN: Towards Real-Time Object Detection with Region Proposal Networks. *IEEE Trans. Pattern Anal. Mach. Intell.* **2017**, *39*, 1137–1149. [CrossRef]
7. Redmon, J.; Divvala, S.; Girshick, R.; Farhadi, A. You only look once: Unified, real-time object detection. In Proceedings of the 29th IEEE Conference on Computer Vision and Pattern Recognition, CVPR 2016, Las Vegas, NV, USA, 26 June–1 July 2016.
8. Liu, W.; Anguelov, D.; Erhan, D.; Szegedy, C.; Reed, S.; Fu, C.Y.; Berg, A.C. SSD: Single shot multibox detector. In Proceedings of the 14th European Conference on Computer Vision, ECCV 2016, Amsterdam, The Netherlands, 8–16 October 2016.
9. Ravi, R.; Habib, A.; Bullock, D. Pothole Mapping and Patching Quantity Estimates using LiDAR-Based Mobile Mapping Systems. *Transp. Res. Rec. J. Transp. Res. Board* **2020**, *2674*, 124–134. [CrossRef]
10. Fakhri, S.A.; Satari Abrovi, M.; Zakeri, H.; Safdarinezhad, A.; Fakhri, A. Pavement crack detection through a deep-learned asymmetric encoder-decoder convolutional neural network. *Int. J. Pavement Eng.* **2023**, *24*, 2255359. [CrossRef]
11. Sami, A.A.; Sakib, S.; Deb, K.; Sarker, I.H. Improved YOLOv5-Based Real-Time Road Pavement Damage Detection in Road Infrastructure Management. *Algorithms* **2023**, *16*, 452. [CrossRef]
12. Hedeya, M.A.; Samir, E.; El-Sayed, E.; El-Sharkawy, A.A.; Abdel-Kader, M.F.; Moussa, A.; Abdel-Kader, R.F. A Low-Cost Multi-sensor Deep Learning System for Pavement Distress Detection and Severity Classification. In Proceedings of the 8th International Conference on Advanced Machine Learning and Technologies and Applications (AMLTA2022), Cairo, Egypt, 5–7 May 2022; pp. 21–33.
13. Matarneh, S.; Elghaish, F.; Al-Ghraibah, A.; Abdellatef, E.; Edwards, D.J. An automatic image processing based on Hough transform algorithm for pavement crack detection and classification. *Smart Sustain. Built Environ.* **2023**. [CrossRef]
14. Redmon, J.; Farhadi, A. YOLO9000: Better, faster, stronger. In Proceedings of the 30th IEEE Conference on Computer Vision and Pattern Recognition, CVPR 2017, Honolulu, HI, USA, 21–26 July 2017.
15. Law, H.; Deng, J. CornerNet: Detecting Objects as Paired Keypoints. *Int. J. Comput. Vis.* **2020**, *128*, 642–656. [CrossRef]
16. Zhang, A.A.; Shang, J.; Li, B.; Hui, B.; Gong, H.; Li, L.; Zhan, Y.; Ai, C.; Niu, H.; Chu, X.; et al. Intelligent pavement condition survey: Overview of current researches and practices. *J. Road Eng.* **2024**, *4*, 257–281. [CrossRef]
17. Wang, S.; Cai, B.; Wang, W.; Li, Z.; Hu, W.; Yan, B.; Liu, X. Automated detection of pavement distress based on enhanced YOLOv8 and synthetic data with textured background modeling. *Transp. Geotech.* **2024**, *48*, 101304. [CrossRef]

18. Yuan, B.; Sun, Z.; Pei, L.; Li, W.; Zhao, K. Shuffle Attention-Based Pavement-Sealed Crack Distress Detection. *Sensors* **2024**, *24*, 5757. [CrossRef] [PubMed]
19. Wang, R.; Wang, C.; Chu, X. Research Progress on Pavement Damage Image Recognition. *J. Jilin Univ. Technol. (Eng. Ed.)* **2002**, *32*, 91–97.
20. Li, L.; Sun, L.; Chen, C. Edge Detection Method for Pavement Damage Image Processing. *J. Tongji Univ. (Nat. Sci. Ed.)* **2011**, *39*, 688–692.
21. Shi, L.; Dang, L.; Yang, L.; Shi, S. Pavement Damage Image Recognition Method Based on Manifold Feature Fusion. *Comput. Appl. Softw.* **2016**, *33*, 150–152 + 96.
22. Zhang, Y.; Li, Q.; Xue, F.; Yu, L. Design of Pavement Crack Detection System Based on Jetson TX2. *Highway* **2023**, *68*, 337–344.
23. Chen, H.; Wang, J. Infrared Asphalt Pavement Crack Detection Method Based on Improved YOLOv5. *Telev. Technol.* **2023**, *47*, 43–50.
24. Zhou, Y.; Zhang, J.; Cao Ji Liu, Y.; Zhang, H. Research on Pavement Pothole Detection Error Compensation Algorithm Based on 3D Laser Technology. *J. Highw. Transp. Technol.* **2023**, *40*, 17–24.
25. Wu, C.; Ti, J.; Ma, J. Digitalization of Asphalt Pavement Maintenance Information Based on Holographic 3D Detection Technology. *Guangdong Highw. Transp.* **2023**, *49*, 1–7.
26. Chen, M.; Zhang, M.; Liu, Z.; Han, Y.; Gu, S. Design and Implementation of a Lightweight Portable Intelligent Pavement Distress Detection System. *Eng. Qual.* **2022**, *40*, 74–79.
27. Chen, X.; Gao, H.; Yang, Z.; Kong, T.; Che, R. Research on Pavement Crack Detection and Recognition Based on Improved Yolov5s. *Softw. Guide* **2014**.
28. Wang, Y.; Zhou, C.; Wang, Y.; Li, W. Digital Research on Pavement Distress Based on Improved YOLOv8 Algorithm. *Highway* **2024**, *69*, 350–356.

infrastructures

MDPI

Article

Cost Efficiency and Effectiveness of Drone Applications in Bridge Condition Monitoring

Taraneh Askarzadeh and Raj Bridgelall *

Department of Transportation and Supply Chain, College of Business, North Dakota State University, P.O. Box 6050, Fargo, ND 58108-6050, USA; taraneh.askarzadeh@ndsu.edu

* Correspondence: raj@bridgelall.com

Abstract: Bridges are an integral and important part of road networks, but monitoring their condition using traditional methods is expensive, dangerous, and laborious. This study examines the rapidly emerging field of drone-based transportation asset monitoring, focusing on analyzing the cost efficiency and effectiveness of drone applications in bridge condition monitoring. This research innovated a multi-dimensional framework that highlights the transformative role of drone technology in enhancing inspection accuracy, safety, and cost savings. Using statistical models and Monte Carlo simulations, the framework provides an extensive cost–benefit analysis to inform drone investment decisions. A case study demonstrates the utility of the framework in quantifying costs and benefits. Furthermore, a sensitivity analysis evaluates how variations in drone costs, driven by technological progress, can potentially influence adoption of the technology.

Keywords: autonomous aircraft; data mining; asset management; drone-based monitoring; cost-efficient bridge inspection; Monte Carlo simulation

1. Introduction

The American Society of Civil Engineers (ASCE) reported that the United States has more than 617,000 bridges, many exceeding 50 years in age, with 42% requiring repairs and 7.5% classified as structurally deficient [1]. This situation highlights the critical need for effective condition monitoring to maintain structural integrity. Many organizations use traditional monitoring methods that are often expensive and time-intensive, increasing interest in more innovative and efficient solutions. Unmanned aerial vehicles (UAVs), or drones, are increasingly adopted across industries for their ability to efficiently cover and assess large infrastructure areas, providing comprehensive visual data often missed in manual inspections [2]. These capabilities accelerate evaluations and offer a cost-effective alternative to traditional approaches.

Drones provide valuable capabilities for bridge monitoring but also have notable limitations. Drone condition monitoring faces challenges, such as battery life constraints, adverse weather conditions (wind, rain, fog), and GPS signal interference, which affect flight stability and data accuracy. Restricted airspace regulations and data processing complexities further hinder deployment. Advancing battery technology, sensor durability, and regulatory frameworks are crucial to enhancing the reliability of inspections [3].

The American Association of State Highway and Transportation Officials (AASHTO) categorized bridge inspections into eight types: initial, routine, damage, in-depth, fracture-critical, underwater, routine wading, and special inspections. Routine inspections, the most common type, primarily rely on visual methods to identify defects. While drones enhance

 https://doi.org/10.3390/infrastructures10030063

routine inspections with their broad coverage, they are less effective at tasks requiring direct physical interaction, such as inspection of internal structural elements. Drones are also unsuited for underwater inspections and cannot detect fractures due to their lack of tactile assessments or advanced nondestructive testing [4].

Drones are particularly effective at inspecting steel and concrete structures. However, they face challenges in navigating tight spaces, capturing specific angles, and detecting subtle color changes in steel bridges. Drones cannot assess the subsurface conditions of composite and timber bridges [5]. These limitations highlight the importance of considering both inspection types and bridge materials when implementing drone technology for monitoring purposes. Hence, the most effective application of drones lies in the routine inspection of concrete and steel bridges, which is the scope of this study's focus.

The main contribution of this study is a framework and methodology to evaluate the costs and benefits of drone-based condition monitoring (D-BCM) relative to traditional-based condition monitoring (T-BCM) of linear transportation assets. The proposed framework aims to help organizations make informed decisions regarding the adoption of drone technology by providing insights into its economic viability, potential benefits, and associated risks.

The rest of this paper proceeds as follows. Section 2 reviews the literature on drone technology, remote sensing technologies, and quantitative modeling. Section 3 presents the data mining workflow and defines all of the variables utilized in the study. Section 4 defines the mathematical models, outlining the methodology used to quantify the costs and benefits of D-BCM. Section 5 presents the results and discusses their implications for stakeholders. Section 6 concludes the research and suggests directions for future work.

2. Literature Review

Recent research highlights the significant benefits of drone technology in bridge inspection and monitoring. Perry et al. (2020) demonstrated the efficiency and accuracy of drones in assessing structural damage, improving both visualization and quantification [6]. Hubbard and Hubbard (2020) investigated the safety advantages of drones, quantifying these benefits through worker compensation rates and survey data from a state Department of Transportation case study [7]. Azari, O'Shea, and Campbell (2022) introduced a sensor-equipped drone prototype that enhances data collection and management processes for bridge inspections [8].

Song, Yoo, and Zatar (2022) have developed iBIRD, a web-based tool for managing drone-assisted bridge inspections, featuring 3D modeling and report generation [9]. Dorafshan and Maguire (2018) demonstrated drones equipped with self-navigation and image processing, enabling the creation of accurate, autonomous 3D models for damage identification [10]. Chen et al. (2019), however, point out the limitations of image-based methods in areas lacking distinct features [11]. Addressing environmental challenges, Aliyari et al. (2021) performed a hazard analysis of drone inspection risks under harsh conditions, focusing on human performance impacts, including drone pilots [12].

Dorafshan et al. (2017) demonstrated the effectiveness of drones in detecting damage on concrete and steel bridges, with results comparable to human inspections and the added advantage of real-time feedback [13]. They argued that current drone technology primarily serves as an assistive tool, improving the speed, cost efficiency, and safety of bridge inspections while eliminating the need for traffic closures [13]. Hubbard et al. (2020) investigated the role of drones in enhancing bridge inspection safety [7]. The study surveyed bridge inspectors and developed a benefit–cost methodology based on worker compensation rates to assess the safety benefits of drones. Although both studies recognize the cost and time efficiencies of drones, they lacked detailed quantification of these benefits.

The present study fills these research gaps by developing a comprehensive framework to quantify the benefits and costs of using drones in bridge condition monitoring.

3. Data

The next subsections discuss the data mining workflow, define the variables in the framework, and explain the methods of quantifying the costs and benefits, data analysis, uncertainty management, and related simulations.

3.1. Data Mining Workflow

Figure 1 illustrates the data mining and analytical workflow for quantifying variables. The workflow starts by identifying the cost and benefit components of D-BCM. It then models these aspects for both drone-based and traditional methods, incorporating direct and indirect factors before formulating the overall cost–benefit structure.

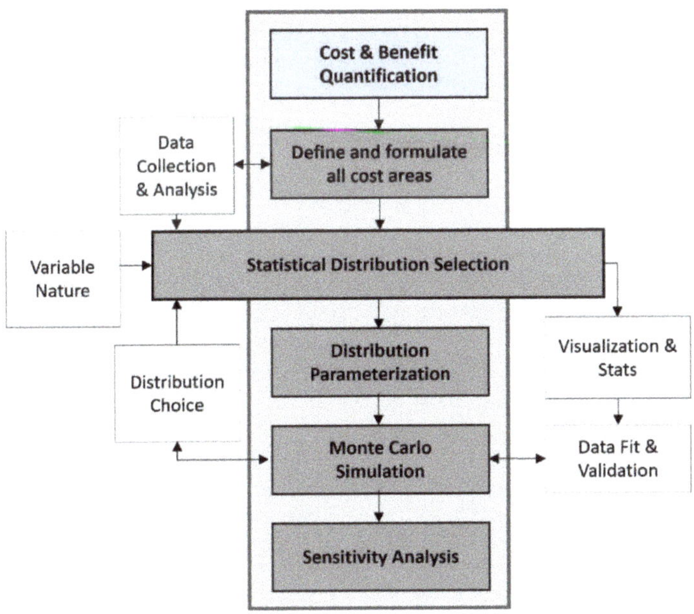

Figure 1. Workflow for quantifying the variables.

Monte Carlo simulation (MCS) serves as a probabilistic forecasting tool, selecting values from a user-defined probability distribution to assess multiple model scenarios [14]. The workflow compiles data by categorizing variables as either stochastic or deterministic. Statistical models and goodness-of-fit tests fit appropriate distributions to the data. Given database uncertainties, stochastic variables undergo MCS, and results are compared with empirical data to validate accuracy. Additionally, the case study utilizes sensitivity analysis and scenario forecasting for future drone pricing trends.

3.2. Variables

Table 1 presents all of the variables used in the cost and benefit models of this study, including their descriptions, categories, and classification as either stochastic or deterministic. It also details the time frame associated with each variable, the cost for deterministic variables, and the corresponding data sources.

Table 1. Variables used in the model.

Cost Components	Variable	Description	Category	Type	Time Frame	Cost	Source
Drone Component Costs (C1)	N_{DP}	Number of drones	Cost	-	-	-	-
	DP	Cost of a drone	Cost	Stochastic	Once	-	Market
	N_{PP}	Number of payloads (e.g., cameras, sensors)	Cost	-	-	-	-
	N_{SBP}	Number of standardized batteries	Cost	-	-	-	-
	PP	Cost of a payload	Cost	Deterministic	Unit	-	Market
	SBP	Cost of a standardized battery	Cost	Stochastic	Hour	-	Market
Ground Infrastructure Costs (C2)	GCS	Cost of the ground control station	Cost	Deterministic	Once	USD 5000	[15]
	GL	Cost of ground landing pads	Cost	Deterministic	Unit	-	Market
Personnel Costs (C3)	T_{BILT}	Time required for the bridge inspection team leader (BILT)	Cost and Benefit	Deterministic	Hour	USD 150	[16]
	T_{ABI}	Time required for assistant bridge inspectors (*ABI*)	Cost and Benefit	Deterministic	Hour	USD 120	[16]
	N_{ABI}	Number of *ABIs*	Cost and Benefit	-	-	-	-
	P	Cost of a drone pilot	Cost	-	-	-	-
	N_P	Number of drone pilots	Cost	-	-	-	-
	T_P	Training costs for personnel	Cost	Deterministic	Once	USD 2575	[17]
Upkeep Costs (C4)	AMC	Annual maintenance costs	Cost	Stochastic	Year	-	[18]
	URC	Unexpected repair costs	Cost	Stochastic	Year	-	[18,19]
IT Infrastructure Costs (C5)	N_{Soft}	Number of software licenses required	Cost	-	-	-	-
	C_{Soft}	Cost of software licenses	Cost	Stochastic	Month	-	Market
	$C_{Storage}$	Cost of data storage	Cost	Deterministic	Month	USD 180 once + USD 100 per month	Market
Data Processing Costs (C6)	PPE	Cost of post-processing engineers	Cost	Deterministic	Hour	USD 120	[16]
	N_{PPE}	Number of post-processing engineers	Cost	-	-	-	-
	T_{PPE}	Time required for post-processing engineers	Cost	Stochastic	Inspection	-	[16]

Table 1. *Cont.*

Cost Components	Variable	Description	Category	Type	Time Frame	Cost	Source
Insurance Costs (C7)	LI	Cost of liability insurance	Cost	Stochastic	Year	-	Market
	IE	Cost of hull insurance	Cost	-	-	-	-
	N_{IE}	Number of insured equipment units	Cost	-	-	-	-
Deployment Costs (C8)	RE	Registration cost	Cost	Deterministic	3 Years	USD 5	[17]
	N_{RE}	Number of registered drones	Cost	-	-	-	-
Crew Time Savings (B1)	ITSP	Inspection time saving percentage	Benefit	Stochastic	%	-	[16]
Operational Vehicle Cost Savings (B2)	$UBIV$	Cost of under-bridge inspection vehicles ($UBIVs$)	Benefit	Stochastic	Daily	-	[16]
	T_{UBIV}	Time required for UIBV	Benefit	-	-	-	[16]
	N_{UBIV}	Number of UIBVs	Benefit	-	-	-	-
	C_{UBIV}	Maintenance costs of UIBV	Benefit	-	-	-	-
	P_{UBIV}	Infrastructure costs associated with UIBV	Benefit	-	-	-	-
Tool Cost Savings (B3)	N_I	Number of inspection tools replaced by drones	Benefit	-	-	-	-
	C_{TM}	Cost of traditional monitoring tools	Benefit	-	-	-	-
Safety Cost Savings (B4)	ISE	Safety equipment savings	Benefit	Deterministic	Inspection/ Person	USD 275	[20]
	$RISK_{FI}$	Risk reduction costs	Benefit	-	-	-	-
	VSL	Value of a statistical life	Benefit	-	-	-	-
	P_F	Probability of fatality during inspections	Benefit	-	-	-	-
	VSI	Value of a statistical injury	Benefit	-	-	-	-
	P_I	Probability of injury during inspections	Benefit	-	-	-	-

Table 1. *Cont.*

Cost Components	Variable	Description	Category	Type	Time Frame	Cost	Source
Reduced Lane Closure and Traffic Costs (B5)	LCC	Lane closure costs	Benefit	Stochastic	Hour	-	[16]
	TTC_{Total}	Total travel time cost savings	Benefit	Deterministic	Hour	USD 6915 at peak hrs. USD 1235 during off-peak hrs.	[21]
	AVOR	Average vehicle occupancy rates	Benefit	-	-	-	-
	VOC	Vehicle operation cost savings	Benefit	Deterministic	Hour	USD 345 during peak times USD 115 during off-peak times	[22]
	OP	Operating cost per mile	Benefit	-	-	-	-
	AMTD	Annual miles traveled	Benefit	-	-	-	-
	$RISK_{Accident}$	Reduced accident risks and associated cost savings	Benefit	-	-	-	-

The framework classified the variables into deterministic and stochastic categories based on the databases and their nature to provide a clear overview of the economic factors considered.

Drone Price. For monitoring linear assets, multicopters are preferred, with quadcopters accounting for the majority and hexacopters making up the remaining 40% [3]. Among drone manufacturers, DJI currently leads the industry, holding over 70% of the global drone market, as reported by CNBC [23]. Aside from any governmental restrictions on manufacturing origin, key considerations when selecting a drone for inspections include flight time, camera quality, weather stability, obstacle detection capabilities, and industrial-grade features [24]. Inspectors commonly use optical cameras, thermal cameras, and LiDAR systems based on specific inspection requirements. The present study focuses on concrete and steel bridge inspections, where cameras are the most suitable payload. Table 2 provides a summary of suitable drones and their payload specifications.

Standardized Battery Price. D-BCM necessitates multiple battery replacements. The price of drone batteries varies based on type (LiPo or Li-ion), the number of batteries required, drone brand, and battery life. While some drones require only one battery, others may need up to six. This variability makes it challenging to directly compare prices. Thus, for a clear comparison of drone batteries, standardization becomes essential. This formula converts the raw battery cost into a standardized cost per flight hour.

$$SBP = \left(\frac{BP \times N_{BP}}{\frac{FT}{60} \times cc} \right) \tag{1}$$

where FT is the battery flight time and CC denotes the estimated number of charge cycles (approximately 300) before the LiPo battery degrades by retaining 80% of its original capacity [25,26]. The lifespan of a Skydio X2 battery is one year or 200 battery cycles [27]. BP is the price of the battery, and N_{BP} is the required number of batteries for each drone.

Post-Processing Time. A significant benefit of D-BCM is that it not only decreases the duration of inspections but also reduces the need for extensive crew and time spent on inspection vehicles. The Minnesota Department of Transportation (MnDOT) provided inspection time data for both traditional and D-BCM [16].

Table 2. Recommended drones and their payload for D-BCM.

Drone	Price	Flight Time (Min)	Built-In Payload	Extra Required Payload	Data Source
DJI Mavic 3 Pro	USD 2199 to USD 3299	43	Hasselblad: 4/3 CMOS, 20 MP Medium Tele: 1/1.3-inch CMOS, 48 MP Tele: 1/2-binch CMOS, 12 MP	-	[28]
DJI Air 3	USD 1099 to USD 1550	46	Wide-Angle: 1/1.3-inch CMOS Effective Pixels: 48 MP Medium Tele: 1/1.3-inch CMOS Effective Pixels: 48 MP	-	
DJI Phantom 4 Pro	USD 1599 to USD 1699	30	1″ CMOS Effective Pixels: 20 MP	-	
DJI Matrice 600 Pro	USD 5000 to USD 6000	32	-	Zenmuse (Z) X3: 1/2.3″ CMOS/12 MP photos and 4K video at 30 fps: USD 500–USD 700 ZX5 and X5R: USD 1400–USD 1600/USD 3000 ZX7: USD 2700–USD 3000 Z30: USD 2500–USD 4000	

Table 2. *Cont.*

Drone	Price	Flight Time (Min)	Built-In Payload	Extra Required Payload	Data Source
DJI Matrice 300 RTK	USD 13,000	55	Infrared Sensing System	ZH20 series: hybrid multi-sensor camera USD 5000 to USD 10,000 ZP1: full-frame sensor camera: USD 8000	[28]
MATRICE 210 RTK V2	USD 10,000 to USD 15,000	34	-	Z30: USD 2500–USD 4000 ZX4S: USD 600–USD 800 ZX5S: USD 1900–USD 2200 ZX7: USD 2700–USD 3000 ZXT2	
Skydio 2+	USD 5000	27	Camera: Sony IMX577 CMOS sensor and Qualcomm RedDragon™ QCS605: 12 MP photos, 4K60 HDR video/45 MP	-	[27]
Skydio X10	USD 15,000	40	Narrow camera: 64 MP 1″ wide camera: 50 MP Radiometric thermal: 640 × 512 px	-	
Parrot Anafi	USD 7000	32	Vertical camera, ultra-sonar/2 × 6-axis IMU, 2 × 3-axis accelerometers, 2 × 3-axis gyroscopes, 4K video, thermal	-	[29]
Yuneec H520E	USD 2500	28	-	E90 Camera: 1-inch CMOS sensor, 20 MP resolution. USD 1299–USD 1499 E50 Camera: USD 1200 CGOET Camera: USD 1900	[30]
Elios 3	USD 5000	12	Visual camera and onboard LED lighting capable of 4K UHD videos. CMOS Effective Pixels: 12.3	-	[31]
DJI Inspire 3	USD 16,500	28	X9-8K Air	-	[32]
AUTEL EVO 2 PRO RTK	USD 1500–USD 3000	40	1-inch CMOS	-	[33]

Annual Maintenance and Unexpected Repair Cost. In alignment with other electronic devices, the values for AMP and URC are 10% and 2% of the drone price [18,19].

Software Cost. Effective D-BCM relies on specialized software for flight planning, photogrammetry, real-time monitoring, and data management. Manufacturers equip many drone models with proprietary software for flight planning and real-time monitoring, often including it in the package at no additional cost. Advanced cataloging systems, incorporating photogrammetric 3D models of bridges, enable precise identification and examination of specific bridge sections within inspection images [34]. Table 3 consolidates data from various DOT reports, showcasing commonly used software solutions and their associated costs in this field.

Table 3. Software cost used by DOTs for D-BCM.

Software	Price Range	DOT
Photogrammetry		
Pix4D V4.8.2	USD 32–USD 291/month	[16,35–38]
Agisoft Metashape V2.1.2	USD 179–USD 3500 (one-time)	[38–40]
AutoCAD V25.0	USD 40/month	[16,36]
ContextCapture V20	USD 3900/year	[16,41]
Data Management		
Airdata UAV V1.34.7	Free to USD 300/year	-
Dronelogbook V10.0.6	USD 10/month	-
Intel Insight V10.1	USD 99/month	[16]

Insurance Cost. According to a DroneDeploy survey, most drone service providers choose liability insurance coverage of USD 1 million [42]. The insurance costs database lists 12 drone insurance companies for USD 1 M coverage.

Inspection Time Saving Percentage. Table 4 details the inspection durations for each method and lists the corresponding bridge sizes in feet [16].

Table 4. Comparison of inspection hours for D-BCM and T-BCM.

T-BCM			D-BCM		BS (Feet)
T_{BILT}	T_{ABI}	T_{UBIV}	T_{BILT}	T_{ABI}	
8	8	0	4	4	505
4	4	0.5	1	1	2740
4	4	0	4	4	45
8	8	0	4	4	1887
24	24	3	20	20	635
8	8	0	2	2	214
8	8	0	6	6	3360
12	12	0	6	6	160
4	4	0.5	3	3	1914
4	4	1	4	4	2100
8	8	0	4	4	2769

Under-Bridge Vehicle Operation. Maintenance inspections for various bridge types require access to the upper or lower bridge areas, often utilizing under-bridge inspection vehicles. These specialized vehicles, equipped with articulated booms (sometimes extendable to three or four booms for enhanced reach), are available in self-propelled, truck-mounted, and trailer-mounted configurations [43]. Organizations typically rent these vehicles daily due to their high costs, often including operator services. This case study evaluates the costs associated with T-BCM by considering distinct vehicle types, their rental costs (including operators), and usage probabilities using data from MnDOT. Table 5 details vehicle categories, costs, and probabilities.

Table 5. Inspection vehicles for T-BCM, their associated costs, and usage probabilities.

Vehicle with Operator	Probability	Daily Rental (USD)	Type of Inspection
Snooper Truck	30%	USD 3000	Under-bridge access.
Bucket Truck	40%	USD 700	Mainly used for overhead inspections where direct access is required at a certain height. Effective for bridge superstructure elements.
Scissor Lift	15%	USD 500	Where vertical elevation is required without the need for lateral movement. Primarily used for low-height under-bridge areas or decks.
Boom Lift	25%	USD 1000	For both vertical and horizontal movement, facilitating access to difficult areas of a bridge, especially for superstructure elements.

Reduced Lane Closure. T-BCM often requires lane or road closures, increasing costs and disruptions. Temporary traffic control measures must comply with the Manual on Uniform Traffic Control Devices (MUTCD) and local standards. Using equipment like *UBIVs* requires lane closures, incurring traffic control costs ranging from USD 500 to USD 2500 daily. Drone inspections reduce reliance on such equipment, minimizing lane closures and associated direct and user costs. Table 6 presents the direct costs associated with various road closure types, based on data sourced from MnDOT [16].

Table 6. Direct costs per day for various road closure types.

Category	Cost (USD)
Misc. Traffic Control (Ped. Only, etc.)	USD 500
Low Speed Lane/Shoulder Closure	USD 2000
Mobile Lane/Shoulder Closure	USD 1500
High Speed Lane/Shoulder Closure	USD 2500

4. Methodology and Modeling

Several cost components collectively shape the economic landscape of D-BCM. As technology progresses, components, such as drone hardware, software algorithms, and operational expenses, experience fluctuations. This study categorizes costs into eight groups: drone components, ground infrastructure, personnel, maintenance, IT infrastructure, data processing, insurance, and deployment. The cost model integrates these elements to quantify total costs (C_{total}), covering acquisition, maintenance, infrastructure setup, personnel expenses, software, data processing, insurance, and regulatory fees. This structured approach ensures transparent and systematic evaluation of the financial feasibility of D-BCM.

The following cohesive financial model consolidates these varied costs:

$$C_{total} = \sum_{i=1}^{8} Ci \tag{2}$$

$$C_1 = N_{DP} \times DP + N_{PP} \times PP + N_{SBP} \times SBP \tag{3}$$

$$C_2 = GCS + GL \tag{4}$$

$$C_3 = T_{BILT} + (T_{ABI} \times N_{ABI}) + (P + N_P\ ) + T_P \tag{5}$$

$$C_4 = AMC + URC \tag{6}$$

$$C_5 = \left(N_{Soft} \times C_{Soft}\right) + C_{Storage} \tag{7}$$

$$C_6 = PPE \times N_{PPE} \times T_{PPE} \tag{8}$$

$$C_7 = LI + IE \times N_{IE} \tag{9}$$

$$C_8 = RE \times N_{RE} \tag{10}$$

The total benefits (B_{total}) are the sum of contributions from various direct and indirect benefits, as follows:

$$B_{total} = \sum_{i=1}^{5} Bi \tag{11}$$

$$B_1 = T_{BILT} + (N_{TABI} \times T_{ABI}) \tag{12}$$

$$B_2 = (UBIV \times T_{UBIV} \times N_{UBIV}) + (C_{UBIV} \times N_{UBIV}) + P_{UBIV} \tag{13}$$

$$B_3 = N_I \times C_{TM} \quad (14)$$

Crew time savings (B_1) result from the reduced workforce required for drone-based inspections compared with traditional methods. High-definition drone cameras minimize the need for roles, such as *BILT* and *ABI*, where costs depend on the number of inspectors and their durations. Operational vehicle cost savings (B_2) stem from decreased reliance on *UBIVs*. These savings include costs for fuel, wages, operational hours, maintenance, and infrastructure expenses for vehicle facilities. Additionally, drones equipped with advanced sensors and cameras replace specialized inspection tools like gauges or ultrasonic testing devices, leading to tool cost savings (B_3).

$$B_4 = ISE + RISK_{FI} \quad (15)$$

$$RISK_{FI} = (VSL \times P_F) + (VSI \times P_I) \quad (16)$$

Drone-based inspections offer significant indirect benefits, including cost savings from enhanced safety (B_4). For example, drones reduce reliance on costly safety equipment like harnesses and scaffolding. Drones also lower the risk of injuries or fatalities, which are more common in traditional methods due to hazardous conditions, fatigue, and extended work hours. This reduced risk translates into lower insurance premiums and fewer injury-related costs. P_F and P_I are, respectively, the likelihood of fatality and injury during inspection. According to the United States Department of Transportation (USDOT), the value of a statistical life primarily finds application in health economics, transport economics, and environmental economics [44].

$$B_5 = LCC + TCC_{Total} + VOC + RISK_{Accident} \quad (17)$$

Drones minimize the need for lane or track closures during inspections (B_5), reducing labor, equipment, and permit costs associated with disruptions. This accounts for direct lane closure costs, travel time costs, vehicle operation costs, and accident risks. By avoiding closures and delays, drones enhance operational efficiency and reduce public inconvenience.

$$TTC = DV \times VTTS \times AVOR \quad (18)$$

$$VOC = OP \times AMTD \quad (19)$$

where *DV* is the average delay. Wisconsin Department of Transportation (WisDOT) data indicate that vehicles on major highways endure an average delay of 10 min (or 0.1667 h) due to lane closures. In contrast, those on minor roads experience a five-minute delay (or 0.0833 h) [45]. The value of travel time savings (*VTTS)* refers to the benefits provided by reductions in the amount of time spent riding in a vehicle [46]. The values are based on USDOT hourly *VTTS* values of USD 17.00/person-hour for personal journeys and USD 31.90/person-hour for business trips. Accounting for the customary distribution of 88.2% personal and 11.8% business travel, the result is a blended rate of USD 18.76/person-hour [22]. Considering the average vehicle occupancy rates (*AVOR*) of 1.48 persons per vehicle during weekday peak times and 1.58 persons per vehicle in off-peak times [22], the *TTC* values become USD 5.10/vehicle and USD 2.75/vehicle, respectively.

Furthermore, data from the FWHA specify an average traffic volume of 1500 vehicles per hour on major highways during peak hours and 500 vehicles per hour on minor roads in off-peak times [21]. This results in an aggregate TTC_{Total} of USD 6915 per hour for major highways at peak times and USD 1235 per hour for minor roads during off-peak hours.

5. Results and Discussion

5.1. Variables and Scenarios

This case study implements MCS using well-defined stochastic models to account for uncertainties in cost and benefit estimation for D-BCM. Because actual probability distributions for these variables may be incomplete due to limited observational data, statistical distribution fitting techniques model their variability. The selection of probability distributions is based on empirical data, industry reports, and expert judgment.

This study tests several statistical models to determine the best-fit distributions for each variable. Right-skewed distributions (such as log-normal, gamma, or Weibull) model cost-related variables based on their potential for extreme values and increasing costs over time. Normal and uniform distributions model benefit-related variables under the assumption of symmetrical variability. Goodness-of-fit tests (such as the Kolmogorov–Smirnov test, the Anderson–Darling test, and the Akaike Information Criterion) validate the distributions, ensuring that the simulated data closely match observed real-world distributions.

After defining the probability distributions, the Monte Carlo model randomly sampled from these distributions over 10,000 iterations, generating a large set of potential outcomes. This process quantified uncertainty and risk in D-BCM implementation, offering a probabilistic assessment instead of a single deterministic estimate. The model assumes independence among most cost and benefit factors, except where empirical data suggest correlation.

To strengthen the MCS, empirical distributions derived from actual data are compared to simulated distributions, ensuring accurate representation of stochastic variability. This enhances the reliability of the cost–benefit analysis for decision making. Figure 2 presents the results of the 10,000 MCS trials, displaying the mean, standard deviation (SD), and probability distribution type for each variable.

As mentioned earlier, AMC and URC, expressed as percentages of the drone price, are represented using mean and standard deviation within a log-normal distribution to capture their skewed nature. However, due to limited data, full probability distributions could not be derived from VO and LCC. Instead, these variables are treated as discrete probabilities rather than continuous distributions, ensuring that the model incorporates uncertainty without overfitting to insufficient or unreliable data. Table 7 presents these three variables along with their statistical details.

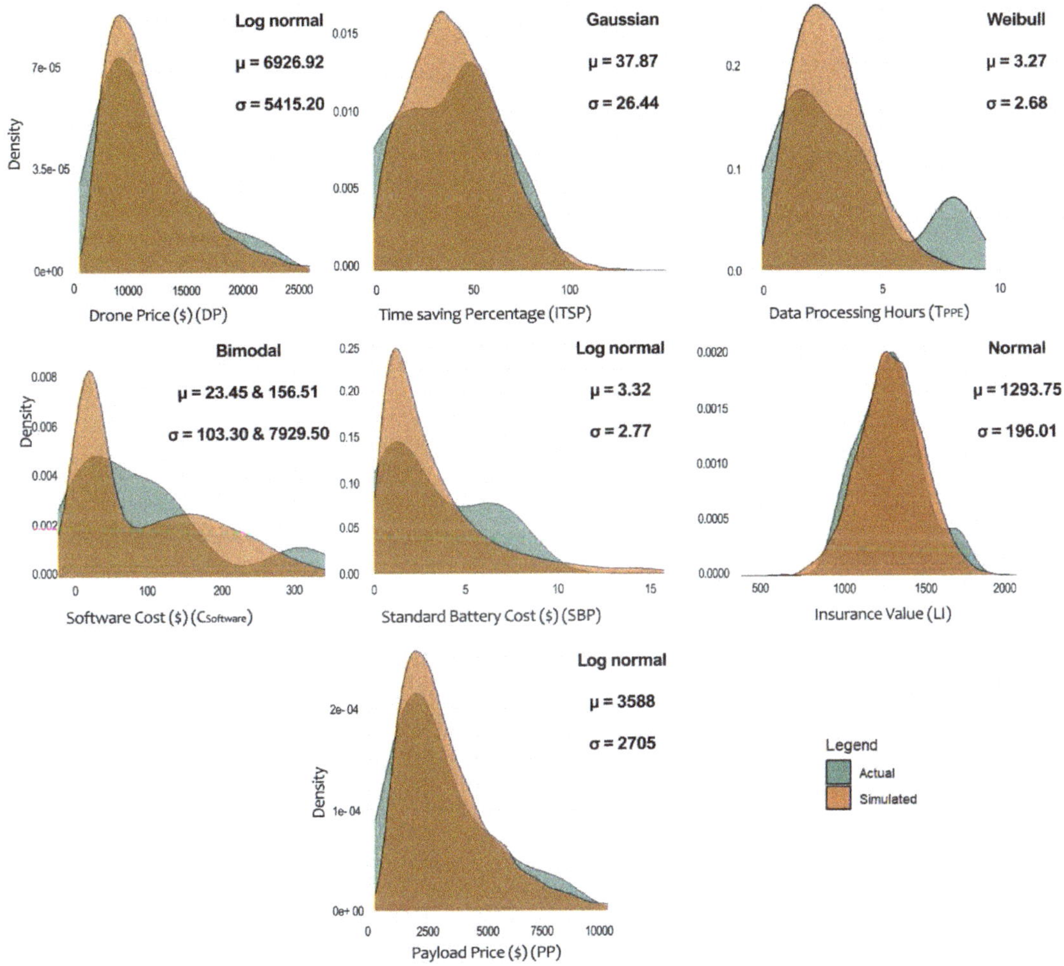

Figure 2. Comparing actual and simulated distributions of stochastic variables and their distribution type, mean, and SD.

Table 7. The stochastic variables of D-BCM and their distribution statistics.

Variable	μ	σ	Distribution
AMC and URC	1202.17	3151.33	Log-Normal
VO	3588	2705	Probability
LCC	1391	1133	Probability

The sensitivity analysis of drone price uses trends in smartphone pricing as a benchmark for predicting drone price evolution. It assumes drones might adopt a similar price trajectory as smartphones, based on technological progress and market penetration. Figure 3 presents historical smartphone pricing trends from the application to smartphones report [47].

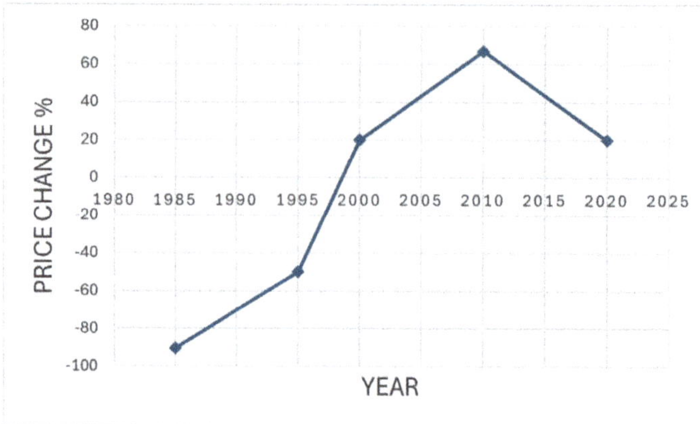

Figure 3. Changes in smartphone price over time.

In the short term (2023–2028), the present study predicts that drones will advance in AI, autonomy, and navigation as FAA regulatory updates enable more beyond visual line of sight operations. Energy storage may improve slightly, and consumer drones will likely become more user-friendly, differentiating them from commercial models. Following smartphone pricing trends, drone prices may increase by a mean of 40% with a 20% variance.

In the midterm (2028–2033), further AI developments and automated fleet management are anticipated, boosting drone adoption for inspections and logistics. This study expects the consumer drone market to stabilize with standardized technology and pricing, leading to a +70% mean price change and a +10% variance. In the long term (2033+), universal regulations and advanced AI autonomy will enable complex operations and broader applications, with the market maturing into diverse offerings. Analysts forecast a mean price change of −20% and a variance of −5%. Figure 4 illustrates these trends, showing higher prices in the short term and midterm and balancing in the long term with stable, diverse options for varied needs.

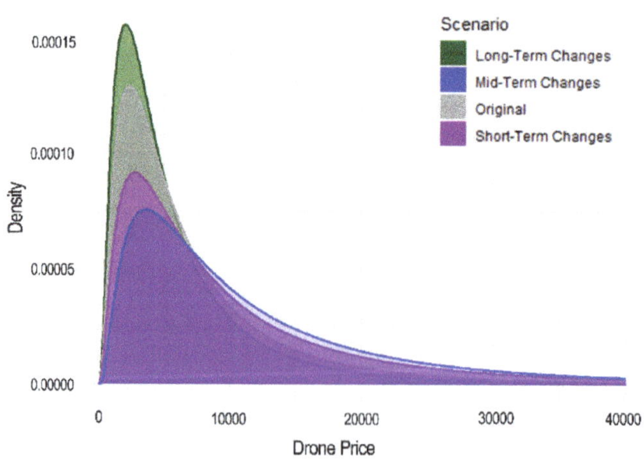

Figure 4. Histograms of sensitivity analysis of DP.

5.2. Stochastic Investment Costs

This case study models the investment cost (D_I) using the following equation:

$$D_I = D_{Log}\left(\mu_{DP}, \sigma^2_{DP}\right) + D_{Log}\left(\mu_{PP}, \sigma^2_{PP}\right) + T_P + C_2 + SD + RE \quad (20)$$

Here, D_I represents the distribution of I, derived from 10,000 MCS. Figure 5 illustrates the empirical distribution of I, revealing a mean cost of USD 18,788.63, a median cost of USD 15,483.40, and a mode of approximately USD 8121.33. The standard deviation of USD 11,214.13 indicates significant cost variation, with half of the values between the 25th percentile (USD 12,272.6) and the 75th percentile (USD 21,362.03). The histogram displays 68% and 95% confidence intervals, highlighting likely cost boundaries. The right-skewed distribution suggests costs are more concentrated near the mode but with occasional spikes in the longer tail. Most costs fall below USD 30,000, aligning with the median and providing stakeholders with reliable cost expectations.

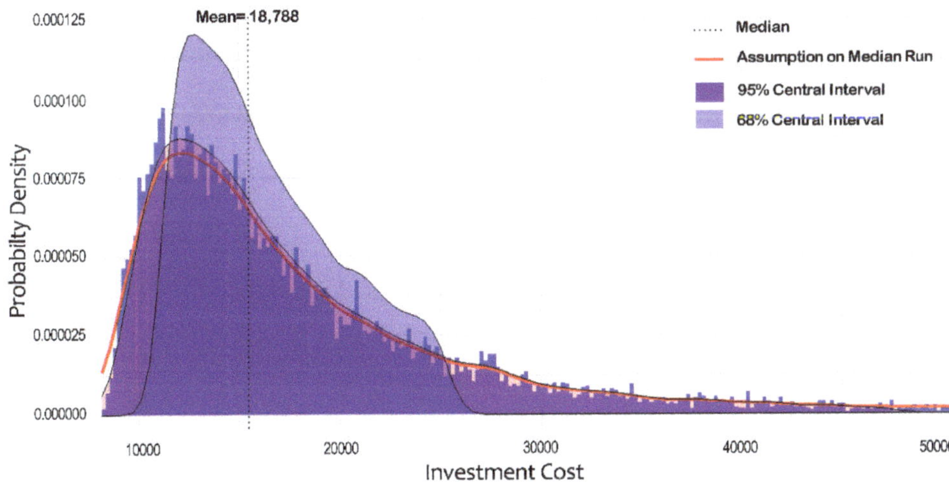

Figure 5. Distribution of investment cost.

5.3. Stochastic Cost and Benefits per Inspection

The present research used stochastic variables to model the costs per inspection for both D-BCM and T-BCM. For D-BCM, inspection costs (D_{CS}) depend on inspection time, data processing, and battery price, with deterministic inputs, such as wages for *BILT*, *ABI*, and *PPE*, calculated as

$$D_{CS} = D_W(\lambda_{SBP}, k_{SBP}) + \left(D_W\left(\lambda_{T_{PPE}}, k_{T_{PPE}}\right) \times PPE\right) + \left(D_G\left(\mu_{ITSP}, \sigma^2_{ITSP}\right) \times (BILT + ABI)\right) \quad (21)$$

For T-BCM, inspection costs (D_B) account for stochastic *UBIV* costs and deterministic variables, calculated as

$$D_B = \sum\nolimits_{i \in \{1,4,5\}} B_i + P\left(\mu_{UBIV}, \sigma^2_{UBIV}\right) \quad (22)$$

The analysis used MCS to compare cost distributions for both methods. Figure 6 shows the results, with D-BCM demonstrating a median cost of USD 1770, significantly lower than T-BCM's USD 7216. Quartile values reinforce this trend; D-BCM costs remain below T-BCM across all percentiles, peaking at USD 4870 compared with USD 9716.

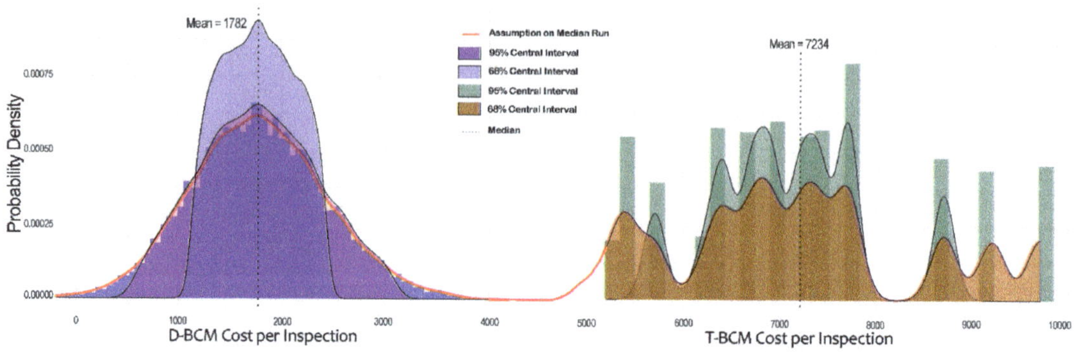

Figure 6. Distribution of D-BCM and T-BCM cost per inspection.

The standard deviation for D-BCM is USD 653.85, indicating more consistent costs than T-BCM with USD 1258.37. Kurtosis values highlight differences in distribution shapes, with D-BCM showing a leptokurtic distribution (3.01), while T-BCM displays a platykurtic one (2.39). The 95% confidence interval further emphasizes the advantage of D-BCM, with costs likely between USD 532.81 and USD 3117.12 compared with between USD 5216.5 and USD 9716.5 for T-BCM.

The findings illustrate the financial superiority of D-BCM, offering lower and more predictable costs. Hence, adopting D-BCM can provide substantial savings, particularly for large-scale or recurring inspections.

5.4. Net Saving

The net savings distribution (D_{NS}) reflects the cost difference between T-BCM and D-BCM, calculated as

$$D_{NS} = D_B - D_{CS} \tag{23}$$

MCS reveal an average net savings of USD 5043 per inspection, with a median of USD 4935. Most outcomes fall between the first quartile (USD 3997) and the third quartile (USD 5983), with a maximum saving of USD 9271. Variability, indicated by a standard deviation of USD 1435.83, remains moderate. The dataset exhibits a skewness of 0.243 and a kurtosis of 2.5, indicating a slightly right-skewed distribution with fewer outliers. The 95% confidence interval, ranging from USD 2512.42 to USD 7912.54, highlights consistent financial advantages of drone-based methods.

5.5. Net Present Value

The net present value (NPV) distribution ($D_{NPV\ of\ First\ Inspection}$) for the first inspection combines investment, cost, and benefit distributions:

$$D_{NPV\ of\ First\ Inspection} = (D_I + D_{CS}) - D_B \tag{24}$$

The first inspection yields a median NPV of $-$USD 11,538, with a significant standard deviation of USD 13,157.72 and a left-skewed distribution (skewness: -4.03). The 95% confidence interval ($-$USD 43,833 to $-$USD 4275) confirms predominantly negative NPVs initially. Annual NPV analysis for 10 inspections shows a shift to positive returns, calculated as

$$D_{NPV} = (D_I + (D_{CS} \times 10) + (D_V \times 12)) - (D_B \times 10) \tag{25}$$

The average annual NPV increases to USD 31,600, with reduced variability (SD: USD 12,849.6) and a symmetric distribution (skewness near zero). The confidence interval (USD

45,481 to USD 64,569) emphasizes the long-term financial benefits of D-BCM, highlighting their cost-effectiveness as cumulative inspections offset initial expenses. Figure 7 compares the distribution of NS, *NPV* of the first inspection, and *NPV* of the 10th inspection.

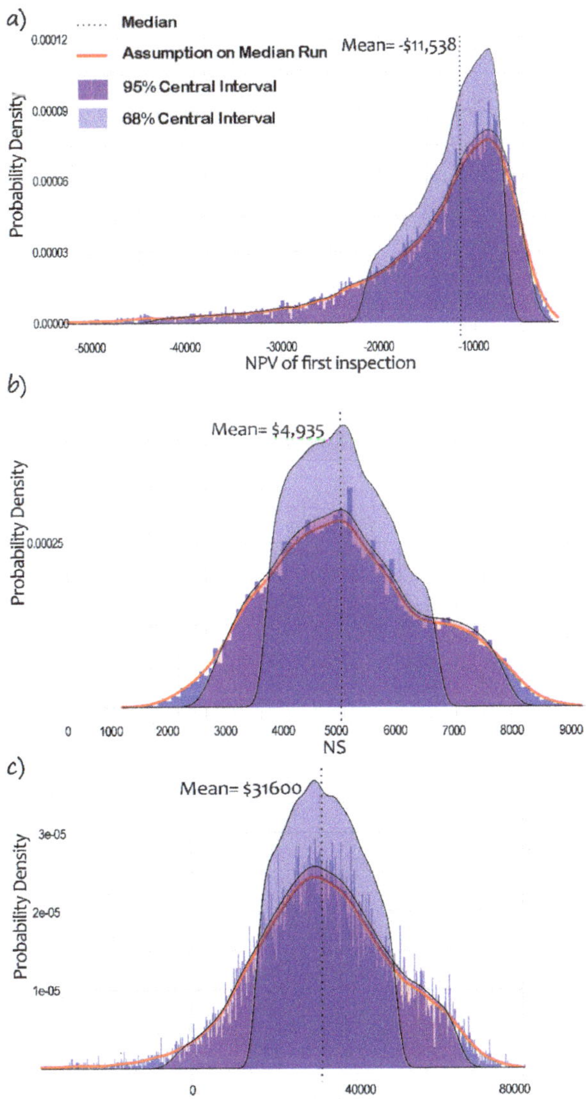

Figure 7. Distribution of (**a**) *NPV* of the first inspection, (**b**) net saving, (**c**), and *NPV* of the 10th inspection.

5.6. Benefit–Cost Ratio

Figure 8 compares the benefit–cost ratio (*BCR*) distribution ($D_{BCR\ of first\ Inspection}$) for the initial and 10th inspections, calculated as

$$D_{BCR\ of\ First\ Inspection} = D_B/(D_I + D_{CS}) \quad (26)$$

$$D_{BCR} = (D_B \times 10)/(D_I + (D_{CS} \times 10) + (D_V \times 12)) \quad (27)$$

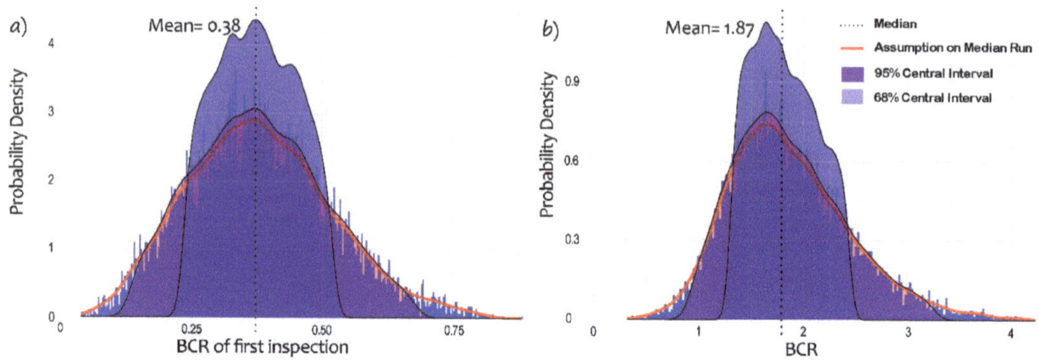

Figure 8. The *BCR* distribution of (**a**) the first and (**b**) the 10th inspections.

In the first inspection, the distribution is unimodal with a slight rightward skew, as the mean (0.38) exceeds the median (0.37), with a skewness of 0.31. Most values cluster near the mean within the 68% interval, while the 95% interval captures broader variability. The model aligns well with observed data, as shown by the red curve matching the median.

The distribution for the 10th inspection shows a similar unimodal shape but with increased skewness (0.89), as the mean (1.87) surpasses the median (1.79). Greater variability is evident, with a standard deviation of 0.60 compared with 0.13 for the first inspection. The 95% interval expands from [0.13, 0.68] to [0.91, 3.28], indicating growing uncertainty over time.

The skewness and variability of the *BCR* increase from the first to the 10th inspection, reflecting greater potential for high or varied *BCR* values. This trend suggests that repeated inspections yield a broader range of benefit–cost outcomes, driven by changes in costs, benefits, or both.

5.7. Cost–Benefit Measures

The *NPV* calculates the difference between current benefits (B_i) and costs (C_i) over a specific timeframe, indicating cost effectiveness. A positive *NPV* signifies profitability, while a negative value suggests otherwise. Among multiple alternatives with positive *NPVs*, the option with the highest value yields the greatest return. The *NPV* formula is

$$NPV = \sum_{i=1}^{Y} \frac{B_i - C_i}{(1+r)^i} \quad (28)$$

Here, r is the discount rate, i is the year, and Y is the total payback period. B_i includes benefits, such as reduced lane closures, improved safety, and cost savings from T-BCM. Costs (C_i) are derived as

$$\sum_{i=1}^{Y} C_i = I + C_S + C_V + C_4 \quad (29)$$

where I is the investment cost ($C_1 + C_2 + T_P + C_{SD}$), C_S is the inspection cost ($(C_3 - T_P) + C_6 + SBP$), and C_V is the monthly deployment cost ($C_V = C_5 + C_7$).

The *BCR* measures the ratio of benefits to costs. A *BCR* above 1 indicates cost effectiveness, with higher values denoting greater returns. For multiple alternatives, the option with the highest *BCR* is preferred. The *BCR* formula is

$$BCR = \frac{\sum_{i=1}^{Y} \frac{B_i}{(1+r)^i}}{\sum_{i=1}^{Y} \frac{C_i}{(1+r)^i}} \quad (30)$$

The USDOT (2023) recommends using a real discount rate of 7% per year to discount monetized benefits and costs to their present value, excluding inflation effects [22]. This case study adopts a 10-year duration, reflecting the anticipated operational lifespan of current drone technologies [48].

5.8. Case Scenario

This scenario assumes an eight-hour traditional inspection time. The simulation identified a 75% reduction in inspection time (ITSP), reducing drone-based inspection time to two hours. Table 8 shows that the initial costs for D-BCM are USD 14,605.99 compared with USD 1528.78 for T-BCM. The *NPV* for the first inspection is negative (−USD 9011.45), and the *BCR* is 0.44, indicating non-cost-effectiveness at the outset. However, the cumulative *NPV* turns positive after the second inspection, as shown in Table 9.

Table 8. *NPV* and *BCR* of the first D-BCM (snapshot: September 2023).

Investment Cost Area	Cost (USD)	Costs Area	Time	Cost (USD)	Benefits Area	Time	Cost (USD)
Drone	5000	*SBP*	2 h	22.62	*TBI*	8 h	960
Payload	1850.99	$Cost_{Soft}$	1 month	260	*TABI*	8 h	1200
Training Pilot	2575	*LI*	1 month	126.16	*UBIV*	1 h	1056.82
GCS	5000	*BILT*	2 h	300	LCC	1 h	1850
Memory Card	180	*ABI*	2 h	240	$Risk_{FI}$	1 day	431.5
Registration Cost	5	*PPE*	4 h	480	ISE	1 inspection	275
		$Cost_{Storage}$	1 month	100	*TTC*	1 h	1235
					VOC	1 h	115
Total	14,605.99			1528.78			7123.32
	NPV				−9011.45		
	BCR				0.441488785		

Table 9. The payback inspection and return rate.

Payback Inspection	Cumulative *NPV*	Return Rate
1	−9011.45	−61.72%
2	−3416.91	−23.40%
3	2177.63	14.92%
4	7772.17	53.23%
5	13,366.71	91.54%
6	19,961.25	136.64%
7	26,555.79	181.75%
8	33,150.33	226.86%
9	39,744.87	271.97%
10	44,755.41	306.51%

The analysis evaluates 10 inspections annually over a 10-year period with 3% and 7% discount rates. The five-year lifespan of drones requires two investments. At both rates, the first-year *NPV* is USD 39,797, with *BCRs* of 2.27 for investment years and 4.23 for subsequent years (Tables 10 and 11).

Table 10. *NPV* and *BCR* analysis for a 10-year time horizon at a 7% discount rate.

Year	Project Year	Discounted Investment Cost at 7%	Discounted Monthly Costs at 7%	Discounted Mainte-nance Costs at 7%	Discounted Costs per Inspection at 7%	Discounted Benefits per Inspection at 7%	Discounted *NPV* at 7%	*BCR*
2023	0	14,605.99	5833.92	570.00	10,426.20	71,233.20	39,797.09	2.26
2024	1	0	5452.26	532.71	9744.11	66,573.08	50,844.00	4.23
2025	2	0	5095.57	497.86	9106.64	62,217.83	47,517.76	4.23
2026	3	0	4762.21	465.28	8510.88	58,147.50	44,409.12	4.23
2027	4	0	4450.66	434.85	7954.09	54,343.46	41,503.85	4.23
2028	5	10,413.86	4159.50	406.40	7433.73	50,788.28	28,374.78	2.26
2029	6	0	3887.38	379.81	6947.41	47,465.68	36,251.07	4.23
2030	7	0	3633.07	354.96	6492.91	44,360.45	33,879.50	4.23
2031	8	0	3395.39	331.74	6068.14	41,458.37	31,663.09	4.23
2032	9	0	3173.26	310.04	5671.16	38,746.14	29,591.67	4.23

Table 11. *NPV* and *BCR* analysis for a 10-year time horizon at a 3% discount rate.

Year	Project Year	Discounted Investment Cost at 3%	Discounted Monthly Costs at 3%	Discounted Mainte-nance Costs at 3%	Discounted Costs per Inspection at 3%	Discounted Benefits per Inspection at 3%	Discounted *NPV* at 3%	*BCR*
2023	0	14,605.99	5833.92	570.00	10,426.20	71,233.20	39,797.09	2.26
2024	1	0	5664.00	553.39	10,122.52	69,158.44	52,818.52	4.23
2025	2	0	5499.02	537.27	9827.69	67,144.12	51,280.12	4.23
2026	3	0	5338.86	521.63	9541.44	65,188.46	49,786.52	4.23
2027	4	0	5183.36	506.43	9263.54	63,289.77	48,336.43	4.23
2028	5	12,599.25	5032.39	491.68	8993.73	61,446.38	34,329.32	2.26
2029	6	0	4885.81	477.36	8731.77	59,656.68	45,561.72	4.23
2030	7	0	4743.51	463.46	8477.45	57,919.11	44,234.68	4.23
2031	8	0	4605.35	449.96	8230.53	56,232.14	42,946.29	4.23
2032	9	0	4471.21	436.85	7990.81	54,594.31	41,695.43	4.23

Figures 9 and 10 highlight the cost advantage of D-BCM, which consistently outperforms traditional methods. The savings grow significantly under a 7% discount rate, demonstrating greater cost efficiency for D-BCM over time.

Based on the 15-year changes in drone prices discussed earlier, this scenario adjusts drone prices in the investment component every five years. Additionally, the model re-simulates all stochastic variables every five years using MCS to generate new random values. However, benefits, costs, and inspection durations—eight hours for T-BCM and two hours for D-BCM—remain constant. Table 12 presents the sensitivity analysis of *NPV* and *BCR* values, highlighting the financial feasibility of drone investments. A negative *NPV* and a *BCR* below one indicate that the initial investment may not yield immediate returns.

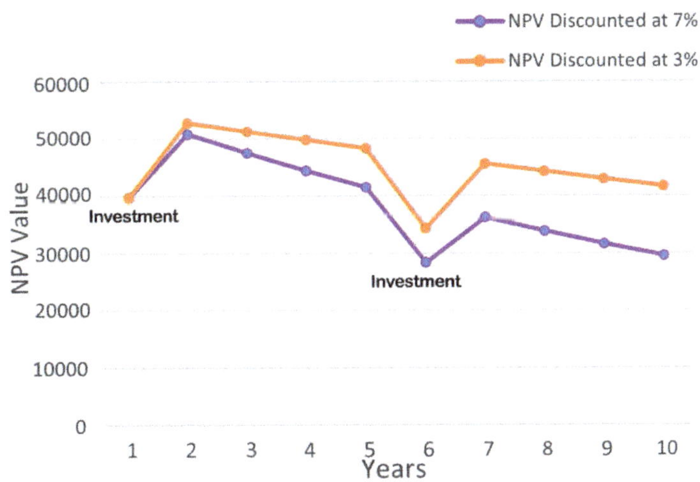

Figure 9. Ten-year *NPV* projection comparing 3% and 7% discount rates.

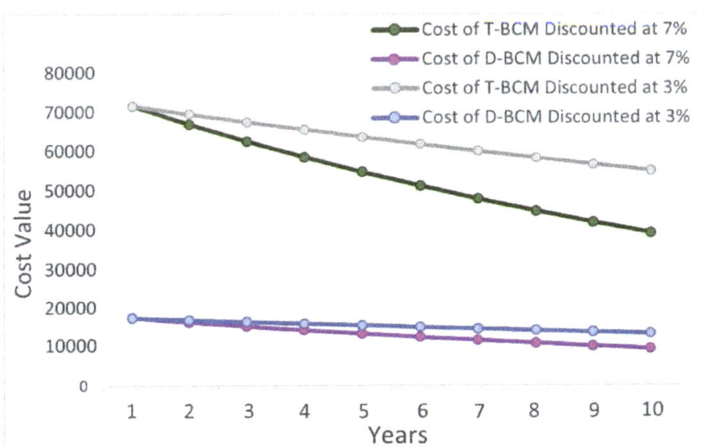

Figure 10. Comparative 10-year cost analysis of T-BCM and D-BCM at 3% and 7%.

Table 12. Statistics of sensitivity analysis (USD).

	DP	*NPV*	*BCR*	Investment Costs	Monthly Costs	Maintenance Costs
Short Term	10,363.19	−15,245.725	0.318445	20,784.19	505.31	380
Midterm	12,365.81	−19,180.20417	0.270812	24,751.81	466.51	851
Long Term	5911	−10,765.20083	0.398206	16,382	468.57	199.5

Table 13 highlights the downward trend of *NPV* over time. This suggests limited gains initially but improved outcomes with sustained drone use. The sixth year shows a significant dip, with the midterm drone purchase yielding a more negative *NPV*, reducing profitability. However, in the long term, the *NPV* improves and the *BCR* approaches 0.4, driven by lower drone prices.

Table 13. *NPV* analysis of DP scenario at a 7% and 3% discount rate (USD).

Calendar Year	*NPV* Discounted at 7%	*NPV* Discounted at 3%
2023	33,209.83	33,209.83
2024	50,461.70	52,421.37
2025	47,160.46	50,894.54
2026	44,075.20	49,412.17
2027	41,191.77	47,972.98
2028	29,636.14	35,855.38
2029	35,937.18	45,167.22
2030	33,586.15	43,851.67
2031	31,388.93	42,574.44
2032	29,335.45	41,334.41
2033	25,592.53	36,802.86
2034	25,920.51	39,414.47
2035	24,224.78	38,266.47
2036	22,639.98	37,151.92

The 15-year sensitivity analysis of drone prices provides insight into the economic impact of adopting drone technology. Although initial investments may not yield immediate profits, consistent benefits and long-term cost reductions can make the initiative economically viable. Hence, balancing upfront costs with long-term gains and strategic opportunities is essential.

6. Conclusions

This study develops a quantitative framework to evaluate the cost efficiency and effectiveness of drone-based condition monitoring (D-BCM) compared to traditional bridge inspection methods (T-BCM). By integrating Monte Carlo simulation (MCS) and probabilistic modeling, the analysis captures the stochastic nature of cost and benefit variables, providing a data-driven foundation for decision making in drone adoption in infrastructure monitoring.

The findings show that while initial investments in drones, software, and training may be substantial, long-term savings in operational costs, inspection time, and safety-related expenses outweigh these costs. D-BCM significantly reduces lane closures, workforce requirements, and reliance on under-bridge inspection vehicles (*UBIVs*), making it a cost-effective alternative for transportation agencies. Monte Carlo results confirm the financial sustainability of drone investments, indicating that technological advancements and market trends will likely lower acquisition and operational costs over time, further enhancing economic feasibility.

Despite these advantages, drones cannot fully replace traditional inspection methods due to their inability to perform tactile, subsurface, or underwater assessments. Additionally, airspace regulations, operator training, and data processing challenges must be addressed for seamless integration into routine infrastructure monitoring.

Future research should explore AI-driven automation, advanced sensor technologies, and regulatory frameworks to enhance drone efficiency and expand applications in bridge, highway, and railway asset management. Further validation through real-world pilot studies and comparative analyses across multiple infrastructure types would strengthen the generalizability of the findings.

By providing a rigorous cost–benefit analysis, this study supports informed decision making for policymakers, transportation agencies, and industry stakeholders, enabling strategic investments in drone technologies to improve the efficiency, safety, and sustainability of infrastructure inspections.

Author Contributions: Conceptualization, T.A. and R.B.; methodology, T.A. and R.B.; software, T.A.; validation, T.A. and R.B.; formal analysis, T.A.; investigation, T.A. and R.B.; resources, R.B.; data curation, T.A.; writing—original draft preparation, T.A.; writing—review and editing, T.A. and R.B.; visualization, T.A. and R.B.; supervision, R.B.; project administration, R.B.; funding acquisition, R.B. All authors have read and agreed to the published version of the manuscript.

Funding: This research received funding from the United States Department of Transportation, Center for Transformative Infrastructure Preservation and Sustainability (CTIPS), Funding Number 69A3552348308.

Data Availability Statement: This article includes the data presented in the study.

Conflicts of Interest: The authors declare no conflicts of interest.

References

1. ASCE. *Report Card for America's Infrastructure*; ASCE: Reston, VA, USA, 2021.
2. Askarzadeh, T.; Bridgelall, R.; Tolliver, D. Drones for Road Condition Monitoring: Applications and Benefits. *J. Transp. Eng. Part B Pavements* **2025**, *151*, 04024055. [CrossRef]
3. Askarzadeh, T.; Bridgelall, R.; Tolliver, D.D. Systematic Literature Review of Drone Utility in Railway Condition Monitoring. *J. Transp. Eng. Part A Syst.* **2023**, *149*, 04023041. [CrossRef]
4. AASHTO. *The Manual for Bridge Evaluation*, 2nd ed.; American Association of State Highway and Transportation Officials: Washington, DC, USA, 2011.
5. Seo, J.; Duque, L.; Wacker, J. Drone-enabled bridge inspection methodology and application. *Civ. Eng.* **2018**, *94*, 112–126. [CrossRef]
6. Perry, B.J.; Guo, Y.; Atadero, R.; van de Lindt, J.W. Streamlined bridge inspection system utilizing unmanned aerial vehicles (UAVs) and machine learning. *Meas. J. Int. Meas. Confed.* **2020**, *164*, 108048. [CrossRef]
7. Hubbard, B.; Hubbard, S. Unmanned Aircraft Systems (UAS) for Bridge Inspection Safety. *Drones* **2020**, *4*, 40. [CrossRef]
8. Azari, H.; O'shea, D.; Campbell, J. Application of Unmanned Aerial Systems for Bridge Inspection. *Transp. Res. Rec.* **2022**, *2676*, 401–407. [CrossRef]
9. Song, H.; Yoo, W.-S.; Zatar, W. Interactive Bridge Inspection Research using Drone. In Proceedings of the 2022 IEEE 46th Annual Computers, Software, and Applications Conference (COMPSAC), Los Alamitos, CA, USA, 27 June–1 July 2022.
10. Dorafshan, S.; Maguire, M. Bridge inspection, human performance, unmanned aerial systems and automation. *J. Civ. Struct. Health Monit.* **2018**, *8*, 443–476. [CrossRef]
11. Chen, S.; Laefer, D.F.; Mangina, E.; Zolanvari, S.M.I.; Byrne, J. UAV Bridge Inspection through Evaluated 3D Reconstructions. *J. Bridge Eng.* **2019**, *24*, 05019001. [CrossRef]
12. Aliyari, M.; Ashrafi, B.; Ayele, Y.Z. Drone-based Bridge Inspection in Harsh Operating Environment: Risks and Safeguards. *Int. J. Transp. Dev. Integr.* **2021**, *5*, 118–135. [CrossRef]
13. Dorafshan, S.; Maguire, M.; Hoffer, N.V.; Coopmans, C. Challenges in bridge inspection using small unmanned aerial systems: Results and lessons learned. In Proceedings of the 2017 International Conference on Unmanned Aircraft Systems (ICUAS), Miami, FL, USA, 13–16 June 2017.
14. Mun, J. *Modeling Risk, Applying Monte Carlo Risk Simulation, Strategic Real Options, Stochastic Forecasting, and Portfolio Optimization*, 2nd ed.; John Wiley & Sons: Hoboken, NJ, USA, 2010.
15. Banh, M.L.; Foina, A.; Li, D.; Lin, Y.; Redondo, X.A.N.; Shong, C.; Zhang, W.-B. *Evaluation of Feasibility of UAV Technologies for Remote Surveying BART Rail Systems*; Bay Area Rapid Transit (BART): Oakland, CA, USA, 2017. Available online: https://escholarship.org/uc/item/3qr9v29d (accessed on 11 March 2025).
16. Wells, J.; Lovelace, B. *Improving the Quality of Bridge Inspections Using Unmanned Aircraft Systems (UAS)*; Minnesota Department of Transportation: Saint Paul, MN, USA, 2018. Available online: https://trid.trb.org/View/1539868 (accessed on 11 March 2025).
17. FAA. 2023. Available online: https://www.faa.gov/faq/how-much-does-it-cost-get-remote-pilot-certificate (accessed on 27 September 2023).
18. Lofsten, H. Measuring maintenance performance in search for a maintenance productivity index. *Int. J. Prod. Econ.* **2000**, *63*, 47–58. [CrossRef]
19. Dunn, R. Advanced maintenance technologies. *Plant Eng.* **1987**, *40*, 80–82.
20. MnDOT. *Bridge Inspectors Reference Manual*; MnDOT: Saint Paul, MN, USA, 2012. Available online: https://www.dot.state.mn.us/bridge/pdf/insp/birm/birmchapt0-cover.pdf (accessed on 11 March 2025).

21. FHWA. *Traffic Data Computation Method Pocket Guide*; Publication No. FHWA-PL-18-027; FHWA: Washington, DC, USA, 2018. Available online: https://www.fhwa.dot.gov/policyinformation/pubs/pl18027_traffic_data_pocket_guide.pdf (accessed on 11 March 2025).
22. USDOT. *Benefit Cost Analysis Guidance for Discretionary Grant Programs*; U.S. Department of Transportation: Washington, DC, USA, 2023. Available online: https://www.transportation.gov/mission/office-secretary/office-policy/transportation-policy/benefit-cost-analysis-guidance (accessed on 11 March 2025).
23. Anwar, N. World's Largest Drone Maker Is Unfazed—Even If It's Blacklisted by the U.S. 7 February 2023. Available online: https://www.cnbc.com/2023/02/08/worlds-largest-drone-maker-dji-is-unfazed-by-challenges-like-us-blacklist.html (accessed on 1 June 2023).
24. Askarzadeh, T.; Bridgelall, R.; Tolliver, D. Monitoring Nodal Transportation Assets with Uncrewed Aerial Vehicles: A Comprehensive Review. *Drones* **2024**, *8*, 233. [CrossRef]
25. Bradbury, T. The Drone Girl. 2023. Available online: https://www.thedronegirl.com/2020/05/24/lipo-batteries-last/ (accessed on 6 June 2024).
26. Castro, J.; Flores, C.; Gonzalez, D.; Quintero, V.; Perez, A. From the Air to the Ground: An Experimental Approach to Assess LiPo Batteries for a Second Life. In Proceedings of the 2022 Prognostics and Health Management Conference (PHM-2022 London), London, UK, 27–29 May 2022.
27. Skydio. 20 July 2023. Available online: https://support.skydio.com/hc/en-us/articles/1260804644430-How-to-charge-and-maintain-your-Skydio-X2-batteries (accessed on 27 September 2023).
28. DJI. Consumer Drones Comparison. 1 August 2023. Available online: https://www.dji.com/products/comparison-consumer-drones (accessed on 12 December 2023).
29. Parrot. ANAFI Ai Photogrammetry. 2023. Available online: https://www.parrot.com/us/drones/anafi-ai/technical-documentation/photogrammetry (accessed on 2 August 2023).
30. Yuneec. Yuneec a Company of ATL Drones. 2023. Available online: https://yuneec.online/e30z/ (accessed on 4 August 2023).
31. RMUS Unmanned Solutions. Flyability Elios 3. 2023. Available online: https://www.flyability.com/elios-3-rad-payload (accessed on 15 July 2023).
32. DJI Inspire. DJI Inspire 3. 2023. Available online: https://www.dji.com/inspire-3 (accessed on 27 August 2023).
33. Autelrobotics. Go Beyond the Boundaries of Aerial Photography. 2023. Available online: https://shop.autelrobotics.com/collections/autel-evo-ii-series (accessed on 29 August 2023).
34. Helmi, W.; Bridgelall, R.; Askarzadeh, T. Remote Sensing and Machine Learning for Safer Railways: A Review. *Appl. Sci.* **2024**, *14*, 3573. [CrossRef]
35. Gillins, D.T.; Parrish, C.; Gillins, M.N.; Simpson, C. *Eyes in the Sky: Bridge Inspections with Unmanned Aerial Vehicles*; Oregon DOT: Washington, DC, USA, 2018.
36. Michael Baker International. *UAS Bridge Inspection Pilot*; Wisconsin Department of Transportation: Madison, WI, USA, 2017. Available online: https://trust.dot.state.wi.us/ftp/dtsd/structures/pub/gurtner/UAS%20Pilot%20Bridge%20Inspection%20Report_final.pdf (accessed on 11 March 2025).
37. O'Neil-Dunne, J. *Unmanned Aircraft Systems for Transportation Decision Support*; U.S. Department of Transportation: Washington, DA, USA, 2016. Available online: https://rosap.ntl.bts.gov/view/dot/32744 (accessed on 11 March 2025).
38. Zajkowski, T.; Snyder, K.; Arnold, E.; Divakaran, D. *Unmanned Aircraft Systems: A New Tool for DOT Inspections*; NCDOT: Charlotte, NC, USA, 2016.
39. Dorafshan, S.; Maguire, H.N.V.; Coopmans, C. *Fatigue Crack Detection Using Unmanned Aerial Systems in Under-Bridge Inspection*; Idaho Transportation Department: Boise, ID, USA, 2017. Available online: https://apps.itd.idaho.gov/apps/research/Completed/RP256.pdf (accessed on 11 March 2025).
40. Bridge, J.A.; Ifju, P.G.; Whitley, T.; Tomiczek, A.P. *Use of Small Unmanned Aerial Vehicles for Structural Inspection*; Florida Department of Transportation: Tallahassee, FL, USA, 2018. Available online: https://rosap.ntl.bts.gov/view/dot/36598 (accessed on 11 March 2025).
41. Burgett, J.M.; Bausman, D.C.; Comert, G. *Unmanned Aircraft Systems (UAS) Impact on Operational Efficiency and Connectivity*; U.S. DOT: Washington, DC, USA, 2019. Available online: https://rosap.ntl.bts.gov/view/dot/53574 (accessed on 11 March 2025).
42. DroneDeploy. Drone Insurance and Liability Coverage: Do You Need It? 17 January 2018. Available online: https://www.dronedeploy.com/blog/drone-insurance-and-liability-coverage-do-you-need (accessed on 2 October 2023).
43. FHWA BIRM. *Bridge Inspector's Reference Manual (BIRM)*; FHWA: Washington, DC, USA, 2023.
44. USDOT. *Departmental Guidance Treatment of the Value of Preventing Fatalities and Injuries in Preparing Economic Analyses*; USDOT: College Station, TX, USA, 2021. Available online: https://tti.tamu.edu/news/tti-publishes-2021-urban-mobility-report/ (accessed on 11 March 2025).

45. WisDOT. *General Lane Closure Impact Analysis*; WisDOT, Wisconsin Department of Transportation: Madison, WI, USA, 2020. Available online: https://transportal.cee.wisc.edu/closures/devel/#:~:text=Welcome,to%20time%20or%20budget%20constraints (accessed on 11 March 2025).
46. Glover, B.J. *2021 Urban Mobility Report*; Texas A&M Transportation Institute, United States Department of Transportation: Washington DC, USA, 2021. Available online: https://www.transportation.gov/sites/dot.gov/files/2021-03/DOT%20VSL%20Guidance%20-%202021%20Update.pdf (accessed on 11 March 2025).
47. Cordella, M.; Alfieri, F.; Sanfelix, J. *Guidance for the Assessment of Material Efficiency: Application to Smartphones*; Publications Office of the European Union: Luxembourg, 2020.
48. Cohn, P.; Green, A.; Langstaff, M.; Roller, M. Commercial Drones Are Here: The Future of Unmanned Aerial Systems. 2017. Available online: https://www.mckinsey.com/industries/travel-logistics-and-infrastructure/our-insights/commercial-drones-are-here-the-future-of-unmanned-aerial-systems#/ (accessed on 1 June 2023).

infrastructures

MDPI

Article

Assessment of Anisotropy in Cold In-Place Recycled Materials Using Shear Wave Velocity and Computed Tomography Analysis

Quentin Lecuru [1,*], Yannic Ethier [1], Alan Carter [1] and Mourad Karray [2]

[1] Department of Construction Engineering, École de Technologie Supérieure, Montréal, QC H3C 1K3, Canada; yannic.ethier@etsmtl.ca (Y.E.); alan.carter@etsmtl.ca (A.C.)

[2] Department of Civil Engineering, Université de Sherbrooke, Sherbrooke, QC J1K 2R1, Canada; mourad.karray.benhassen@usherbrooke.ca

* Correspondence: quentin.lecuru.1@ens.etsmtl.ca

Abstract: Pavement materials like hot mix asphalt (HMA) and cold recycled mixes (CRMs) are typically considered isotropic. This study evaluates the anisotropy of a cold in-place recycled (CIR) material using the shear wave velocity (V_s) parameter. The piezoelectric ring actuator technique (P-RAT) is utilized to assess the V_s parameter in three directions in CIR slabs. Similarly, the ultrasonic pulse velocity (UPV) technique is employed to measure P-wave velocities. Both methods evaluate mechanical properties in multiple directions. Complex modulus tests are conducted to link velocities results to $|E^*|$ modulus. Finally, computed tomography (CT) scans are performed on the specimens in order to evaluate anisotropy resulting from aggregate alignment. The V_s obtained using P-RAT and the V_p from UPV indicate anisotropy, as the wave velocities differ across the three directions. Differences range from 0.6 to 11.6% in V_s, influenced by measurement location. UPV results are analysed in relation to the $|E^*|$ modulus master curves, demonstrating that the first peak arrival time for the P-wave corresponds with the master curve. CT scan analysis reveals that the aggregates tend to be more aligned in the direction of the compacting wheel's displacement, which also highlights anisotropy.

Keywords: anisotropy; shear wave velocity (V_s); cold recycled asphalt material; frequency analysis; CT scan

1. Introduction

Cold recycled materials (CRMs) have been employed for several years in the rehabilitation of pavement. They offer an economical and ecological way of renewing upper layers through cold in-place recycling (CIR) techniques or of correcting deeper default by using full-depth reclamation (FDR). However, the evolutive mechanical properties of CRMs make their behavior difficult to precisely analyze over time. For instance, CIR materials are composed of 100% reclaimed asphalt pavement (RAP), to which a binder in the form of asphalt emulsion or foamed asphalt is added to enhance mechanical resistance. At early ages, considering actual humidity and added water during mixing, CIR mixtures contain a great amount of water. While curing, the water content in the mixture decreases, resulting in an increase in the stiffness of the material. The cure brings the CIR behavior closer to that observed in hot mix asphalt (HMA). The characterization of CRMs is of interest and is covered by many researchers. It appears that, as with HMA, CRMs are typically regarded as isotropic. While a few studies have evaluated the validity of this assumption in HMA, as of the time of writing this article, no studies have specifically addressed this hypothesis for CRMs.

However, the compaction process may result in a stress-dependent anisotropy in the material, as the aggregates are rearranged in a specific manner to reach the targeted void ratio. This anisotropic behavior observed in both HMA and CRM could impact the methodology and interpretation of laboratory testing in academic research and quality control. Indeed, some measurements depend on what is being measured and the nature of the test being performed, as well as the way it is performed, thus, the characterisation of anisotropy has practical implications.

2. Objectives

The main objective of this study is to demonstrate the applicability of the shear wave velocity (V_S) parameter to evaluate the anisotropic behavior of a CIR material using the P-RAT. It is compared with the UPV technique. Also, the relationship between the anisotropic behavior and the orientation of the aggregate obtained by 3D image analysis is evaluated. The final objective here is to link the wave velocity results with the complex modulus.

3. Background

Numerous studies have been carried out on CRMs to improve the global knowledge, either to better understand compaction and mixing temperature [1–3] or to evaluate the influence of components on the mechanical behavior of the mix [4–6]. These studies use mechanical testing methods. As such, the complex modulus test (E*) is used to evaluate the dynamic modulus, for both CRMs [5,7,8] and HMA [9–11]. These previous studies have focused specifically on the anisotropic behavior of HMA. However, pavement is often regarded as an isotropic material, implying that the mechanical properties of the material are identical in all directions.

In the field of geotechnical engineering, two types of anisotropy are generally considered [12,13]. Inherent anisotropy is directly linked to the materials and the manner of their constitution (gravity, natural deposition, for instance). Induced anisotropy, on the over hand, is stress-dependent and related to external actions. According to Masad et al. [14], Underwood et al. [15], Bhasin et al. [16], and Alanazi et al. [17], the two types of anisotropy exist in bituminous materials, as they are bound granular materials. The inherent anisotropy would be related to aggregate orientation, aggregate contacts, and air void distribution and orientation. The use of CT scanning provides means to evaluate this anisotropy in HMA [16–19] and granular materials [20]. The orientation and distribution of aggregates in HMA has been evaluated by Hassan et al. [18] and Huang et al. [19] and an orientation perpendicular to the direction of compaction was observed. Quinteros et al. [20] quantified the anisotropy with the particles orientation.

Bhasin et al. [16] showed that bitumen mastic also presents a perpendicular orientation to compaction direction and that the addition of coarse aggregate tends to modify this preferential orientation. This would possibly lead to induced compaction anisotropy. Alanazi et al. [17] showed that anisotropy decreases as compaction increases and that the compaction method influences the anisotropy.

The induced anisotropy would be considered as a damaged-induced anisotropy, as the accumulation of damage modifies the structure of the specimen. This anisotropy occurs when tested under confinement pressure, for instance. Underwood et al. [15] and Alanazi et al. [17] show that as the confinement pressure increases, the anisotropy evolves, as the air void and aggregates contacts are modified. However, in these studies the anisotropy was evaluated by a ratio of radial and axial strains, and the deformations were found to be quite high and, therefore, away from the small strain domain.

Underwood et al. [15] show that HMA exhibits directionally dependent properties, especially in the case of compression tests. The testing temperature influences the anisotropic behavior. Alanazi et al. [17] confirm these findings, and also show that a greater nominal maximal aggregate size increases the anisotropy. Moreover, for a similar mixture, a greater air void ratio results in a more pronounced anisotropy. The following studies did not described anisotropy as inherent nor induced. Pham et al. [9] investigated the behavior of HMA and warm mix containing RAP. They determined that complex Poisson ratio can be considered as transverse isotropic. Di Benedetto et al. [10] studied the complex modulus and Poisson ratio in three dimensions of HMA. The modulus in the three directions exhibited a difference of up to 25%. With regard to the complex Poisson ratio, the difference could reach up to 100%, but the authors mentioned that this measurement is delicate. They concluded that the specimens can be considered as orthotropic. Nguyen et al. [11] conducted E* tests on specimens cored from a slab in three directions. This allowed them to compare E* and ν^* in these three directions. The 2S2P1D model for the three directions gave identical parameters, with exception of E_{00}, E_0 ν_{00}, and ν_0. Differences in the value of E* and ν^* in the three directions show anisotropy in the mix. The normalized curve of E* and nu* can be considered as identical. Benedetto et al. [21] highlighted that specimens compacted with a slab compactor and with a shear gyratory compactor (SGC) exhibited anisotropic behaviors, but with inverted characteristics. A wave-based method was employed to evaluate the compressive velocity (V_p) in three directions. In considering the axis along which the compaction energy is applied (Z), the transverse axes (X and Y), and the compressive velocity (V_p axis) in the aforementioned direction, it can be observed that for the cylinder specimen, V_pX and V_pY are equal and lower than V_pZ. In contrast, for the slab specimen, V_pX and V_pY remain equal but higher than V_pZ.

3.1. Wave-Based Methods

Wave-based methods are non-destructive and, for most of them, non-intrusive. These methods have been used to characterize various construction materials, such as concrete, soils, or bituminous materials. Mechanical wave-based methods employ compression or shear waves for testing. The ultrasonic pulse velocity (UPV) method, such as the portable ultrasonic non-destructive digital indicating tester (PUNDIT), is extensively used and primarily relies on compression waves, although shear waves may be used occasionally [22]. In this method, sensors are placed on opposite ends of the material being tested. These sensors transmit and receive signals that propagate through the material. By measuring the travel distance and flight time of the signals, the velocity of the waves can be determined. This method operates in the temporal domain.

To obtain the time of flight, the arrival time of the received signal is compared to a reference time of the emitted signal. However, there is no consensus regarding the precise method for obtaining this time of flight [23,24]. Indeed, one can choose to set the reference time of the emitted signal at the first arrival or at the maximum peak of it. As well, one can choose to set the arrival time of the received signal at the first arrival, first peak, or maximum amplitude peak [22,25–29]. Figure 1 shows an example of such a dilemma. Depending on which of the precedent is chosen, the time of flight drastically changes, along with the wave velocity. Moreover, the near-field effect can occur when using S-wave [30].

Wave-based methods require signal processing, so time-domain methods such as UPV provide a straightforward and convenient solution, rendering them attractive as initial approximations. However, when it comes to characterizing materials and monitoring the evolution of mechanical parameters, the requisite level of precision may warrant the utilization of more intricate techniques, such as frequency-domain methods.

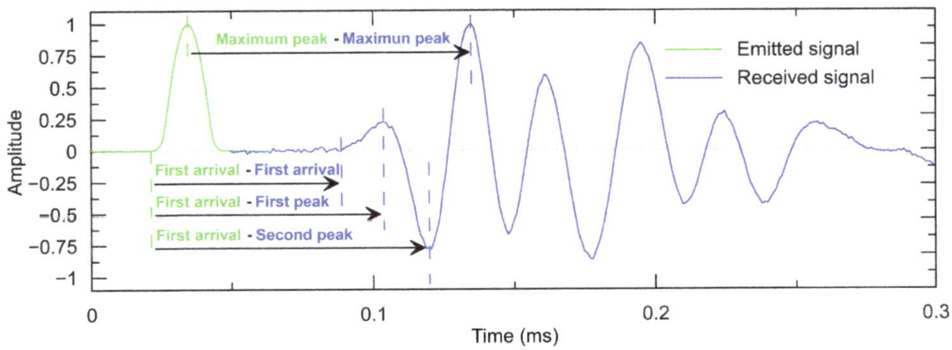

Figure 1. Example of input signal and received signal.

Unlike time-domain methods that concentrate on analyzing temporal signals, frequency-domain methods involve decomposing the signal into multiple harmonics using discrete Fourier transform. This decomposition allows for extracting valuable information such as the frequency spectrum, resonant frequency, and even the phase shift [22,30,31]. Although these methods entail more calculations, their signal processing is objective and offers means of verification.

3.2. The P-RAT

The piezoelectric ring actuator technique (P-RAT) method, developed at Sherbrooke University, was initially used for geotechnical applications [30,32]. It is based on the cross-correlation method, which determines the shear wave velocity by comparing the emitted and received signal at different frequencies. The P-RAT method uses spectral analysis of elastic shear waves propagating in a specimen. Back calculations provide the shear wave velocity V_s instead of the phase velocity V_{ph}, which is obtained via the cross-correlation method and correction of phase-shift produced by emitter–receiver sensors. Unlike the phase velocity, which varies with frequency, the shear wave velocity is an intrinsic material parameter that should remain constant irrespective of the frequency.

In the P-RAT configuration, piezoelectric transducers are used. They are composed of a metal cylinder fitted inside a piezoelectric ring using epoxy. The piezoelectric ring is welded on each side to a shield wire, which transfers the electrical pulses of varying forms and parameters to the ring. Due to the piezoelectric nature of the ring's material, these pulses induce radial deformation of the rings. This enables the generation of shear waves. Two transducers are used at a time, with one serving as emitter and the other as the receiver. The signals are analyzed with a dedicated Labview program, and the shear wave velocity (V_s) is obtained. Details may be found in Karray et al. [30] and Lecuru et al. [33,34].

P-RAT has been successfully used to characterize CIR material at young age [33,34], due to its use of mechanical shear waves that do not propagate into water. P-RAT appears to be suitable for such utilization. Moreover, P-RAT has been employed to characterize soils [30,32] and concretes at an early age [35–37]. Elbeggo et al. [32] demonstrated the robustness of the P-RAT in clays, by using different actuators and different set-ups in two distinct laboratories. Finally, the use of P-RAT focuses on very small strain, allowing the operator to consider the test as being performed in the elastic domain [38]. The P-RAT's transducers characteristics and the methodology of the test allows measuring V_s in different directions on a single specimen, if the shape of the specimen permits it.

4. Materials and Methods

The study is composed of three different parts. The first part consists of V_s measurement in sections of a slab in three directions with the P-RAT and UPV techniques. The second part consists of conducting complex modulus tests on cylinders cored from these slab sections. Finally, the third part consists of performing CT scans on the same cylinders to evaluate the aggregates orientation. Figure 2 displays a schematic representation of the experimental campaign, with a detailed description of each part in the subsequent sections.

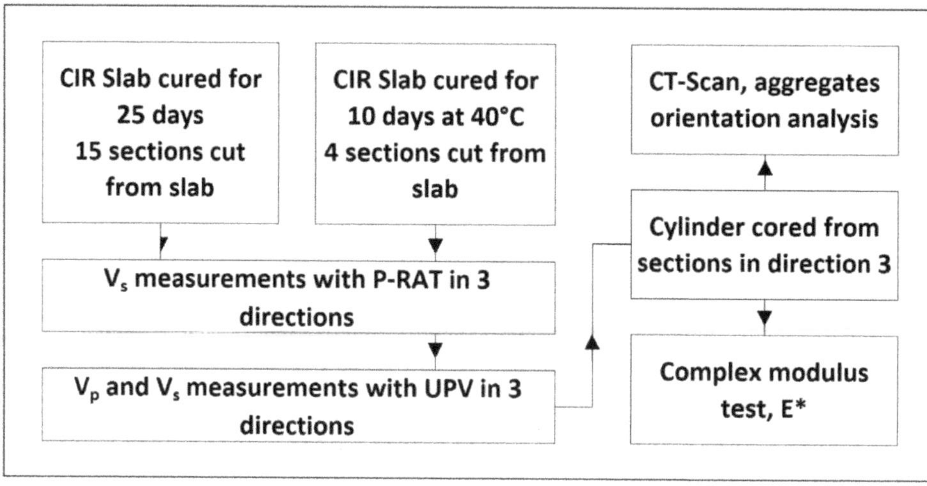

Figure 2. Schematic of the experimental campaign.

4.1. Materials Used and Specimens

The tested specimens are two CIR materials slabs. The first one is composed of 100% RAP material (0–10 mm) with 2% added bitumen in the form of asphalt emulsion, 6% added water, and 1% cement. The second slab is composed of 100% RAP material (0–20 mm) with 2% added bitumen in the form of asphalt emulsion (CSS-1, from McAsphalt, Toronto, ON, Canada) and 2% added water. All the mentioned percentages are related to the mass of RAP. The components were stored at room temperature and the RAP was not oven dried.

The slabs were compacted with a LPC French wheel compactor, following the procedure, although in the case of CIR materials, no metal wheel was used. The slab dimensions are 600 × 400 × 130 mm and 500 × 180 × 110 mm for slab 1 and slab 2, respectively. Slab 1 and slab 2 were cured in an oven at 40 °C for 25 and 10 days, respectively. After curing, the slabs were left in a freezer for 24 h to facilitate the cutting process. The slabs were cut in 15 sections of 105 × 105 × 130 mm and 4 sections of 110 × 140 × 110 mm blocs for slab 1 and slab 2, respectively. Figures 3A and 3B present schematic views of slab 1 and slab 2, respectively.

The sections are referred to by the name of the slab and the number of the section, according to Figure 3. Section 4 from slab 2 is referred to as Slab2_4, for instance. Some sections were also covered with a bitumen mastic on each side to assess the improvement in contact and its influence on the P-RAT measurement. The bitumen mastic was made of $\frac{1}{4}$ bitumen (PG 58S-28) for $\frac{3}{4}$ sand (concrete sand), by mass.

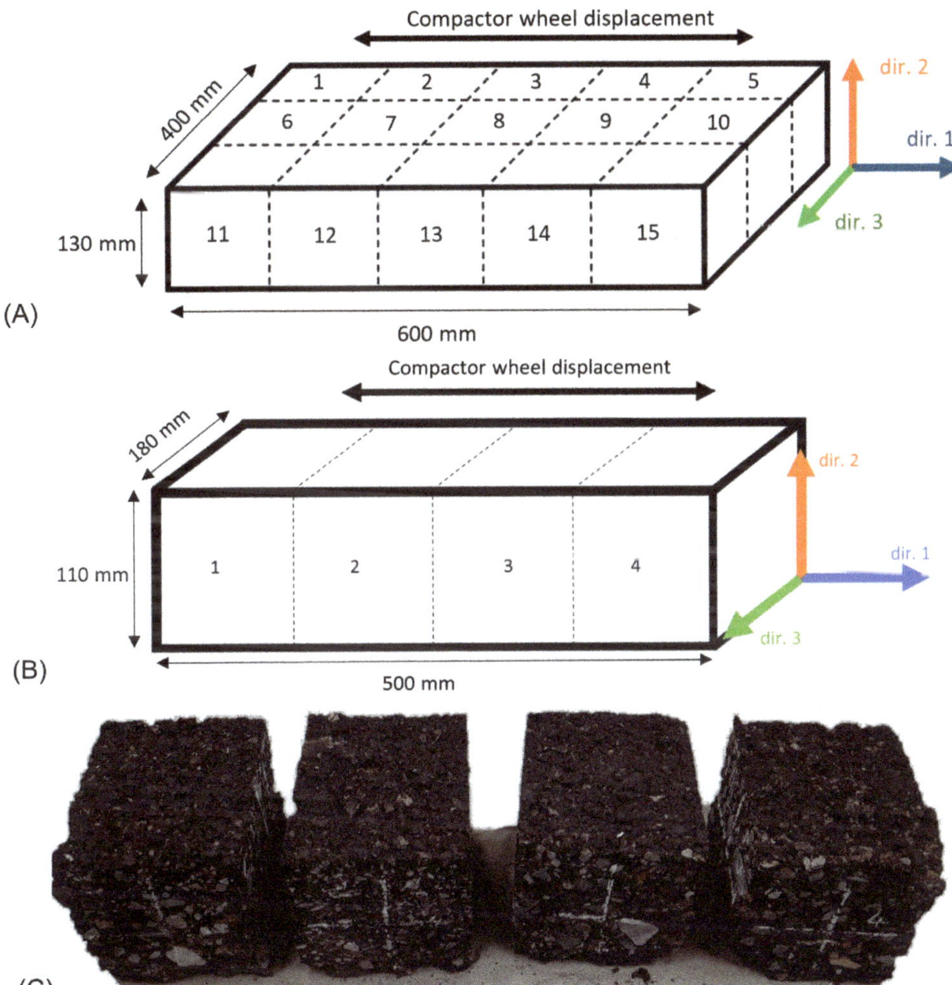

Figure 3. Schematic view of: (**A**) slab 1; (**B**) slab 2; (**C**) real specimens after curing and cutting, slab 2. The sections are identified by their numbers.

Four cylinders of diameter 75 mm were cored from the sections of slab 2 in direction 3, as E* tests are usually performed with cylindrical specimens with a $\frac{1}{2}$ diameter to length ratio. The four cylinders are designated specimens Slab2_1C to C_Slab2_4C, according to the section from which they were cored from.

Slab 1 was compacted to a targeted void ratio of 20%.

The bulk density was calculated according to LC-26-040 [39] by volumetric means due to the void ratio being larger than 10%. There is a lack of materials on the edge of the sections/cylinders due to cutting process. Therefore, the bulk densities are underestimated, and the resulting air void ratios are overestimated. The 3D analysis provides the air void ratio, without considering the edges of the specimens. The bulk density of specimen Slab2_2 to Slab2_4, their estimated void ratio, and calculated void ratio with 3D analysis are presented in Table 1. Due to damages on specimen Slab2_1C, no 3D analysis was performed.

Table 1. Information on sections of slab 2.

Method	Specimen	Slab2_1C	Slab2_2C	Slab2_3C	Slab2_4C
LC 26-040	Volumetric bulk density (g·cm^{-3})	1.910	1.940	1.985	1.900
	Air void ratio (%)	24.1	22.9	21.1	24.5
Three-dimensional analysis	Air void ratio (%)	N.A.	17.2	16.5	17.1
	Calculated bulk density (g·cm^{-3})	N.A.	2.084	2.102	2.086

4.2. P-RAT Anisotropy Measurements

The tests were conducted uniformly across all tested sections. The objective of the study is to evaluate the anisotropic behavior of the CIR material. To this end, each block was tested in 3 different directions (dir. 1, dir. 2, and dir. 3), according to Figure 3.

Different pairs of transducers were used in this study. As shown in Lecuru et al. [33], the dimensions of the transducer do not influence the results significantly, although for smaller transducer size, wave diffraction can occur if the aggregate diameter exceeds the transducer wavelength [37]. Still, three main different transducer pairs were used in this study for control purposes and to ensure reproducibility. The designation of the transducers refers to the outer–inner diameter, namely 19–16 mm, 26–22 mm, and 36–31 mm. The inner stones of the transducers are made of four-quarter aluminum cylinders. Other transducers were used in the case of slab 1, mostly for testing newly made transducers or enhanced former transducers. Some of the transducers are encased in a plastic capsule, which mitigates the stress on the transducer during testing.

Transducers are positioned with care, ensuring that they are correctly aligned with one another. One transducer, acting as the emitter, is connected to a signal generator, Handyscope HS3 100 MHz (TiePie Engineering, Sneek, The Netherlands). The second transducer, acting as the receiver, is connected to another entry of the same apparatus, which also act as an oscilloscope. An amplifier, 7602M from Krohn-Hite (Brockton, MA, USA), is used on the generator side. The Handyscope is operated by a personal computer on which one can observe the emission and the reception signals. The form, amplitude, and frequency of the input signal are selected on the computer.

A single pair of transducers was used at a time, a rubber band was used to hold the transducer onto the specimens. Each pair was positioned in the same location, at the center of the surface. If the contact at this spot is insufficient to ensure optimal signal quality, the transducers are repositioned a few millimeters away. For repeatability and control concerns, signals of different shapes and frequencies are used when testing the specimen with a given transducer pair. To remove potential noise in an individual signal, the signals are sent eight times and stacked. Every specimen was tested at room temperature, and under the same conditions. A total of 1334 and 703 signals, without considering the stacking steps mentioned earlier, were collected for slab 1 and slab 2, respectively, considering all pairs of transducers used. Finally, these signals are then analyzed with a dedicated P-RAT software (version 2) developed in LabView (National Instruments, Austin, TX, USA) with the aforementioned concept obtained from Sherbrooke University [30].

4.3. UPV

The PUNDIT PL-200 device (ScreeningEagle, Zurich, Switzerland) is a device mainly employed in the monitoring of concrete materials via UPV. The Pundit device was used to monitor the P-wave and the S-wave in the tested specimens. To ensure good contact between the specimens and the sensors, coupling media was used as recommended by the PUNDIT supplier. The sensors used for the generation and reception of P-waves and

S-waves have a resonant frequency of 54 and 250 kHz, respectively. Each sensor was manually held in place to ensure a sufficient contact with the tested specimen, and the sensors were kept aligned. The results have been analyzed in the Pundit software (PL-Link, version 3).

To obtain the S-wave flight time, one has to compare the received P-wave signal with the received S-wave. Indeed, even with a dedicated S-wave sensor, a weaker P-wave is generated. The following study investigates three different configurations regarding the arrival time of the received signal: first arrival, first peak, and second peak. There is no information on the nature of the input signal generated by the Pundit PL-200 device. Therefore, the flight time of either the P-wave or the S-wave is taken as shown in Figure 1, as instructed by the Pundit method.

4.4. Complex Modulus E*

The linear viscoelastic behavior of the cored cylinder was evaluated by dynamic tension–compression tests. Sinusoidal cyclic axial loads were applied with a hydraulic press (MTS) in a controlled deformation mode. The amplitude of the deformation was selected to be around 50 μm/m. The cylinders were glued to metallic caps using epoxy and equipped with three extensometers of 50 micrometer in the axial direction and temperature probes.

Specimens Slab2_3C and Slab2_4C were tested at 1, 0.3, 0.1, 0.03, and 0.01 Hz at 40, 30, 20, 10, 0, −10, and, −20 °C; and 25, 15, 5, −5, −15, −25, and −35 °C for specimen Slab2_3C and with the last array of temperature for Slab2_4C. The experimental results were analysed using the 2S2P1D model, which is a linear viscoelastic model composed of two springs, two parabolic elements, and one dashpot [9–11,40,41]. More details can be found in the mentioned references. The following equation is used to calculate the complex modulus E*:

$$E^*_{2S2P1D}(\omega) = E_{00} + \frac{E_0 - E_{00}}{1 + \delta(j\omega\tau_E)^{-k} + (j\omega\tau_E)^{-h} + (j\omega\beta\tau_E)^{-1}} \quad (1)$$

where ω is the angular frequency; j is the imaginary unit; k, h, δ are constant ($0 < k < h < 1$); E_{00} is the static modulus, when ω tends to 0; E_0 the glassy modulus, when ω tends to infinite; β, parameter linked with η, the dynamic viscosity of the dashpot, $\eta = (E_0 - E_{00})\beta\tau_E$; τ_E, the characteristic time values. The characteristic time values vary with the chosen temperature T and the shift factor a_T, $\tau_E(T) = a_T(T) \times \tau_{0E}$.

4.5. Computed Tomography Scan (CT-SCAN) and 3D Image Analysis

CT scans were performed to assess any specific particles orientation in the mix. The XTH-225 device from Nikon (Tokyo, Japan) was used for specimen Slab2_3C while a FF35 from Comet Yxlon was utilized for specimens Slab2_2C and Slab2_4C. CT scans were conducted with a voltage of 210 kV and a current of 210 μA for specimen Slab2_3C, and a voltage of 209 kV and a current of 108 μA for specimens Slab2_2C and Slab2_4C. Two distinct scanning devices were utilized for availability concerns. The 3D specimens were analysed with Dragonfly software, version 2023 (Comet Technologies Canada Inc., Montréal, QC, Canada). The cubic voxel size was 68.63 μm for specimen Slab2_3C and 30.75 μm for specimens Slab2_2C and Slab2_4C.

The 3D images data set size was first reduced to remove parts affected by artefacts, such as cone beam artefacts at the upper and lower parts of the data set. Then the data set was filtered to ease the segmentation process. Mainly, a smoothing filter was applied to reduce noise within the images, and a radial basis function (RBF) filter was used to mitigate greyscale variation across the images due to beam hardening and the cylindrical shape of the specimens. Granular materials were segmented and labelled individually with the tools offered by Dragonfly software, version 2023. During the segmentation process, some errors can be introduced, mostly due to the spatial and the contrast resolution of the voxels. For instance, a voxel can be labelled as "aggregate" though it could be "bitumen mastic". But, in the present study, it does not impact the aggregate orientation as aggregates composed of few voxels were not considered in the final analysis.

The orientation of the longest axis of a labelled particle was evaluated with 2 parameters, Φ and θ angles, θ being the angle between the X axis and the projection of the longest axis in the X–Y plane; and Φ the angle between the longest axis and Z axis, as shown in Figure 4.

Figure 4. Parameter Φ and θ schematic definition, ORS Dragonfly.

Air voids were also segmented, in order to calculate the air void ratio of the tested specimens by comparing the labeled void voxel to the global voxel count. A contrast-limited adaptive histogram equalization (CLAHE) filter was applied to enhance the contrast in the data set, when required.

5. Results and Analysis

In this section, the results concerning anisotropy are presented first. V_s results obtained with the P-RAT in the slabs are analysed and the particles orientation of aggregates is presented. UPV results are discussed. Afterward, the results of the complex modulus are presented, and a comparison with wave velocities is made.

5.1. P-RAT Shear Wave Velocity (V_s) Results

Figure 5 illustrates the effects of bitumen mastic on two signals. In both cases, the presence of a P-wave is observed, but it is more pronounced in the case of the mastic interface. However, the signal is clearer. The P-RAT transducers are designed to generate mainly shear waves. Due to the size of the specimen and the quality of the contact, in order to obtain a signal at the receiver end, it is sometimes necessary to amplify the emitted signal, generating larger deformation of the piezoelectric ring. Such amplification tends to generate more P-wave. It has been observed that in the case of the bitumen interface, the

P-wave amplitude is higher than the S-wave one compared to the case without mastic. The arrival times of the wave, calculated from the P-RAT V_s measurement are 0.294 and 0.309 ms for the signals with and without bitumen mastic, respectively. Figure 5C,D display an example of the P-RAT analysis. See Karray et al. [30] for more details.

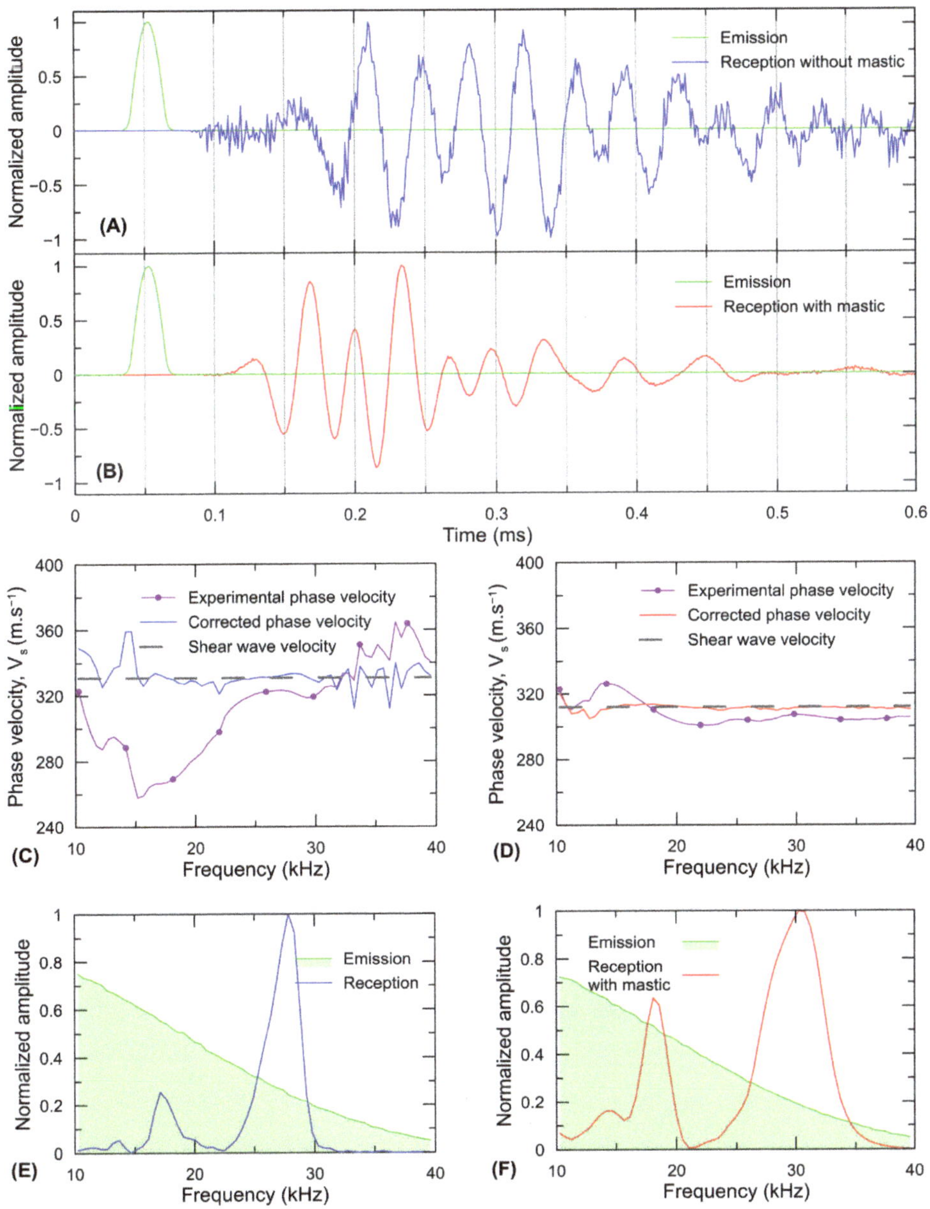

Figure 5. V_s measurements using P-RAT and influence of the bitumen mastic interface; temporal signal (**A**,**B**); P-RAT analysis (**C**,**D**); frequency decomposition (**E**,**F**) without and with bitumen mastic, respectively. Section Slab2_3, dir. 2, 26–22 mm transducers, DSA-10 kHz-12 V signal.

The difference between V_s values in section Slab2_3 with and without mastic is 1.5, 2.7, and 1.3% for directions 1, 2, and 3, respectively. It was more convenient to lay the same thickness of mastic on the ends of Section 3 in directions 1 and 3 (2 mm on each face) than in direction 2. Indeed, the upper part surface of the slab is uneven, resulting in a thicker layer of mastic. This explains the higher difference observed in direction 2. P-RAT testing with various pairs of transducers have demonstrated that the mastic used to improve the interface gives V_s values of approximately 300 $m{\cdot}s^{-1}$. In the case of a mastic interface, the S-wave propagates slower in the mastic layer, reducing the overall V_s. The actual V_s of the specimen can be calculated as the thickness of the mastic layer and the V_s values of the mastic used are known. In the following Figures, V_s depicted as "with mastic" display the corrected V_s values.

Regarding slab 2, for direction 2, only section Slab2_1 was monitored with the 36–31 mm transducer pair. The surface of the three other sections were too rough to allow for a proper contact of this larger size of transducer. The same situation occurred for specimens of Slab 1, preventing the acquisition of V_s results in its direction 2. V_s measurements in the tested section of slab 1 and in the four sections of the slab 2 are presented in Figures 6 and 7, respectively. Each point represents the mean value of all measured signals. As mentioned in the P-RAT measurement section, several signals of various shapes and frequencies were sent for repeatability and control issues. For example, "V_s values in Direction 1 19–16 mm" is the mean value of all five treated signals (after stacking) for the 19–16 mm transducer pair in direction 1.

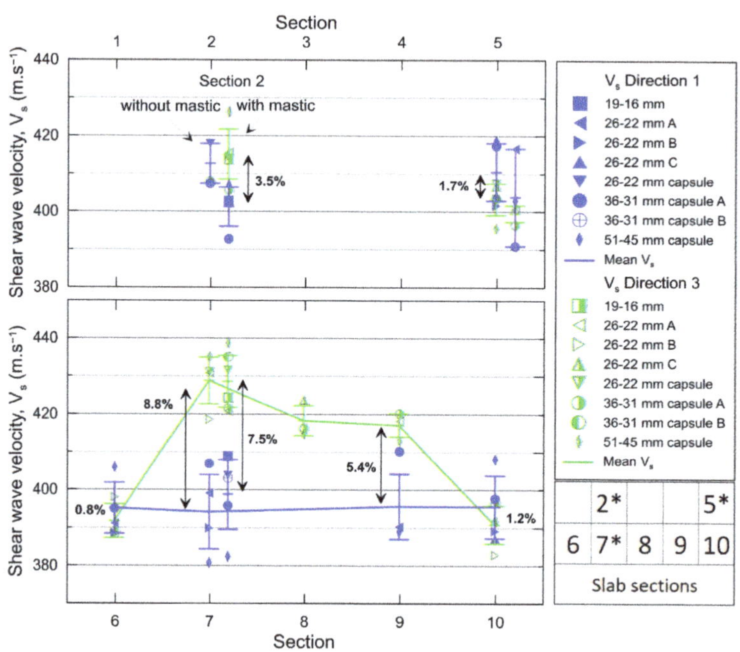

Figure 6. Shear wave velocity with respect to direction for tested sections of slab 1. Sections with * have also been tested with a bitumen mastic interface. The relative differences between mean V_s are displayed.

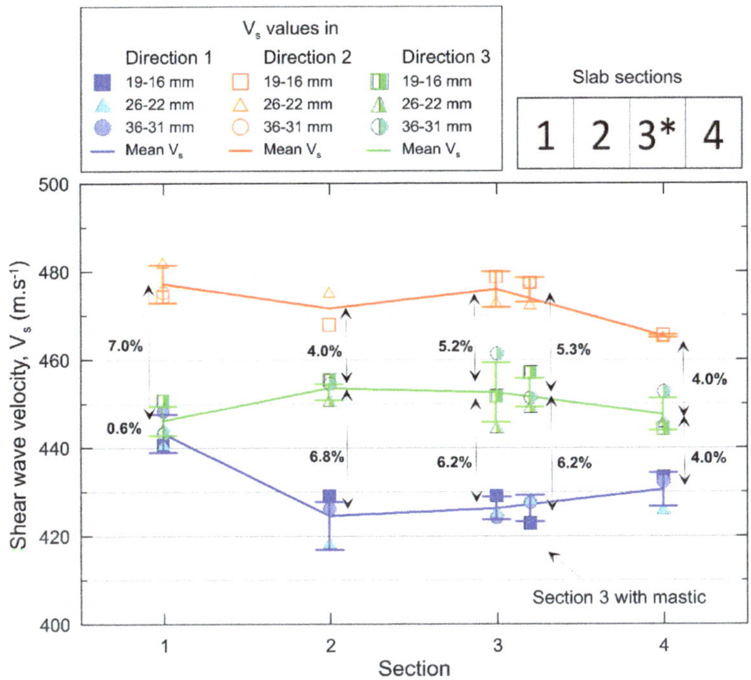

Figure 7. Shear wave velocity with respect to direction for tested sections of slab 2. Sections with * have also been tested with a bitumen mastic interface. The relative differences between mean V_s are displayed.

To simplify the text, V_s values for directions 1, 2, and 3 are referred to as V_{s1}, V_{s2}, and V_{s3} respectively. Tables 2 and 3 presents the relative standard deviation between the pairs of transducers for a single direction in the same section for slab 1 and slab 2, respectively. The maximum standard deviation in regard to the mean V_s values are 3.1 and 1.8% in slab 1 and slab 2, respectively. In a previous study, the mean standard deviation among different pair of transducers was approximately 2% [34]. It shows good reproducibility of the measure. More transducer pairs were utilized to evaluate V_s in slab 1's sections, which explains the higher standard deviation in slab 1 compared to slab 2. As mentioned in Lecuru et al. [33], the different sizes of actuators do not influence significantly the values, as shown by the low differences presented in Tables 2 and 3. Naji et al. [37] showed that the size of aggregates can generate wave diffraction, depending on the size of the P-RAT transducer used. Wave diffraction influences the propagation of shear waves in the tested specimen, therefore, it influences the shear wave velocity. The results obtained herein using transducers of various dimensions do not demonstrate systematically higher or lower values of V_s, depending on the dimension of the transducer, compared with the average values.

A trend emerges from the curves of Figures 6 and 7. Especially away from the end of the slabs, V_{s2} values are systematically greater than V_{s3} and V_{s1}, and V_{s3} is greater than V_{s1}.

In Figure 6, there is an axial symmetry relative to the middle tendency in the V_s values for sections Slab1_6–Slab1_7 and Slab1_9–Slab1_10. The same symmetry is observed in Figure 7, in V_s values for sections Slab2_2 and Slab2_3, and somewhat for sections Slab2_1 and Slab2_4. It is expected, as the tested specimens are pieces of a slab, that section Slab2_2 and Slab2_3 should be compacted in the same state, as should section Slab2_1 and Slab2_4.

Table 2. Minimal V_s values, maximal V_s values, and relative standard deviation (σ%) for the pairs of transducers in slab 1.

Section	V_{s1} (m·s^{-1})			V_{s3} (m·s^{-1})		
	Min.	**Max.**	**σ%**	**Min**	**Max**	**σ%**
Slab1_2	407.5	417.9	1.8%	408.0	408.0	N.A.
Slab1_2 mastic	388.4	401.9	1.4%	401.4	420.9	1.7%
Slab1_5	401.6	418.8	2.1%	395.6	407.4	1.2%
Slab1_5 mastic	387.2	411.7	3.1%	393.8	398.5	0.6%
Slab1_6	388.5	406.0	1.9%	388.4	398.1	1.4%
Slab1_7	380.8	406.9	2.9%	418.6	435.0	1.6%
Slab1_7 mastic	378.9	403.9	2.5%	415.8	432.7	1.7%
Slab1_8	N.A.	N.A.	N.A.	415.0	423.9	1.1%
Slab1_9	389.0	410.3	2.5%	413.1	420.3	0.9%
Slab1_10	387.3	408.2	2.4%	383.0	396.6	1.5%

Table 3. Mean V_s values and relative standard deviation (σ%) among the three pairs of transducers in slab 2.

Section	V_{s1} (m·s^{-1})			V_{s2} (m·s^{-1})			V_{s3} (m·s^{-1})		
	19–16	**26–22**	**36–31**	**19–16**	**26–22**	**36–31**	**19–16**	**26–22**	**36–31**
Slab2_1	440.7	440.9	448.3	474.3	482.2	475.2	450.8	444.5	443.2
	σ%		1.0%	σ%		0.9%	σ%		0.9%
Slab2_2	429.1	418.5	426.2	467.9	475.4	N.A.	455.4	450.9	454.5
	σ%		1.3%	σ%		1.1%	σ%		0.5%
Slab2_3	429.2	425.6	424.2	478.8	473.1	N.A.	451.8	444.8	461.3
	σ%		0.6%	σ%		0.8%	σ%		1.8%
Slab2_3 mastic	417.0	422.2	421.4	465.6	461.2	465.6	450.3	443.2	445.0
	σ%		0.7%	σ%		0.5%	σ%		0.8%
Slab2_4	433.2	426.2	432.3	465.6	465.1	N.A.	444.3	446.0	452.8
	σ%		0.9%	σ%		0.1%	σ%		1.0%

V_{s1} and V_{s3} are higher in Slab1_5 than in Slab1_10. This may be because Slab1_5 is located at the corner of the sample. V_{s1} is higher in Slab1_2 and Slab1_5 than in Slab1_6 to 10. Slab1_1 to 5 are along the side of the sample. It can be assumed that the edges of the mold create a confinement in the materials during compaction. The confinement might induce a better aggregate contact, resulting in a higher V_s value. This confinement is greater in direction 3 than in direction 1, due to the dimension of the mold, which is smaller in direction 3 than in direction 1. This confinement becomes more homogeneous in sections Slab2_1, Slab2_4, and Slab1_5, which explains the closeness of V_{s1} and V_{s3} values in these sections.

The difference between V_{s1} and V_{s3} values could also be explained by the nature of the S-waves and the way the slabs are compacted. The compacting wheel moves in direction 1 and, therefore, the aggregates tend to move along. This movement can result in friction between the aggregate in a perpendicular direction, i.e., direction 3. In S-waves, the local displacement of matter is perpendicular to the direction of wave propagation.

The relative differences between mean V_s values in the different direction are presented in Figures 6 and 7. Concerning slab 2, the maximum relative differences are 7.6, 11.1, 11.6, and 8.1% for sections Slab2_1, Slab2_2, Slab2_3, and Slab2_4, respectively. In order to compare these values with the literature, Equation (2) is used to estimate the equivalent modulus [22]:

$$V_s = \frac{1}{\cos\left(\frac{\varphi}{2}\right)}\sqrt{\frac{|E^*|}{2\rho(1+\nu)}} \tag{2}$$

With the phase angle ϕ, the bulk density ρ (kg·m^{-3}), the Poisson ratio ν, V_s the shear wave velocity (m·s^{-1}), and E^* the complex modulus (Pa).

The maximum relative differences in modulus for the aforementioned specimens are 15.9, 23.4, 24.6, and 16.8%, respectively. The relative differences values observed in the literature for HMA range from 10 to 50% [10,11,42]. As explained earlier, the anisotropy is influenced by many parameters, which explains the wide array of values in the literature.

The maximum value always appears between V_{s2} and V_{s1}, and the symmetric trend is respected. Although no analysis of variance was performed, it can be assumed that the V_s values in the three directions are different in each section, except for V_{s1} and V_{s3} in sections Slab2_1, Slab1_5, Slab1_6, and Slab1_10. The symmetry of the results demonstrates the consistency of the results, and the systematic differences observed implies anisotropy in the tested CIR.

5.2. UPV Test Results

The results of UPV tests are presented in Figure 8 for the P-wave. Although the choice of arrival time configuration is not obvious, the first peak (bold line in Figure 8) tends to depict a similar behavior to what is observed in P-RAT test results (Figure 7). A trend emerges among the sections monitored with P-wave. The axial symmetry relative to the middle observed for P-RAT analysis occurs here as well. For sections Slab2_2 and Slab2_3 (i.e., in the middle of the slab), V_p in dir. 2 is higher than in the other two directions. The opposite happens in sections Slab2_1 and Slab2_4. The relative difference between V_p values in a same section are displayed in Figure 8. The difference between V_p values can reach up to 12.8% in the case of section Slab2_3. The relative differences are lower in sections Slab2_1 and Slab2_4, for every arrival time setup. This reduction in relative difference between the edges of the slab and the inner parts was also observed in V_s values obtained with P-RAT. The boundaries conditions applied by the metallic mold in the outer sections of the slab may impact the compaction of the mix and may cause anisotropy.

The analysis of the UPV S-wave is more difficult. As mentioned earlier, to obtain the S-wave flight time, one must compare the received P-wave signal with the received S-wave. In the tested sections, the distinction between the P-wave and the S-wave was not readily apparent, and this can can lead to erroneous assumptions regarding the location of the first and second peaks. It was almost impossible to determine where the first arrival of the S-wave was. Figure 9 presents examples of P- and S-wave signals to illustrate the difficulties mentioned before. In the case of section Slab2_4 in direction 3, two different S-wave signals give different results. Due to these erratic results, and their inconstancy even for a same specimen in the same direction, S-wave UPV results are reported as Supplementary Materials.

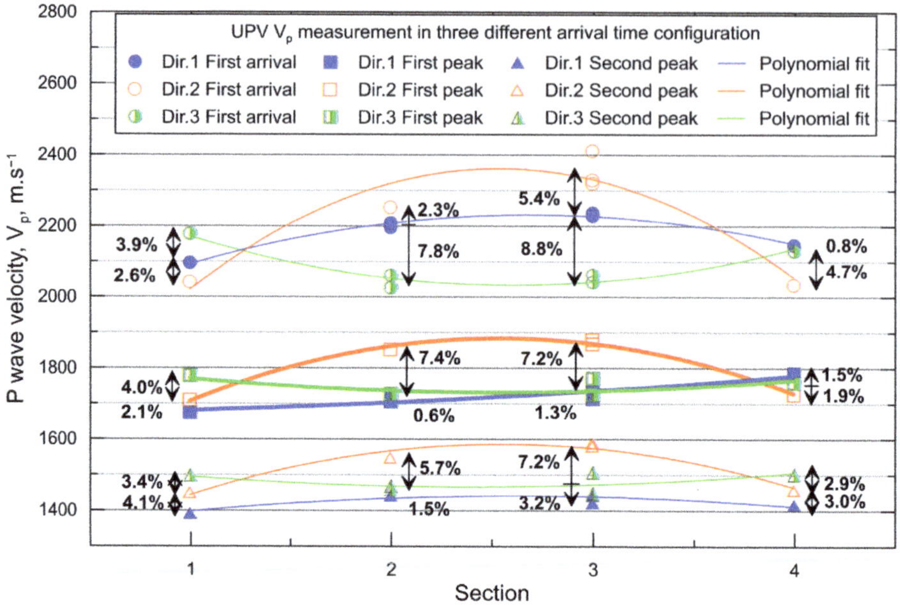

Figure 8. PUNDIT P-wave results in three directions for each section of slab 2 in three arrival setups. The relative differences between mean V_S values are displayed.

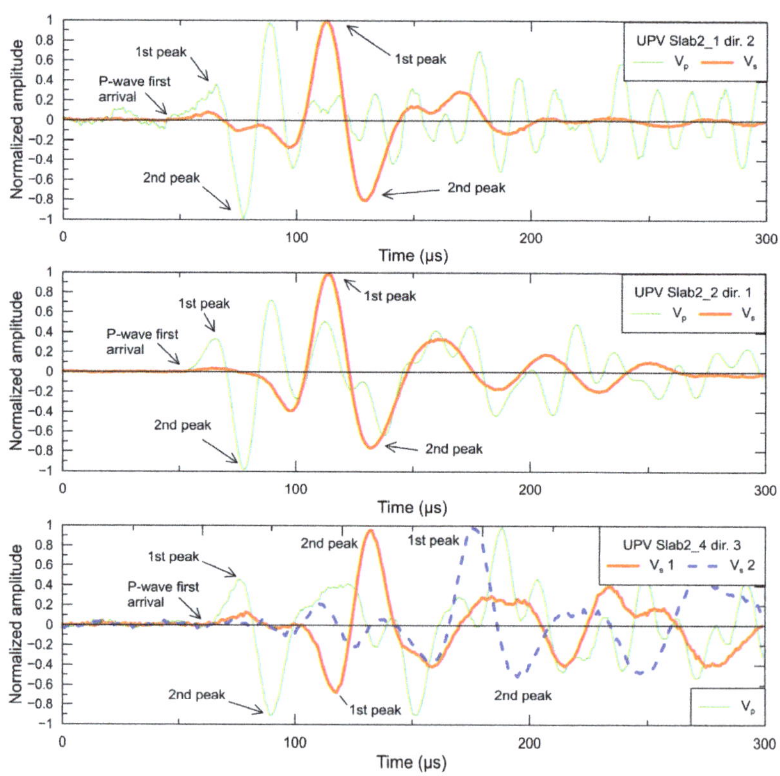

Figure 9. UPV P- and S-wave signals comparison in different directions and sections of slab 2.

5.3. Complex Modulus

This section presents the results of the complex modulus measurements. It should be noted that specimen Slab2_3C has been tested twice. The master curves of specimens Slab2_3C 1 and 2 and Slab2_4C are plotted in Figure 10. Table 4 compiles the 2S2P1D parameter used for the analysis of the experimental results.

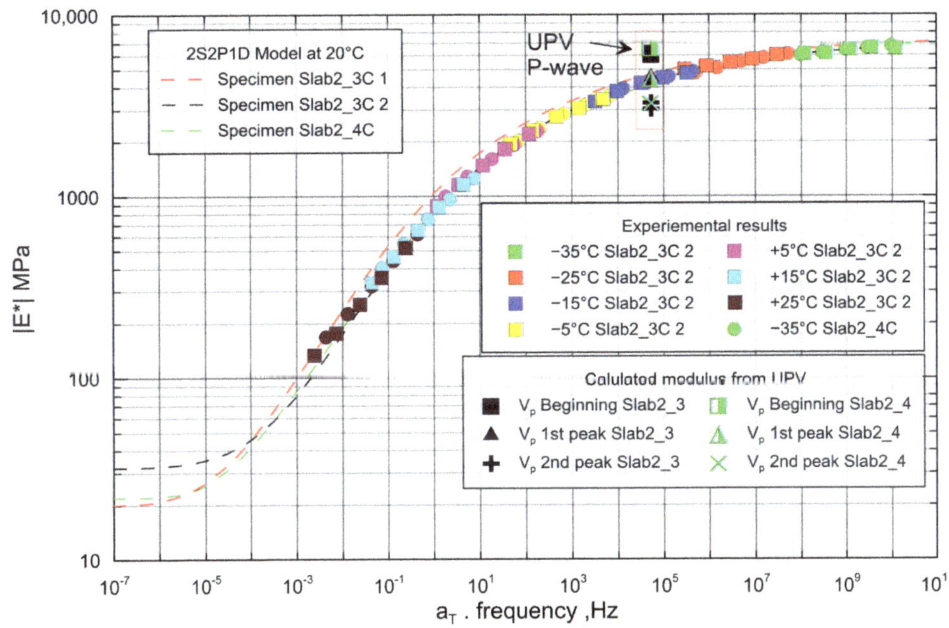

Figure 10. Master curve of cylinder cored in direction 3 for section Slab2_3 and Slab2_4 at a reference temperature of 20 °C. The calculated complex modulus from UPV tests are also plotted.

Table 4. The 2S2P1D parameters.

	E_{00}	E_0	ν_{00}	ν_0	k	H	δ	τ	β	C1	C2
	MPa	MPa						(T_{ref})			
Slab2_3C 1	20	7500	/	/	0.170	0.475	2.56	0.2 (10.93)	5000	20.00	159.12
Slab2_3C 2	32	7200	/	/	0.170	0.490	2.69	1.1 (4.96)	5000	24.02	158.55
Slab2_4C	22	7675	/	/	0.161	0.450	2.85	1.1 (4.96)	5000	23.47	159.08

The specimens were equipped with a chain to measure radial deformations, but the results obtained were erratic and did not follow the usual trends that appear in the literature. It is assumed that the applied strain was too small to obtain enough amplitude in the transverse direction. The strain measured by the chain was mainly noise, rather than actual displacement. The results are not presented for this reason.

It appears that there is a reduction in the modulus for specimen Slab2_3C when performing the second tension–compression test. Although the strain amplitude was small, some damage can be induced to the specimen.

The results from UPV testing were converted into modulus, with the following equation [22]:

$$V_p = \frac{1}{\cos\left(\frac{\varphi}{2}\right)}\sqrt{\frac{|E^*|(1-\nu)}{\rho(1+\nu)(1-2\nu)}} \quad (3)$$

and considering the following parameters: the phase angle ϕ = 20° (from E* tests), the bulk density ρ = 2000 kg·m^{-3} (from measurement), and the Poisson ratio ν = 0.3 (assumed). Their $|E^*|$ moduli are plotted along the master curve, at a reference temperature of 20 °C, considering the frequency of the UPV sensor used, 50 kHz for P-wave. The moduli for the different arrival time setups, i.e., first arrival of the signal, first peak, and second peak (see Figure 1) are plotted. The relative differences between the experimental $|E^*|$ modulus and the UPV modulus calculated from the aforementioned arrival time configurations are 35, 2, and 30%. It appears that the first peak arrival time configuration of P-wave fits the master curve.

Mounier et al. [22] and Larcher et al. [26] observed differences between the UPV modulus and 2S2P1D model, considering the arrival of the wave at the first arrival of the signal, from 5 up to 30% and 12%, respectively. Both studies point out the importance of accuracy in measuring the arrival time of the wave.

Mean P-RAT V_s values in dir. 3 for Section 3 and 4 are 452.6 and 447.7 m·s^{-1}, respectively. Their equivalent $|E^*|$ modulus considering ϕ = 20°, ν = 0.3, and ρ presented in Table 1 are 1057 and 990 MPa for sections Slab2_3 and Slab2_4, respectively. These moduli come from a group shear wave velocity and, therefore, are not bound to frequency. Therefore, these moduli cannot be plotted in the master curve. In a previous study by Lecuru et al. [34], moduli calculated from V_s results obtained with P-RAT tests were compared with indirect tensile strength modulus (ITSM) test results. They demonstrated consistency, for a similar CRM mix.

For V_p, the first peak arrival time configuration appears to be the one to consider. It was mentioned earlier that V_p results obtained with this particular configuration show a similar trend to the one observed for P-RAT V_s results. This gives greater confidence regarding anisotropic results in the mix.

5.4. Three-Dimensional Image Analysis Results

For the analysis of the 3D images, the particles were segmented and then individually labelled, as shown in Figure 11. Among all labelled particles, any larger than 1 mm^3 were selected for analysis according to the Dragonfly analysis, corresponding to 6778, 4698, and 4364 particles in specimen Slab2_2C, Slab2_3C, and Slab2_4C, respectively. It appears that smaller labelled particles tend to be spherical, and the analysis attributes arbitrary values of phi and theta angle of 0° or 90° to them. Specimen Slab2_1C was destroyed during a tension compression test and was not scanned.

Figure 12A–C presents the results of the angular orientation analysis for specimens Slab2_2C, Slab_3C, and Slab2_4C, respectively. The axis systems in Figure 12 are the same as that presented in Figure 3.

In each specimen, the particles tend to be oriented in the dir. 1 dir. 2 plane rather than the dir. 3 axis (phi values) as 48.4, 52.3, and 49.8% of the aggregates are enclosed in the 60 to 90° range for specimens Slab2_2C, Slab2_3C, and Slab2_4C, respectively.

For the theta analysis, i.e., the orientation in the dir. 1 dir. 2 plan, three sets of data are considered and are represented in Figure 12A–C. The aggregates aligned with dir. 1 are enclosed in the bins from 60 to 120° plus 240 to 300° (orange dashes); the aggregates aligned with dir. 2 are enclosed in the bins from 330 to 30° plus 150 to 210° (orange triangles); the rest are considered as the cross direction (orange dots).

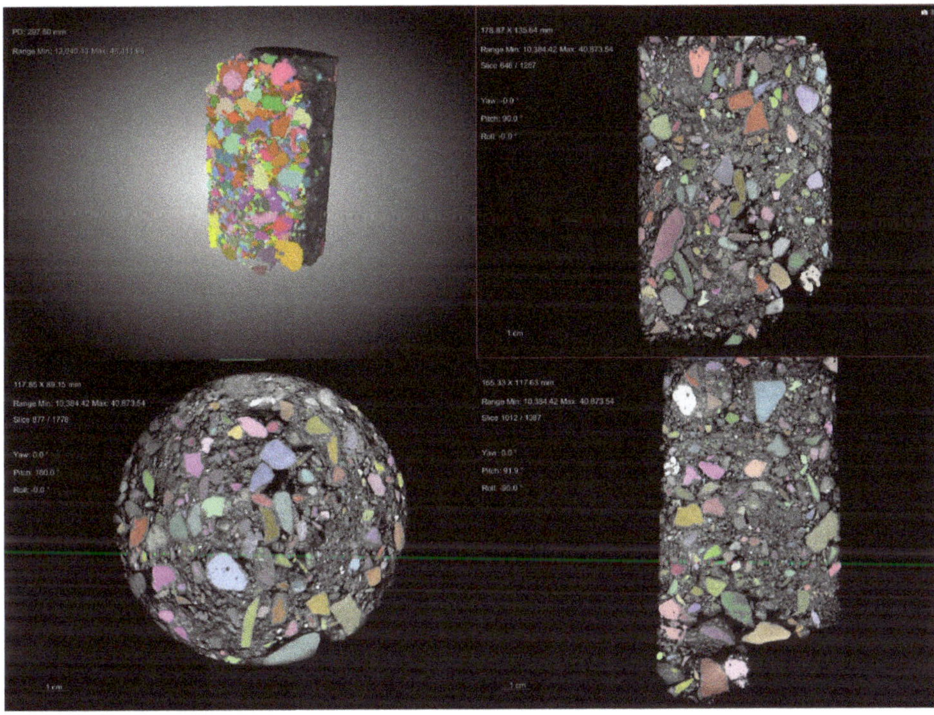

Figure 11. Individually labelled particles bigger than 1 mm^3 (each color represents a particle), specimen Slab2_3C, ORS Dragonfly.

For each specimen, the aggregates tend to be more oriented in dir. 1 than in dir. 2. As the compactor wheel is moving along the dir. 1 axis, it is expected that aggregates would be more aligned in this direction. However, a non-negligible part of the aggregates is aligned in the dir. 2 axis and in a cross direction of dir. 1 and dir. 2 axes.

The phi and theta angles are bound to the axis system of the Dragonfly software. An axis swap enables the user to evaluate the orientation of aggregates from different perspectives. Figure 13 presents the results of phi and theta angle analysis in, according to Dragonfly, the XZY (Y and Z axis swapped) and YZX configuration. The default configuration (XYZ) is presented in Figure 12B.

In Figure 12B, the particles tend to be oriented in the dir. 1 dir. 2 plane rather than dir. 3 (phi values) and more aligned in dir. 1 (theta values). In Figure 13, view (A), the particular orientation is not as straightforward. Considering the phi values, more aggregates are enclosed in the 30° to 60° area and less in the 60° to 90° area, compared to views (A) and (C). There is a slight increase in the 0° to 30° area, but not as much as expected if we consider a preferential orientation of the aggregates in dir. 1. Regarding the theta values, there is no particular orientation in dir. 2, dir. 3, nor in the cross direction. In Figure 13, view (B), particles tend to be oriented in the dir. 1 dir. 3 plane rather than dir. 2 and more aligned in dir. 1.

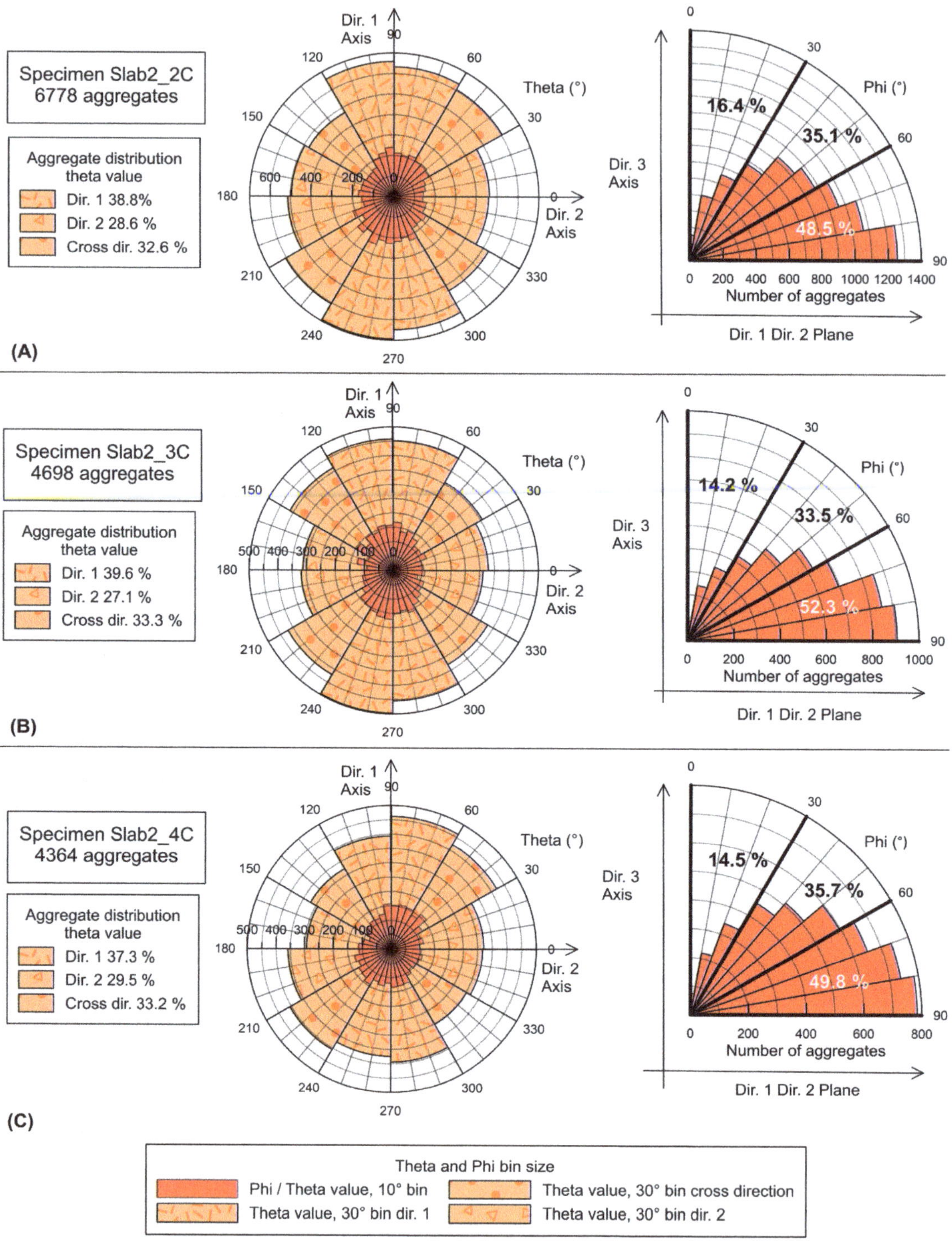

Figure 12. Particles orientation in specimens Slab2_2C (**A**), Slab2_3C (**B**), and Slab2_4C (**C**). Phi and theta values.

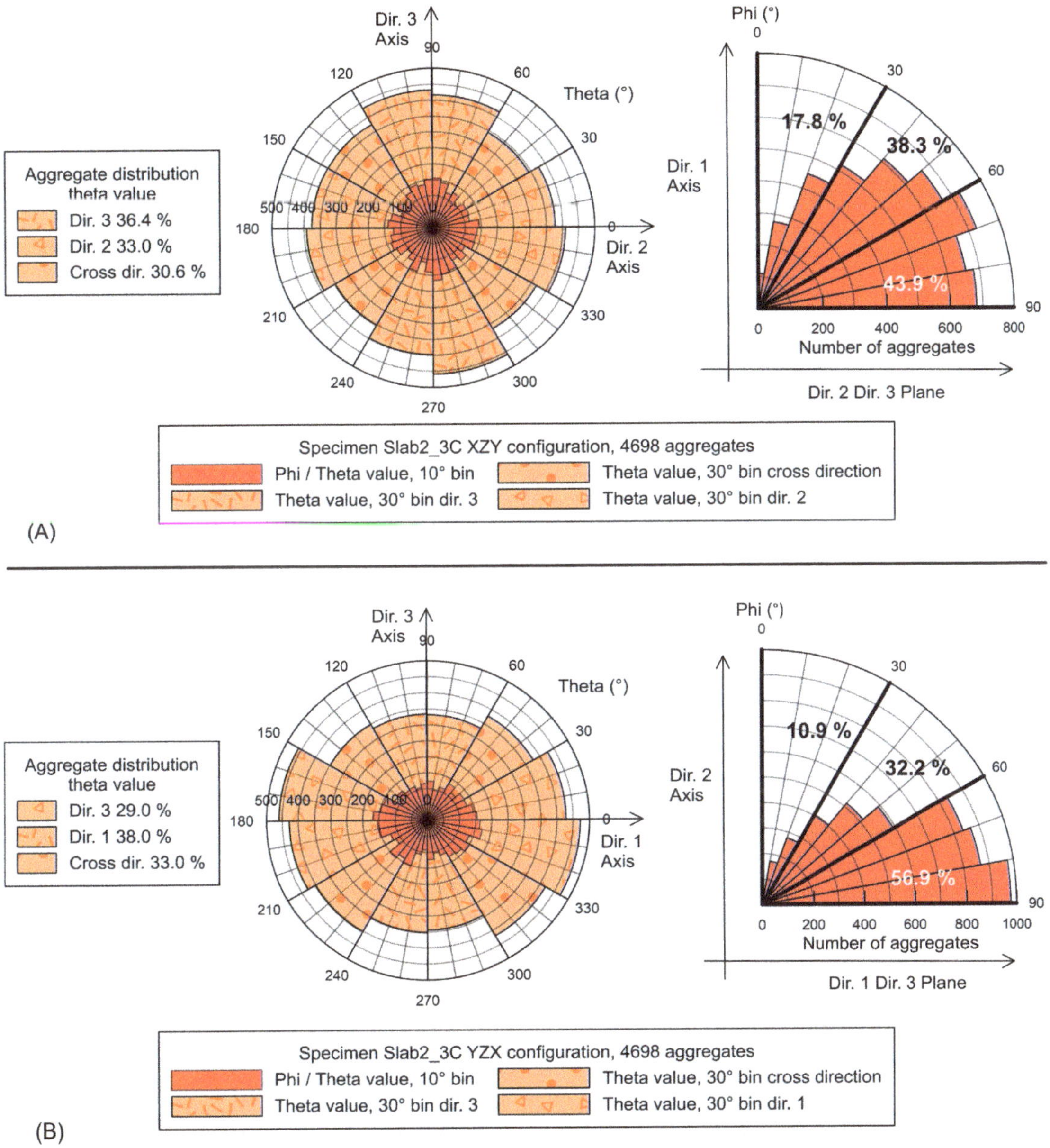

Figure 13. Particles orientation in specimen Slab2_2C for 2 configurations; (**A**) XZY, phi calculated regarding dir. 1 axis; (**B**) YZX, phi calculated regarding dir. 2 axis.

Displaying 3D orientation in a 2D representation has always been challenging. However, the results indicate that the aggregates in the three tested specimens tend to be aligned in dir. 1, which corresponds to the direction of the compacting wheel. However, the different points of view for specimen Slab2_3C confirm that a non-negligeable portion of the aggregates are aligned in dir. 2 and dir. 3. This preferential orientation shows that compaction has induced anisotropy in the tested CIR.

In the case of clays, V_s is higher when propagating in the direction of the main axis orientation of particles [13,43]. The same phenomenon can be observed in granular materials such as micas [12]. This is the result of inherent anisotropy. However, clays and micas are characterized by their flat and elongated geometry, which is not the case of the aggregates composing a CIR mix.

In the case of the tested CIR, V_s is higher in dir. 2, followed by dir. 3, then dir. 1. Although the preferential orientation tends to be in dir. 1, V_{s1} is the smallest. It can be assumed that compaction induces inherent anisotropy and that induced anisotropy, as defined in the background section, is not relevant in the case of wave propagation where strains are small [14,15,17], although the compaction generates some confinement due to the edges of the mold, which then become the inherent (natural) state of the material after unmolding. Wave propagation is, as mentioned earlier, impacted by the orientation of particles, but also by the contacts between them. However, in HMA and CIR mixes, the bitumen mastic plays a role in the stiffness of the material and, therefore, the alignment of aggregate alone is not sufficient to fully understand the differences in the measured V_s in the three directions. Finally, the air voids orientation and distribution could also provide useful information, although it has not been evaluated in this study.

6. Conclusions

A non-destructive method using shear wave velocity was used in the study. The P-RAT was used to measure V_s in the three directions of a CIR slab cut in sections. For comparison purposes, UPV measurements were also performed. Cylinders cored from the section in direction 3 were tested by dynamic tension–compression test to evaluate the complex modulus. Finally, CT scans were performed on the cylinders to evaluate the orientation of the aggregates. Anisotropy was witnessed for both approaches.

- V_s measurements with P-RAT show a systematic anisotropy in the two tested slabs. Different sizes of P-RAT transducers were used and the variability was low. V_s in the direction of compaction is higher than V_s in the orthogonal direction of displacement of the compacting wheel, which is higher than V_s in the direction of the compacting wheel; $V_{s2} > V_{s3} > V_{s1}$. Differences between V_s values range from 0.6 to 11.6%. V_s values on the outer sections of the slabs tend to be closer to one another compared to the middle section. The difference in wave velocities is explained by the compaction method and the dimensions of the mold, which generate uneven confinement in the mix.
- UPV results with PUNDIT also show anisotropy. UPV V_p values are different in every direction, and the same axial symmetry relative to the middle of the slab was observed as in P-RAT results. V_s values from UPV were trickier to obtain, as the arrival time of the wave is hard to determine. Various setups of arrival time were then tested. V_s from P-RAT and V_p from UPV results are consistent one with each other.
- E* results are consistent in the tested specimens, and they were modelized with the 2S2P1D model. Moduli from UPV were plotted in the master curve and the proper arrival time setup were determined to be the first peak for P-wave.
- CT scans highlighted a preferential orientation of the aggregate in the direction of the movement of the compacting wheel, dir. 1. However, a non-negligible part of the aggregates is oriented in the two other directions.

This study was focused on a two CIR mixes with a single method of compaction due to the high quantity of signal collected. Future work should focus on determining the influence of the degree of compaction as well as the method of compaction, such as SGC and slab compactor, on the anisotropy of CIR materials. V_s measurements would benefit from the analysis of contacts between aggregates. Finally, the evaluation of the anisotropy in CIR mixes while curing and in the long term could be achieved with the P-RAT and V_s.

Supplementary Materials: The following supporting information can be downloaded at: https://www.mdpi.com/article/10.3390/infrastructures10050115/s1, Figure S1: PUNDIT S-Wave results in 3 directions for each section of slab 2; Table S1: Standard deviation of UPV V_s values in each direction for every section of Slab 2.

Author Contributions: Conceptualization, Q.L., Y.E. and A.C.; methodology, Q.L.; software, Y.E. and M.K.; validation Q.L., Y.E. and A.C.; formal analysis, Q.L.; investigation, Q.L.; writing—original draft preparation, Q.L.; writing—review and editing, Q.L., Y.E., A.C. and M.K.; visualization, Q.L.; supervision, Y.E. and A.C.; funding acquisition, Y.E. and A.C. All authors have read and agreed to the published version of the manuscript.

Funding: This research was funded by Mitacs, grant number MITACS: FR25581.

Data Availability Statement: The date supporting the findings of this study are available within the article.

Acknowledgments: Author would like to acknowledge Vladimir Brailovski for the provision of the XTH-225 and the FF35 CT-Scan devices and Salah Eddine Brika for the assistance in scanning the specimens. Authors would like to acknowledge the CRSNG and Geowave Inc.

Conflicts of Interest: The authors declare no conflicts of interest.

References

1. Raschia, S.; Perraton, D.; Graziani, A.; Carter, A. Influence of low production temperatures on compactability and mechanical properties of cold recycled mixtures. *Constr. Build. Mater.* **2020**, *232*, 117169. [CrossRef]
2. Sangiorgi, C.; Tataranni, P.; Simone, A.; Vignali, V.; Lantieri, C.; Dondi, G. A laboratory and filed evaluation of Cold Recycled Mixture for base layer entirely made with Reclaimed Asphalt Pavement. *Constr. Build. Mater.* **2017**, *138*, 232–239. [CrossRef]
3. Orosa, P.; Orozco, G.; Carret, J.C.; Carter, A.; Pasadín, A.R. Compactablility and mechanical properties of cold recycled mixes prepared with different nominal maximum sizes of RAP. *Constr. Build. Mater.* **2022**, *339*, 127689. [CrossRef]
4. Li, Y.; Lyv, Y.; Fang, L.; Zhang, Y. Effects of Cement and Emulsified Asphalt on Properties of Mastics and 100% Cold Recycled Asphalt Mixtures. *Materials* **2019**, *12*, 754. [CrossRef]
5. Kuchiishi, A.K.; Vasconcelos, K.; Bernucci, L.L.B. Effect of mixture composition on the mechanical behavior of cold recycled asphalt mixtures. *Int. J. Pavement Eng.* **2021**, *22*, 984–994. [CrossRef]
6. Raschia, S.; Graziani, A.; Carter, A.; Perraton, D. Laboratory mechanical characterization of cold recycled mixtures produced with different RAP sources. *Road Mater. Pavement Des.* **2019**, *20*, 233–246. [CrossRef]
7. Diefenderfer, B.K.; Bowers, B.F.; Shwartz, C.W.; Farzaneh, A.; Zhang, Z. Dynamic Modulus of Recycled Pavement Mixtures. *J. Transp. Res. Board* **2016**, *2575*, 19–26. [CrossRef]
8. Nguyen, L.N.; Nguyen, M.H.; Dao, D.V.; Nguyen, P.Q.; Nguyen, T.M.T.; Tran, T.D. Effect of curing regimes on the dynamic modulus of entirely RAP cold recycled asphalt mixture. *J. Mater. Des. Appl.* **2023**, *237*, 1975–1989. [CrossRef]
9. Pham, N.H.; Sauzéat, C.; Di Benedetto, H.; González-Leön, J.A.; Barreto, G.; Nicolaï, A.; Jakibowski, M. Analysis and modeling of 3D complex modulus tests on hot and warm bituminous mixtures. *Mech. Time-Depend Mater.* **2015**, *19*, 167–186. [CrossRef]
10. Di Benedetto, H.; Sauzéat, C.; Clec'h, P. Anisotropy of bituminous in the linear viscoelastic domain. *Mech. Time-Depend Mater.* **2016**, *20*, 281–297. [CrossRef]
11. Nguyen, Q.T.; Pham, N.H.; Di Benedetto, H.; Sauzéat, C. Anisotropic Behavior of Bituminous Mixtures in Road Pavement Structures. *J. Test. Eval.* **2020**, *48*, 178–188. [CrossRef]
12. Santamarina, J.C.; Klein, K.A.; Fam, M.A. *Soils and Waves: Particulate Materials Behavior, Characterization and Process Monitoring*, 1st ed.; John Wiley & Sons Ltd.: Chichester, UK, 2001; ISBN 0-471-49058-X.
13. Nguyen, T.L.; Szymkiewicz, F.; Reiffsteck, P.; Bourgeois, E.; Mestat, P. Caractérisation et effet de l'anisotropie sur le comportement de sols reconstitués. *Rev. Can. De Géotechnique* **2011**, *48*, 1520–1536. [CrossRef]
14. Massad, E.; Tashman, L.; Somedavan, N.; Little, D. Micromechanics-Based Analysis of Stiffness Anisotropy in Asphalt Mixtures. *J. Mater. Civ. Eng.* **2002**, *14*, 374–383. [CrossRef]
15. Underwood, S.; Heidari, A.H.; Guddati, M.; Kim, R. Experimental Investigation of Anisotropy in Asphalt Concrete. *J. Transp.* **2005**, *1929*, 238–247. [CrossRef]

16. Bhasin, A.; Izadi, A.; Bedgaker, S. Three dimensional distribution of the mastic in asphalt composites. *Constr. Build. Mater.* **2011**, *25*, 4079–4087. [CrossRef]
17. Alanazi, N.; Kassem, E.; Jung, S.J. Evaluation of the Anisotropy of Asphalt Mixtures. *J. Transp. Eng. Part B Pavements* **2018**, *144*, 04018022. [CrossRef]
18. Hassan, N.A.; Airey, G.D.; Khan, R.; Collop, A.C. Nondestructive characterization of the effect of asphalt mixture compaction on aggregate orientation and segregation using X-ray computed tomography. *Int. J. Pavement Res. Technol.* **2012**, *5*, 84–92.
19. Huang, J.; Pei, J.; Li, Y.; Yang, H.; Li, R.; Zhang, J.; Wen, Y. Investigation on aggregate particles migration characteristics of porous asphalt concrete (PAC) during vibration compaction process. *Constr. Build. Mater.* **2020**, *243*, 118153. [CrossRef]
20. Quinteros, V.S.; Carraro, J.A.H. The initial fabric of undisturbed and reconstituted fluvial sand. *Geotechnique* **2021**, *73*, 1–15. [CrossRef]
21. Di Benedetto, H.; Sauzéat, C.; Sohm, J. Stiffness of Bituminous Mixtures Using Ultrasonic Wave Propagation. *Road Mater. Pavement Des.* **2009**, *10*, 789–814. [CrossRef]
22. Mounier, D.; Di Benedetto, H.; Sauzéat, C. Determination of bituminous mixtures linear properties using ultrasonic wave propagation. *Constr. Build. Mater.* **2012**, *36*, 638–647. [CrossRef]
23. Arulnathan, R.; Boulanger, R.W.; Riemer, M.F. Analysis of benderelement tests. *ASTM Geotech. Test. J.* **1998**, *21*, 120–131. [CrossRef]
24. Brandenberg, S.J.; Kutter, B.L.; Wilson, D.W. Fast Stacking and Phase Corrections of Shear Wave Signals in a Noisy Environment. *ASCE J. Geotech. Geoenviron. Eng.* **2008**, *134*, 1154–1165. [CrossRef]
25. Abd Elhafeez, T.; Amer, R.; Saad, A.; El Kady, H.; Madi, M. Evaluation of Flexible Pavement Mixtures Using Conventional Test and Ultrasonic Wave Propagation. *Adv. Civ. Eng. Mater.* **2014**, *3*, 1–20. [CrossRef]
26. Larcher, N.; Takarli, M.; Angellier, N.; Petit, C.; Sebbah, H. High frequency shear modulus of bitumen by ultrasonic measurements. In Proceedings of the 5th European Asphalt Technology Association Conference, Braunschweig, Germany, 3–5 June 2013; p. 13.
27. Mounier, D.; Di Benedetto, H.; Sauzéat, C.; Bilodeau, K. Observation of Fatigue of Bituminous Mixtures Using Wave Propagation. *J. Mater. Civ. Eng.* **2016**, *28*, 04015083. [CrossRef]
28. Zargar, M.; Banerjee, S.; Bullen, F.; Ayers, R. An investigation into the Use of Ultrasonic Wave Transmission Techniques to Evaluate Air Voids in Asphalt, GCEC 2017. In Proceedings of the 1st Global Civil Engineering Conference, Kuala Lumpur, Malaysia, 25–28 July 2017; pp. 1427–1439. [CrossRef]
29. Hou, S.; Deng, Y.; Jin, R.; Shi, X.; Luo, X. Relationships between Physical, Mechanical and Acoustic Properties of Asphalt Mixtures Using Ultrasonic Testing. *Buildings* **2022**, *12*, 306. [CrossRef]
30. Karray, M.; Ben Romdhan, M.; Hussien, M.H.; Ethier, Y. Measuring shear wave velocity of granular material using the piezoelectric ring-actuator technique(P-RAT). *Can. Geotech. J.* **2015**, *52*, 1302–1317. [CrossRef]
31. Tavassoti-Kheiry, P.; Solaimanian, M.; Qiu, T. Characterization of High RAP/RAS Asphalt Mixtures Using Resonant Column Tests. *J. Mater. Civ. Eng.* **2016**, *28*, 04016143. [CrossRef]
32. Elbeggo, D.; Hussien, M.N.; Ethier, Y.; Karray, M. Robustness of the P-RAT in the Shear-wave Velocity Measurement of Soft Clays. *J. Geotech. Geoenviron. Eng.* **2019**, *145*, 04019014. [CrossRef]
33. Lecuru, Q.; Ethier, Y.; Carter, A.; Karray, M. Characterization of Cold In-Place Recycled Materials at Young Age Using Shear Wave Velocity. *Adv. Civ. Eng. Mater.* **2019**, *8*, 336–354. [CrossRef]
34. Lecuru, Q.; Ethier, Y.; Carter, A.; Karray, M. Early-age stiffening of Cold Recycled Bituminous Materials using shear wave velocity. *J. Test. Eval.* 2025, *accepted*.
35. Soliman, N.A.; Khayat, K.H.; Karray, M.; Omran, A.F. Piezoelectric ring actuator technique to monitor early-age properties of cement-based materials. *Cem. Concr. Compos.* **2015**, *63*, 84–95. [CrossRef]
36. Naji, S.; Khayat, K.H.; Karray, M. Assessment of Static Stability of Concrete Using Shear Wave Velocity Approach. *ACI Mater. J.* **2017**, *114*, 105–115. [CrossRef]
37. Naji, S.; Khayat, K.H.; Karray, M. Effect of piezoeletric ring sensor size on early-age property monitoring of self-consolidating concrete. *ACI Mater. J.* **2018**, *115*, 813–824. [CrossRef]
38. Mhenni, A.; Hussien, M.N.; Karray, M. Improvement of the Piezo-Electric Ring Actuator Technique (P-RAT) Using 3D Numerical Simulations. In Proceedings of the 68th Canadian Geotechnical Conference, Québec, QC, Canada, 20–23 September 2015; p. 7.
39. Ministère des Transports du Québec (MTQ). *Détermination de la Densité Brute et de la Masse Volumique des Enrobés à Chaud Compactés, LC 26-040*; Ministère des Transports du Québec: Québec, QC, Canada, 2023.
40. Olard, F.; Di Benedetto, H. General "2S2P1D" Model and Relation Between the Linear Viscoelastic Behaviours of Bituminous Binders and Mixes. *Road Mater. Pavement Des.* **2003**, *4*, 184–224. [CrossRef]
41. Perraton, D.; Di Benedetto, H.; Sauzéat, C.; Nguyen, Q.T.; Pouget, S. Three-Dimensional Linear Viscoelastic Properties of Two Bituminous Mixtures Made with the Same Binder. *J. Mater. Civ. Eng.* **2018**, *30*, 04018305. [CrossRef]

42. Lin, M.; Hu, C.; Guan, H.; Easa, S.M.; Jiang, Z. Impacts of Material Orthotropy on Mechanical Behaviors of Asphalt Pavements. *Appl. Sci.* **2021**, *11*, 5481. [CrossRef]
43. Pennington, D.S.; Nash, D.F.T.; Lings, M.L. Anisotropy of G_0 shear stiffness in Gault Clay. *Géotechnique* **1997**, *47*, 391–398. [CrossRef]

infrastructures

Article

Characterization and Evaluation of Agar as a Bio-Based Asphalt Binder Alternative

Melissa R. Frey [1], Sarah L. Williams [1], Wil V. Srubar III [1,2] and Cristina Torres-Machi [1,*]

[1] Department of Civil, Environmental, and Architectural Engineering, University of Colorado Boulder, Boulder, CO 80309, USA; melissa.frey@colorado.edu (M.R.F.); sarah.l.williams@colorado.edu (S.L.W.); wsrubar@colorado.edu (W.V.S.III)

[2] Materials Science and Engineering Program, University of Colorado Boulder, Boulder, CO 80309, USA

* Correspondence: cristina.torresmachi@colorado.edu

Abstract: Over 90% of roads in the United States are surfaced with asphaltic materials that use petroleum-based asphalt binders, a material with high negative environmental impacts and costs. Biopolymers are a sustainable alternative, as they are sourced from renewable materials and offer the potential to reduce carbon footprint. However, their performance and durability in construction applications remain insufficiently understood. This study analyzes the potential of agar, a biopolymer extracted from red seaweed, to serve as a direct and sustainable replacement for asphalt binders. The study characterizes the rheological properties and durability of agar-based binders and the mechanical and microstructural properties of composites. The study found that agar-based binders exhibited resistance to fungal deterioration, adequate stiffness to resist rutting at temperatures up to 80 °C, and potential for energy efficiencies associated with lower mixing and compacting temperatures. Results indicate that agar-based composites illustrate many properties in line with those of traditional engineering materials. Overall, these results suggest that agar-based materials exhibit promising fresh-state and biodeterioration resistance properties to serve as a sustainable alternative to traditional, petroleum-based asphalt binders.

Keywords: pavements; asphalt; agar; biopolymers

1. Introduction

Within the built environment, petroleum acts as the primary material source for a vast spectrum of materials, including plastics such as polyvinyl chloride and high-density polyethylene, adhesives (e.g., polyurethane, epoxies, poly(vinyl acetate), silicones), and binders used for asphalt for roadways, roofing, coatings, and waterproofing. Over 90% of the roads in the United States are surfaced with asphaltic materials comprising aggregate and a petroleum-based binder [1]. The sourcing and widespread use of non-renewable petroleum-based materials is problematic in multiple facets. The accelerated depletion of non-renewable crude oil has led to limited material availability, with the literature estimating that current petroleum reserves will only last for approximately 46 more years [2]. Petroleum-based materials are also associated with negative environmental impacts due to petroleum sourcing, refining, and manufacturing.

With continued shifts toward more sustainable practices, the construction industry has begun investigating bio-based asphalt binders created from renewable biomass sources. Bio-oils have been produced from various organic materials, including swine manure [3] and oils from waste cooking, vegetables and wood [4,5]. These studies, however, have mainly focused on modifying traditional asphalt binders (i.e., <10% replacement). To

 https://doi.org/10.3390/infrastructures10090223

meet the global environmental need for sustainable and resilient construction materials, the transportation industry must begin to shift to direct alternatives (i.e., 100% replacement) for petroleum-based materials.

One algae-derived biopolymer that may be well suited as an alternative material for asphalt binder is agar. Agar is a linear polysaccharide composed of agarose and agaropectin extracted from red seaweed belonging to the *Rhodophyceae* class. Agar gels form via hydrogen bonding between agarose molecules, resulting in a transparent, thermoreversible material that exhibits hysteresis, gelling at temperatures between 38 and 45 °C and melting between 85 and 90 °C [6].

Agar has been investigated as a stabilizing material to improve the engineering properties of geotechnical materials. Smitha et al. [7] investigated the behavior of silty sand treated with 0.5, 1, and 2% agar biopolymer at multiple curing time periods and noted a significant increase in cohesion with an increase in biopolymer content and curing time. Chang et al. [8] evaluated the use of agar and gellan gum biopolymer with both clayey and sandy soil and noted significant improvements in composite compressive strength as compared to unmodified soil samples. Authors note that this is due to the ability of the biopolymers to coat aggregate surfaces, fill pore spaces, and improve particle to particle contact. Khatami and O'Kelly [9] evaluated sand combined with 1, 2, and 4% agar by weight and noted an increase in compressive strength from 150 kPa to approximately 500 kPa. Work completed by Fatehi et al. [10] evaluated agar biopolymer combined with both sand and soil and found that biopolymer use can improve the compressive strength by 115%. Verma et al. [11] evaluated the use of xanthan gum and agar to stabilize municipal solid waste fines and noted that agar produced a denser profile than xanthan gum and produced better mechanical performance. Kantesaria et al. [12] investigated the use of 2% agar in expansive soil. In SEM testing, agar biopolymers coated soil particles and connected aggregate that was not initially in contact.

Although biopolymers like agar are promising material alternatives for petroleum-based materials, they are largely underutilized in civil engineering applications due to uncertainty regarding long term performance and durability [13]. Additionally, there is a lack of systematic methodology available for predicting performance, reporting material characteristics, and ultimately incorporating biopolymer materials into pavement design [14].

The objective of this work is to study the suitability of high concentration agar gels as a direct replacement material for traditional petroleum-based materials used in the construction industry. To achieve this goal, this research follows a three-step approach. First, this study used technical testing standards of asphalt to measure the rheological properties of agar-based binder samples. Second, the study analyzed the durability of agar as an alternative binder. Finally, we evaluated the microstructure and mechanical properties of agar-based composites for use in pavement structures.

2. Materials and Methods

Agar powder, reagent-grade limestone (calcium carbonate, $CaCO_3$) with a particle size < 250 μm, lab-grade glycerol, and Ottawa sand were supplied by Sigma Aldrich (St. Louis, MO, USA), Research Products International (Mt Prospect, IL, USA), Fisher Scientific (Waltham, MA, USA), and Gilson (Middleton, WI, USA), respectively. Fungal cultures were supplied by the American Type Culture Collection (ATCC, Manassas, VA, USA).

The aggregate used in agar-based composites mixtures was supplied by the Colorado Department of Transportation (CDOT). This aggregate was chosen for this analysis because it represents a commonly used and accepted aggregate for civil engineering applications. This aggregate was collected as a "belt cut" from a Coloradan asphalt manufacturing plant.

In this sampling method, the aggregate was collected randomly from the conveyer belt of an asphalt plant prior to being mixed with any asphalt binder [15].

Table 1 summarizes the methods used to characterize: (i) the rheological properties and (ii) durability of agar-based binder samples; and (iii) the microstructure and mechanical properties of agar-based composites.

Table 1. Methods used to characterize agar-based binders and composites.

Characteristics of Interest	Methods
Rheological properties of agar-based binder	Rotational Viscosity (ASTM D4402) Dynamic Shear Rheometry (ASTM D7175) Penetration (ASTM D5)
Durability of agar-based binder	Thermogravimetric Analysis (TGA) Biodeterioration Resistance (ASTM G21) Moisture Sorption Characteristics (ASTM D570)
Mechanical and microstructural characterization of agar-based composites	X-Ray Tomography Testing Compressive Strength Testing Resilient and Elastic Moduli Characterization

2.1. Sample Preparation

Two sets of samples were prepared for this study. One set of 12 samples, described below as "binder testing samples", were used to characterize the rheological properties and durability of agar-based binder. This testing informed the design of the experimental plan used to characterize the mechanical and microstructural characterization of agar-based composites. This plan, referred to as "composite testing samples", explored eight different sample types. The characteristics of each of these sets are described below. Several replicates were tested for each sample type, as described in each of the methods used and the results section.

2.1.1. Binder Testing Samples

A total of twelve samples were prepared to characterize the rheological properties and durability of agar-based binder (Table 2). The number of replicates used in each of the tests is specified in the test description and results section.

Table 2. Sample nomenclature and mixture formulations for binder testing.

Sample	Deionized Water (mL)	Agar (g)	Glycerol (mL)	Limestone (g)
A5-C	200	10	-	-
A5-G	200	10	3.2	-
A5-L	200	10	-	30
A5-GL	200	10	3.2	30
A7.5-C	200	15	-	-
A7.5-G	200	15	4.8	-
A7.5-L	200	15	-	30
A7.5-GL	200	15	4.8	30
A10-C	200	20	-	-
A10-G	200	20	6.4	-
A10-L	200	20	-	30
A10-GL	200	20	6.4	30

Samples were prepared at three agar concentrations (i.e., 5, 7.5, and 10% w/w relative to deionized water) without and with glycerol and without and with limestone additives.

As these are relatively high agar concentrations and agar materials have a high degree of crystallinity and rigidity [16], glycerol was investigated due to reported plasticizing effects on agar [17]. In order to increase the dimensional stability of agar-based materials, ground limestone was investigated as a filler material. Limestone was explored as a filler material because it is widely available and already utilized in the construction industry (i.e., crushed for use as a subbase material, ground for use as a raw material in cement production).

A subset of samples contained glycerol (40% *w*/*w* of agar), limestone (15% *w*/*w*), or both glycerol and limestone, resulting in a total of twelve samples. Mixture formulations are summarized in Table 2. The sample naming convention is the letter A (for agar) followed by the %*w*/*w* agar-to-water concentration. Sample additives are denoted by a C (for control formulations without glycerol or limestone), G (for glycerol), and L (for limestone). For example, sample A7.5-GL indicates a 7.5% agar-to-water concentration with glycerol and ground limestone.

To prepare samples, 200 mL of deionized water was added to 500 mL media bottles. The appropriate amount of agar powder (i.e., 10, 15, or 20 g) was dissolved at ambient laboratory conditions (22 ± 2 °C) under continuous agitation using a magnetic stir bar. The appropriate amount of glycerol or limestone was then added and the sample was agitated until all constituent materials were well dispersed. The media bottles were then placed in an SK101C Yamato autoclave (Santa Clara, CA, USA) and heated to 100 °C for 20 min. After autoclaving, individual media bottles were placed in a heated water bath at 90 ± 5 °C until sample molding or testing.

2.1.2. Composite Testing Samples

The properties of composites were evaluated in samples comprising 5% *w*/*w* agar combined with a typical aggregate mixture specified for use in Coloradan pavement materials. The amount of agar (i.e., 5% *w*/*w* agar) used in composite testing was informed by the results obtained in the binder testing. Composite properties were evaluated at two binder ratios (i.e., mass ratios of 0.2 and 0.5) with and without two filler materials (i.e., limestone powder, Ottawa sand). This resulted in a total set of eight composite sample types.

Table 3 outlines the composite formulations of these eight samples, designed by mass ratios. The mass ratio of graded aggregate and reinforcing materials (i.e., cumulative mass of aggregate, limestone, and Ottawa sand) was kept constant at a ratio of 1.0 for all composites to maintain overall sample ratios and allow for better comparability between sample types. Sample subsets reinforced with limestone or sand fillers were prepared at constant mass ratios of 0.05 and 0.2, respectively, and the mass ratio of graded aggregate varied from 0.75 to 1.0. Composites were prepared at two mass ratios of 5% *w*/*w* agar binder, namely 0.2 and 0.5. The sample naming convention is as follows: C for control samples (i.e., formulations with only aggregate), L for limestone-containing samples, S for Ottawa sand-containing samples, and LS for limestone- and Ottawa sand-containing samples. The letter designation is followed by the mass ratio of agar binder (i.e., 0.2 or 0.5).

After autoclaving, the media bottles were removed and the appropriate amount of gelled agar was weighed out to reflect the mass ratios (i.e., 0.2 or 0.5) outlined in Composites prepared with limestone or Ottawa sand filler were mixed by hand into the agar for approximately 1 min until filler materials were well dispersed. The appropriate amount of aggregate (i.e., 0.75, 0.80, 0.95, or 1.0 mass ratio) was weighed and mixed with agar-based materials in Hobart mechanical mixer for approximately 1 min. Sample mixtures were compacted into 2 inches cube molds in three layers and tamped by hand after the addition of each layer. Samples were allowed to gel for approximately 30 min and then removed from molds.

Table 3. Mass ratios of constituent mix components for agar-based composites.

Subset	Sample Name	5% *w/w* Agar Binder	Aggregate	Limestone	Sand
Control	C-0.2	0.20	1.00	-	-
	C-0.5	0.50	1.00	-	-
Limestone filler	L-0.2	0.20	0.95	0.05	-
	L-0.5	0.50	0.95	0.05	-
Sand filler	S-0.2	0.20	0.80	-	0.20
	S-0.5	0.50	0.80	-	0.20
Limestone and sand filler	LS-0.2	0.20	0.75	0.05	0.20
	LS-0.5	0.50	0.75	0.05	0.20

Composites were acclimated at ambient laboratory conditions (i.e., 22 ± 2 °C) until reaching constant mass, defined as less than 1% change in composite mass in a 24 h period. Since the mechanical properties of agar-based composites have been shown to improve with dehydration to a dry state [13], the composites were acclimated to constant mass to replicate the best conditions for loadbearing or structural applications and to ensure all samples were at a consistent, reproducible moisture state.

2.2. Rheological Properties of Agar-Based Binder

Standard asphalt binder testing methods were implemented to compare the engineering properties of agar with those of asphalt. This methodological approach was chosen to allow for direct comparison to a reference petroleum-based material used in civil engineering applications. Rotational viscosity (RV), dynamic shear rheometry (DSR), and penetration testing were conducted on agar-based materials. RV and DSR results were subsequently compared to performance-grade (PG) specifications [18] and penetration results were compared to penetration-grade specifications [19].

2.2.1. Rotational Viscosity

The rotational viscosity (RV) of fresh-state agar-based materials was measured according to a modified ASTM D4402 procedure [20]. The procedure was modified by using a rotational rheometer instead of a Brookfield viscometer and lowering the testing temperature from 135 °C to 80 °C. The performance-grade (PG) specification [18] utilizes an RV testing temperature of 135 °C to mimic the conditions of asphalt during construction. However, prolonged incubation of fresh-state agar at temperatures above 80 °C has been shown to impact structural and mechanical properties of agar gels [21]. Thus, RV was measured at 80 °C in order to represent agar temperature compatibility and preparation temperatures more accurately.

Fresh-state samples were immediately loaded into an MCR 301 rotational rheometer (Anton Paar, Graz, Austria) with 25 mm diameter stainless steel parallel plate geometry and a sample gap of 1.0 mm. The top plate was cross-hatched to minimize slip. After equilibration at 80 °C for 15 min, a constant shear rate of 20 s^{-1} was applied for 3 min. Three consecutive RV measurements were recorded at 1 min intervals for each sample. These data were averaged to yield final reportable RV values. Each sample was tested in triplicate.

The RV of asphalt binders is typically characterized to predict the workability at specific handling, mixing, and application temperatures to ensure that the mixtures can be properly mixed and compacted to the required pavement density and smoothness [20]. To this end, asphalt mixtures with inadequate workability are often difficult to compact properly, resulting in a lower pavement strength, higher air void content, reduced moisture resistance [22], and loss of service life [23]. Results were compared to PG specifications that require a RV < 3 Pa·s [18].

2.2.2. Dynamic Shear Rheometry

A dynamic shear rheometer (DSR) was used to determine the high-temperature rheological properties and rutting behavior of agar-based materials. The complex modulus (G^*), phase angle (δ), and rutting factor were determined using an MCR 301 rotational rheometer (Anton Paar, Graz, Austria) with 25 mm diameter stainless steel parallel plates. The rutting factor was calculated at various testing temperatures as denoted in Equation (1). The units of the rutting factor will be given by G^*, which in the case of this study was measured in kPa.

$$Rutting\ Factor = \frac{G^*}{sin(\delta)} \quad (1)$$

Cross-hatched top and bottom plates were used to minimize slip. The testing temperature ranged from 10 to 80 °C using a heating rate of 2 °C/min.

Freshly prepared agar samples were poured into 55 mm diameter molds to create disk-shaped samples with a depth of approximately 5 mm. After gelling for 30 min at ambient laboratory conditions (22 ± 2 °C), samples measuring 25 mm in diameter and 1 mm in thickness were taken from the center for testing. Beginning at the lowest testing temperature (i.e., 10 °C), samples were acclimated at the testing temperature for 20 min and then subjected to oscillatory shear flow at 12% strain amplitude and a rotational frequency of 10 rad/s. These parameters were chosen in order to reflect standard unaged asphalt binder rheological testing protocol (i.e., ASTM D7175) [24]. A constant axial force of 0.5 N was applied to account for thermal expansion and contraction. Each sample was tested in duplicate.

To compare the performance of agar-based samples with traditional asphalt binders, the rutting factor obtained from the DSR testing was compared with the requirements in asphalt performance-grade specifications [18]. The rutting factor, calculated from the complex modulus (G*) and the lower phase angle (δ) (i.e., G*/sin(δ)), measures the ability of binders to be both stiff and elastic to maintain shape and recover in response to repeated traffic loading [25]. A higher complex modulus (G*) and a lower phase angle (δ) are advantageous, resulting in the maximization of the rutting factor. To ensure adequate rutting performance, specifications [18] require the rutting factor to exceed 1.00 kPa at the upper pavement design temperature for unaged binders. Agar-based materials were compared to this specification.

2.2.3. Penetration

Penetration was determined according to ASTM D5 [26] using a Universal Penetrometer (Humboldt, Elgin, IL, USA). Fresh samples were poured into seamless tin cups measuring 80 mm in diameter and 50 mm in depth, covered, and stored for less than 24 h in refrigeration at 4 ± 2 °C until testing in order to prevent desiccation. Samples were then submerged in a 25 °C water bath for 30 min prior to testing, and they remained immersed in 25 °C deionized water for the duration of testing to ensure temperature stability. Results were compared to penetration grading standards (ASTM D946), which specify acceptable penetration grades in bins that range from 40 to 300 for unaged asphalt binders [19]. Triplicate penetration readings were collected for each sample.

Penetration is a measurement of binder consistency, and lower penetration values typically correspond to stiffer material behavior while higher penetration values typically correspond to softer material behavior [19]. Traditional asphalt binders with softer consistency characterized by high penetration values (i.e., 200 to 300) are typically used in cold climates to combat pavement cracking. Conversely, binders with low penetration (i.e., stiffer consistency) are typically used in warmer climates where a binder is expected to resist permanent deformation at high temperatures [27].

2.3. Durability of Agar-Based Binder

Durability was evaluated in terms of thermal stability (i.e., thermogravimetric analysis), fungal biodeterioration, and moisture sorption characteristics. These durability metrics were chosen based on the common limitations of biopolymers and bio-based materials noted in the literature. Thermal, moisture, and biodeterioration durability are some of the leading durability concerns that impose challenges in accepting and implementing bio-based materials in widespread industrial applications [28]. Relatedly, the undesirable hydrophilic behavior of biopolymers can provide pathways of moisture for the introduction of harmful microorganisms leading to biodeterioration [28]. The aforementioned mechanisms can negatively impact the appearance and mechanical properties of bio-based materials as well as impact human health and the service life of built infrastructure these materials are integrated into.

2.3.1. Thermogravimetric Analysis

Thermogravimetric analysis (TGA) was performed with a TA Discovery 5500 (TA Instruments, New Castle, DE, USA). Agar samples were poured into 55 mm diameter molds to create disk-shaped samples with a depth of approximately 10 mm. After gelling for 30 min at ambient laboratory conditions (22 ± 2 °C), samples were placed in incubation at 30 °C until samples reached constant mass (i.e., <1% change in mass in 1 h). A 10–20 mg sample was collected from the center of the dehydrated agar sample and placed in a platinum pan. Testing was conducted using continuous nitrogen gas flow at a rate of 10 mL/min. Samples were acclimated at 30 °C and then heated to 950 °C using a heating rate of 10 °C/min. The temperature and sample weight loss were recorded for analysis. Each sample was tested in duplicate.

2.3.2. Fungal Biodeterioration

Resistance to fungal biodeterioration was determined according to ASTM G21 [29]. The following fungal strains were used to create the testing spore solution at a concentration of 1,000,000 ± 200,000 spores/mL: *Aspergillus brasiliensis* (ATCC 9642), *Penicillium funiculosum* (ATCC 11797), *Chaetomium globosum* (ATCC 6205), *Trichoderma virens* (ATCC 9645), and *Aureobasidium pullulans* (ATCC 15233).

Nutrient-salt agar solutions were prepared according to ASTM G21. Solutions were poured into sterile Petri dishes measuring 150 mm in diameter and 15 mm in depth and allowed to gel. Agar samples to be tested for fungal resistance (summarized in Table 2) were poured into 55 mm diameter sample molds to create disk-shaped samples with a depth of approximately 10 mm. Samples were allowed to gel for approximately 30 min at ambient conditions (22 ± 2 °C) and then placed on the top surface of the nutrient-salts agar plates. Each nutrient-salts agar and agar test sample were inoculated by spreading a spore solution over the entire exposed surface with a sterile cell spreader. A negative control (i.e., blank nutrient-salts agar plates without a sample, denoted B-) and a positive control (i.e., nutrient-salts agar plates with sterilized cellulose paper, denoted C+) were also prepared to validate the experiment. Samples were sealed with parafilm and incubated at 30 °C for 28 days. Photos and observations were recorded every 7 days.

Biodegradation is a biochemical process where microorganisms metabolically degrade complex materials into natural compounds such as water, CO_2, and biomass. The ASTM G21 [29] procedure is designed to provide an environment where heterotrophic microorganisms are provided with all the necessary components for growth (e.g., moisture, salts, minerals) except for excess organic carbon. At the end of the 28-day testing period, samples were visually examined and evaluated by using a rating of 0–4 based on the percentage of sample area covered in fungal growth. This rating system is shown in Table 4. All samples were tested in triplicate.

Table 4. Rating system used for the evaluation of agar-based material fungal resistance.

Rating	Observed Growth on Specimens
0	None
1	Traces of growth (<than 10% of sample area)
2	Light growth (10–30% of sample area)
3	Medium growth (30–60% of sample area)
4	Heavy growth (60% to complete coverage)

2.3.3. Moisture Sorption

The moisture sorption properties of agar-based materials were determined according to a modified ASTM D570 procedure [30]. The procedure was modified minimally by lowering the conditioning temperature from 50 °C to 30 °C. Fresh agar samples were poured into 55 mm diameter molds to create disk-shaped samples with a depth of approximately 10 mm. After gelling for 30 min at ambient laboratory conditions, samples were weighed to determine the fresh sample mass and then conditioned in incubation at 30 °C until reaching constant mass, defined as <1% decrease in mass in a 24 h period.

Moisture sorption testing was conducted in 20 ± 2 °C distilled water. At various intervals, samples were removed from the distilled water, the surface moisture was removed, and the mass of each sample was recorded. Sample masses were recorded every 24 h until the sample reached constant mass. The moisture content was calculated according to:

$$Moisture\ Content\ (\%) = \frac{m_i - m_c}{m_c} \cdot 100\% \tag{2}$$

where m_i is the incremental sample mass and m_c is the conditioned sample mass.

After reaching constant mass in immersion testing, samples were reconditioned in incubation at 30 °C until again reaching constant mass, cooled for approximately 1 h, and reweighed. The percentage of matter lost, M_{Loss}, during immersion was calculated for each sample according to:

$$M_{Loss}\ (\%) = \frac{m_c - m_{rc}}{m_c} \cdot 100\% \tag{3}$$

where m_{rc} is the sample mass after reconditioning. Each sample was tested in triplicate.

2.4. *Mechanical and Microstructural Characterization of Agar-Based Composites*

This work investigates the properties of composites comprising 5% *w*/*w* agar combined with a typical aggregate mixture specified for use in Coloradan pavement materials. The microstructure (i.e., porosity, pore network characteristics) of agar-based composites were evaluated by non-destructive X-ray tomography testing, a procedure used by other researchers to investigate composites in pavement applications [31,32]. Mechanical properties were evaluated through unconfined compressive strength testing and data was further processed to obtain sample stress–strain curves and calculate the modulus of elasticity and resilience. Results are compared to traditional engineering materials and other bio-based composites in the literature.

2.4.1. X-Ray Tomography

The porosity of agar-based materials was not evaluated using typical water displacement methods (i.e., ASTM D7063 [33], ASTM C642 [34]) due to the hydrophilicity of agar-based materials that could alter the sample properties. X-ray tomography was chosen as a non-destructive alternative to evaluate the aggregate packing structure and sample porosity. Three-dimensional visualization of the internal structure of agar-based composites was obtained through X-ray tomography testing using a ZEISS Xradia 520 Versa system (ZEISS, Jena, Germany). The source voltage was set to 140 kV and the power to 10 W. An optical magnification of 0.4× and an HE3 filter were used to produce images. Agar-based composites prepared for X-ray tomography analysis are shown in Figure 1.

Figure 1. Representative images of samples prepared for X-ray tomography testing: (**A**) control samples; (**B**) limestone-reinforced samples; (**C**) sand-reinforced samples; and (**D**) limestone- and sand-reinforced samples.

X-ray tomography is conducted by projecting a cone beam of X-rays onto a sample and capturing projections on a detector. The sample is mounted between the X-ray source and detector and is rotated throughout testing to create renderings of 2D "slices" of the sample [35]. Upon completion of X-ray tomography testing, 2D slices are "stacked" and processed to create a 3D reconstruction of the sample. The X-ray density captured by the detector is related to the physical density of the sample, as denser objects appear brighter in readings. Due to this premise, X-ray tomography is frequently applied to identify pore spaces in samples and to quantitatively analyze the spatial distribution and size of pores. This method has been applied to cement [36], concrete [37], and asphalt [31,32].

X-ray tomography analysis of pores requires processing to segment images at delineation thresholds. Dragonfly 2022.2 software was used to create 3D reconstructions and segment images of agar-based composites to produce volumetric data. The volume of air voids (i.e., pores) in samples was determined by splitting the image pixel intensity at the Otsu threshold [38] two times. The first split segmented data into dense particles (i.e., aggregate) and less dense components (i.e., binder, filler, and voids). The second split, conducted on the foreground (i.e., other or binder, filler, and voids) segmented data into binder and filler versus pores. The segmentation process is illustrated in Figure 2.

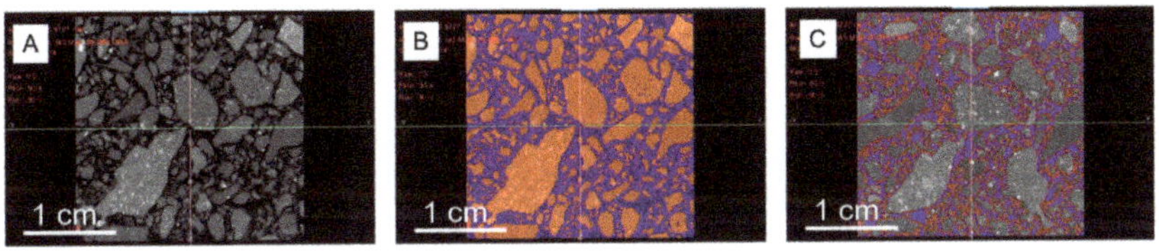

Figure 2. Segmentation processing implemented to analyze the porosity of agar-based composites: (**A**) X-ray tomography of the sample; (**B**) sample split at Otsu to segment particles (orange) from less dense components (blue); and (**C**) binder and air voids split at Otsu to segment binder and filler (orange) from voids (blue). Scale bar = 1 cm.

The volume percentage of pores in samples was determined by comparing the volume of segmented air voids against the total sample volume. Segmented sample data were further processed to create volume thickness maps of the pore space in order to analyze the pore sizes and distribution. Volume thickness mapping is an image processing technique based on a sphere fitting method that creates a "map" by identifying the diameter of the largest sphere that can be bounded within the region of interest and assigning distance values with color values [39]. This technique was applied to the region of pores segmented from the internal structure of agar-based composites to provide a colorized map of the internal pore network.

X-ray tomography testing requires the determination of a representative volume element (RVE), namely the minimum volume required for scanning that represents the bulk properties of the sample. In order to understand impacts from sample size scaling, the volumetric porosity was measured from the reconstruction of the entire tested composite (i.e., 50 mm × 50 mm × 25 mm sample) as well as from a smaller internal cross-section sampled from the center of the tested composite (i.e., 25 mm × 25 mm × 15 mm). Volume thickness maps were created based on reconstructions of the smaller internal cross-section.

2.4.2. Unconfined Compressive Strength

The unconfined compressive strength of agar-based composites was obtained using an Instron Universal Testing Machine and a constant axial extension rate of 0.1 mm/s. Prior to compressive strength testing, the average height (i.e., as an average of all 4 sides) and surface area (i.e., top and bottom) were measured using calipers. The compressive strength was calculated by dividing the maximum compressive load by the average sample test area (A_{Avg}). Stress–strain curves were calculated from compressive strength data using the following equations:

$$Stress\ (\sigma) = \frac{F}{A_{Avg}} \tag{4}$$

$$Strain\ (\varepsilon) = \frac{\Delta L}{L_0} \tag{5}$$

where F represents the compressive force applied during testing (kN in this study), L_0 represents the average height of the sample prior to testing (m in this study), and ΔL represents the change in sample length during testing in the linear elastic regime.

2.4.3. Modulus of Elasticity and Resilience

Stress–strain curves derived using Equations (4) and (5) were used to calculate the modulus of resilience (U_r) and modulus of elasticity, or Young's modulus, (E) for agar-based composites. The modulus of elasticity is a material parameter of stiffness that quantifies the amount of stress a material can withstand per unit of strain before deforming plastically. E

can be found by calculating the ratio of stress to strain in the linear elastic portion of the stress–strain curve (i.e., the slope of the linear region) and is quantified as:

$$E = \sigma/\varepsilon \quad (6)$$

The modulus of resilience (U_r) characterizes the strain-energy per unit volume of a material within the elastic range and quantifies the amount of energy a material can absorb without deforming plastically (i.e., permanently). The term "resilience" is used to indicate the ability of a material to absorb and release the energy in this range [40]. Generally, U_r (measured in J/m^3) is defined as the area under the elastic portion of the stress strain curve. For uniaxial stress, this can be quantified according to:

$$U_r = \frac{1}{2}\sigma\,\varepsilon \quad (7)$$

The elastic portion of each stress–strain curve was found by fitting a line through two points: The data point representing 40% of the maximum stress and the elastic limit, defined as 80% of the maximum stress. Using data from a linear fit (red line in Figure 3) corrects for the initial curved portion of the stress–strain curve that forms due to surface irregularities [41]. An illustration of this method is shown in Figure 3, with the modulus of resilience shown in gray.

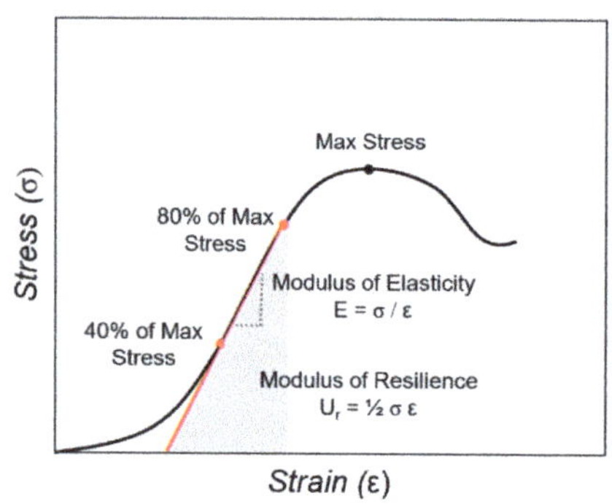

Figure 3. Illustration of methodology used to calculate the modulus of elasticity (E) and modulus of resilience (U_r—shown in gray).

Mechanical characterization results were analyzed through Scheirer Ray Hare (SRH) statistical techniques using a 95% confidence interval. SRH is a non-parametric equivalent of two-way ANOVA. This statistical methodology was chosen because the assumptions of parametric testing through ANOVA could not be met (i.e., normality). The two factors considered were (i) binder content (i.e., 0.2 or 0.5) and (ii) aggregate and reinforcement material composition (i.e., control, limestone-reinforced, sand-reinforced, or limestone- and sand-reinforced). Sample data determined to be statistically significant (i.e., p value ≤ 0.05) were further analyzed in post hoc testing to highlight statistically significant pairwise comparisons. A Dunn's multiple comparison test was applied post hoc to compare pairwise differences.

3. Results and Discussion

3.1. Rheological Properties of Agar-Based Binder

3.1.1. Rotational Viscosity

The results of rotational viscosity (RV) measurements for the agar-based materials are displayed in Figure 4. The PG specification [18] dictates an RV threshold of <3 Pa·s for asphalt binders, represented by a dashed line in. The RV of A5, A7.5, and A10 samples ranged from 1.03 to 2.01 Pa·s, 8.08 to 10.64 Pa·s, and 23.16 to 32.23 Pa·s, respectively. All A5 agar-based materials passed RV standards set by the PG specification for asphaltic binders, while all A7.5 and A10 agar-based materials exceeded current RV standards. Typically, higher RV asphalt binders are used in warmer climatic regions and lower RV binders are used in colder climatic regions [25].

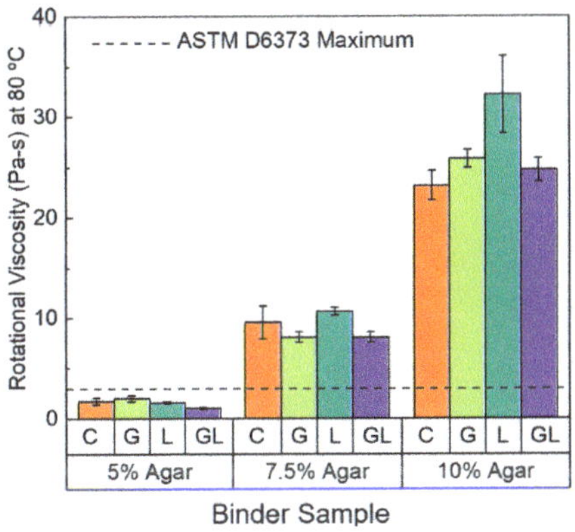

Figure 4. Rotational viscosities (RV) of agar-based materials with no additives (C), glycerol (G), and limestone (L). Error represents ± one standard deviation of triplicate measurements.

Results obtained in this study suggest that agar-based materials, if used in flexible pavement applications, may be better suited for warm-climate applications where a stiff material response is beneficial. However, workability of an asphalt mix is not merely a function of asphalt binder RV. Additional factors such as binder lubricity and the type, shape, texture, gradation, and porosity of aggregate in a mixture have been shown to greatly influence the workability of an asphalt mixture [42].

As expected, RV of agar-based materials increased as the agar concentration increased. Similar trends in increasing viscosity of higher-concentration agar were noted in work by Fernandez et al. [43] and Yu et al. [44] using a parallel plate rheometer and a rotational viscometer, respectively. Notably, the inclusion of glycerol (40% w/w of agar) and limestone (15% w/w of agar) additives did not significantly impact the RV of agar-based binders. No significant trend in RV measurements was observed relative to the addition of each additive and the values of the glycerol- and limestone-containing samples are similar to their respective controls.

Glycerol is one of the most widely used plasticizers for biopolymers due to excellent compatibility with biopolymer chain structures [17]. Glycerol molecules occupy intermolecular spaces between biopolymer chains, which in turn decreases attractive intermolecular forces and increases chain mobility [45]. However, it might be possible that a higher addition of glycerol is needed to observe a measurable plasticizing effect in the agar-based binder formulations investigated herein. For example, Yang and Paulson reported that the lowest effective glycerol concentration was 60% w/w for gellan films [46]. As further discussed in the fungal resistance and moisture sorption results, higher amounts of glycerol, however, would likely have a greater negative effect on the ability of agar-based materials to resist fungal biodeterioration and hydrolytic degradation, indicating a trade-off in fresh-state workability performance and long-term durability.

The RV results suggest that applications of agar-based materials would necessitate minor modifications to traditional placement protocol for asphalt binders. Typical hot mix asphalt is mixed at temperatures between 140 and 180 °C before being transported to a construction site, placed, and compacted. The PG specification utilizes an RV testing temperature of 135 °C to mimic the conditions during construction. The lower temperature compatibility (i.e., 80 °C) of agar-based materials indicates the possibility of mixing and compacting agar mixes (i.e., agar-based binders and aggregate) at a lower temperature than typical hot mix asphalt. Implementation of lower placement temperatures corresponds to a reduction in material embodied energy as well as improved environmental safety. A mixing temperature reduction of 40 to 60 °C has been shown to reduce energy consumption of pavement construction by up to 40% [47]. Additionally, a lower placement temperature allows for quicker reopening of newly paved infrastructure to traffic and a longer paving season [48].

3.1.2. Dynamic Shear Rheometry

Dynamic shear rheometry (DSR) was performed to determine the complex modulus (G^*) and phase angle (δ) for agar-based materials at temperatures ranging from 10 to 80 °C. These results are shown in Figure 5, with solid lines depicting complex modulus values and dotted lines representing phase angles.

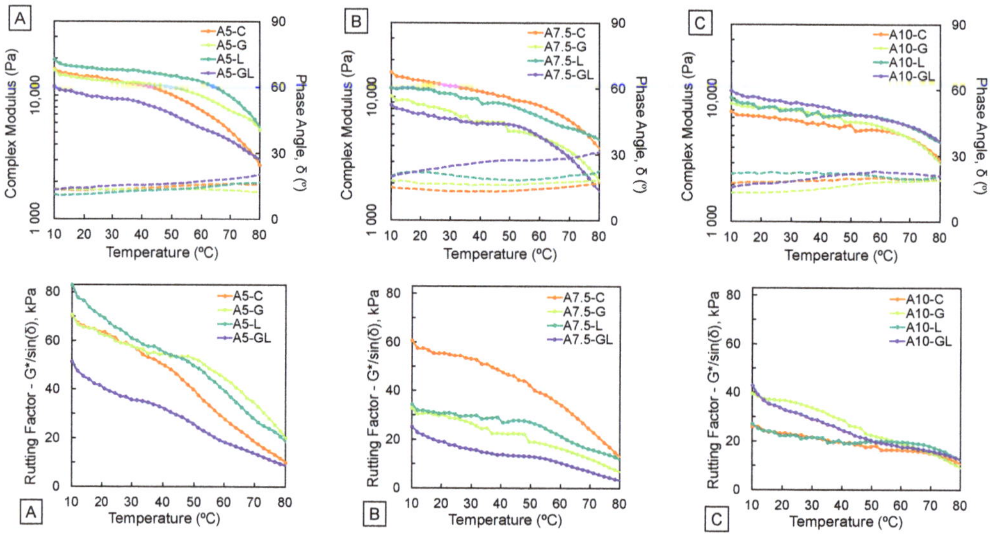

Figure 5. Representative results from dynamic shear rheometry (DSR) temperature sweeps (top) and rutting factor (bottom) for (**A**) 5% agar; (**B**) 7.5% agar; and (**C**) 10% agar specimens. Each sample was tested in duplicate.

Agar-based materials retained high stiffness (i.e., $G^* > 7000$ Pa) at test temperatures and the phase angle was consistently between 15° and 30° for the formulations and temperatures investigated herein. These results align with previous rheological studies on 1% and 1.5% agar and agar blends [49,50]. Agar gels retain strength at high temperatures (i.e., 70–80 °C) analyzed in this study because it is below the temperature necessary (i.e., ~85 to 90 °C) for agar to attain random coil formation to significantly alter the tight 3-D gel network [51].

In general, as the testing temperature increased, the complex moduli decreased while the phase angle increased. Similar behavior has been observed in agar [52] and asphaltic binders [53]. Compared to petroleum-based asphalt binders, agar-based materials exhibited a higher complex modulus and a lower phase angle at PG testing temperatures. The phase angle of petroleum-based asphalt binders is typically between 80 and 90 °C [54] indicating a highly viscous material response. The rheological response of agar-based materials corresponds to stiffer material behavior, which may indicate superior binder resistance to permanent deformation as compared to traditional binders.

Previous work with asphalt mastics has reported an increase in material complex moduli with the use of filler materials like fly ash, stone and brick dust, and mineral fillers [55]. In this work, no significant difference was observed relative to the addition of limestone filler. Similarly to results obtained in rotational viscosity (RV) testing, no significant trend was observed relative to the addition of glycerol or limestone. DSR results for samples with glycerol validate the conclusions of RV data. A greater concentration of glycerol and limestone may be necessary to significantly impact the rheological properties of agar-based materials.

To compare the performance of agar-based samples with traditional asphalt binders, rutting factor results for 60 °C and 80 °C are shown in Figure 6. These temperatures were chosen as they closely resemble common upper pavement design temperatures for high-temperature PG grading (i.e., 58 °C and 82 °C). In the current study, the rutting factor calculated for agar-based binders at 60 °C and 80 °C ranged from 14.0 to 37.1 kPa and 6.9 to 15.6 kPa, respectively. These results suggest that all agar-based binder formulations possessed adequate stiffness to resist early-age rutting at upper pavement design temperatures less than 80 °C. Although these results point towards advantageous early-age rutting properties, the relatively high magnitude of complex moduli and rutting factors exhibited by agar-based binders may also indicate the possibility of low-temperature and fatigue-induced crack propagation.

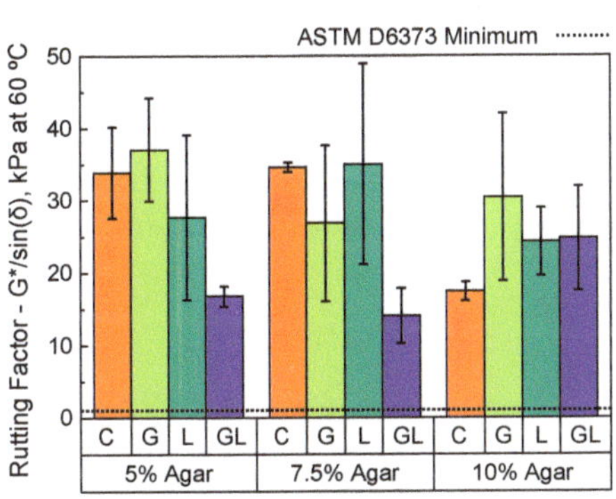

(**A**)

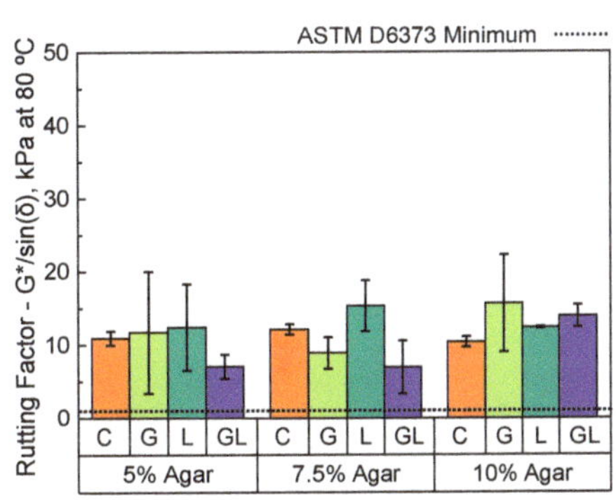

(**B**)

Figure 6. Mean rutting factor (i.e., G*/sin(δ)) for agar-binders at (**A**) 60 °C and (**B**) 80 °C. Error bars represent the high and low results for duplicate measurements.

3.1.3. Penetration

The results from penetration testing of agar-based materials binders are shown in Table 5. All four of the A5 sample formulations (i.e., A5-C, A5-G, A5-L, A5-GL) and both A7.5 sample formulations without limestone (i.e., A7.5-C, A7.5-G) exhibited a penetration greater than 350. Due to limitations based on the geometry of the penetration samples and apparatus, penetration measurements greater than 350 could not be accurately measured. The lowest penetration measurement, 162, was observed in both the A7.5-GL and A10-GL samples. Penetration measurements generally decreased as the agar concentration increased.

Table 5. Penetration results for agar-based binders in 0.1 mm. Error represents one standard deviation of triplicate measurements.

Agar-to-Water Concentration	C	G	L	GL
5% Agar	>350	>350	>350	>350
7.5% Agar	>350	>350	225 ± 8	162 ± 12
10% Agar	211 ± 12	217 ± 2	167 ± 3	162 ± 8

The addition of glycerol minimally impacted penetration measurements for 10% agar samples. The impact of glycerol was more pronounced in 7.5% agar samples with limestone (i.e., A7.5-L as compared to A7.5-GL). The addition of limestone generally resulted in a decrease in penetration, particularly for the A7.5 and A10 samples. These results were expected as previous literature has reported a similar decrease in the penetration of asphalt binders with the use of fillers like corn stalk fiber [56] and crayfish shell powder [57].

Traditional asphalt binders with high penetration values (i.e., softer consistency) are typically used in cold climates to combat pavement cracking. Conversely, binders with low penetration (i.e., stiffer consistency) are typically used in warmer climates where a binder is expected to resist permanent deformation at high temperatures. Asphalt binders with a penetration between 60 and 100 are mostly used in road construction, while asphalt binders with a penetration between 120 and 150 are primarily used in pavements with lighter traffic loading. Asphalt binders with penetration values between 200 and 300 are used less frequently for seal coating or arctic applications [27].

Compared to traditional asphalt binders, agar-based materials exhibited relatively high penetration values. However, these results illustrate a spectrum of binder consistency that might be tailorable with agar concentration and the use of mineral fillers, like ground limestone. Additionally, these measurements were obtained under the most conservative conditions given that the agar samples were unaged and submerged in a 25 °C water bath at the time of testing. Just as petroleum-based binders are known to stiffen with age due to volatilization and oxidation [25], agar-based materials are expected to stiffen with age due to dehydration of the hydrogel structure at ambient conditions.

3.2. Durability of Agar-Based Binder

3.2.1. Thermogravimetric Analysis

All agar-based material formulations were thermally stable up to 210 °C, followed by a steep mass loss due to decomposition of agar and glycerol (Figure 7).

Similar thermal behavior for agar composite materials has been reported previously [58,59]. For the agar-based formulations containing limestone, a second steep mass loss was observed beginning at 650 °C due to the decarbonation of limestone. Although the thermal stability of agar-based materials appeared to be minimally affected by agar concentration, the addition of both glycerol and limestone substantially impacted thermal stability. Specifically, glycerol addition resulted in a decrease in thermal stability and limestone addition resulted in an increase in thermal stability. The 5%, 7.5%, and 10% agar-based binder control formulations retained 29.60%, 29.54%, and 30.59% of their original mass, while limestone formulations retained 43.55%, 44.33%, and 39.45% of their original mass, respectively.

TGA characterization enables direct comparison between traditional petroleum-based construction materials and agar-based materials. Commonly used plastics including polyethylene, polypropylene, polystyrene, and polyethylene terephthalate begin to thermally degrade between 250 and 450 °C, with the maximum thermal degradation taking place between 420 and 490 °C [60]. Petroleum-based asphaltic binders for roadway or roofing applications begin to degrade between 300 and 350 °C [61] and degradation continues until the temperature reaches 500 °C, where typically less than 20% of the original mass is retained [62]. Agar-based materials exhibit partial thermal degradation at lower temperatures than petroleum-based materials and binders commonly utilized in the construction industry (i.e., beginning at 210 °C versus 250–300 °C). However, these data suggest that the addition of limestone filler can improve the thermal stability of agar-based materials allowing for a tailor ability of thermal durability. Previous work on asphalt mastics with additives including aluminum hydroxide, magnesium hydroxide, and limestone filler have shown similar improvements in thermal stability with the addition of a material that is inert until higher temperatures [61].

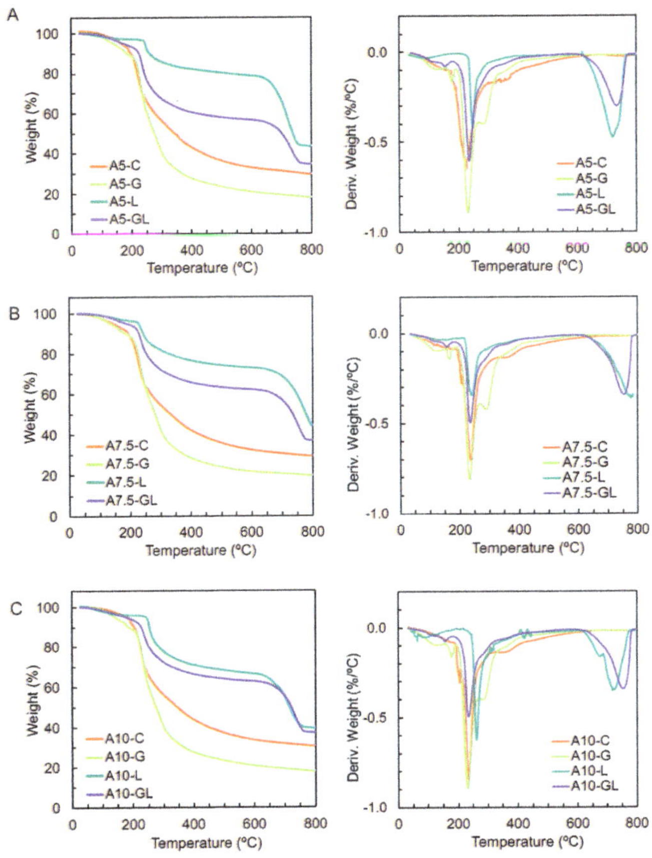

Figure 7. Representative results from thermogravimetric analysis for (**A**) 5% agar; (**B**) 7.5% agar; and (**C**) 10% agar specimens. Each sample was tested in duplicate.

3.2.2. Fungal Biodeterioration

Photographs of a representative sample for each agar-based material formulation after fungal resistance testing are displayed in Figure 8. The control agar samples exhibited good fungal resistance and only traces of fungal growth (i.e., <10% of sample area) were observed after 28 days. The fungal resistance of samples increased with the addition of limestone whereas the addition of glycerol decreased fungal resistance.

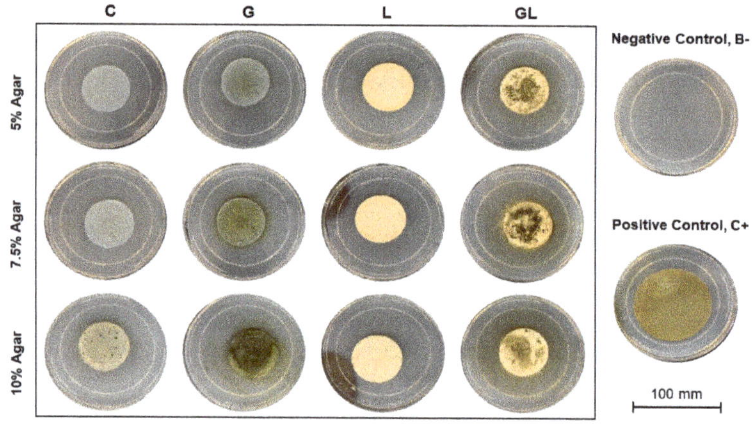

Figure 8. Representative photographs of agar-based materials subjected to ASTM G21 biodeterioration testing. Photos were taken at the end of a 28-day incubation period at 30 °C.

At 28 days, most samples without glycerol addition (i.e., C and L samples) were rated as 1 (traces of growth). Contrastingly, the samples containing glycerol were rated in the range of 2 (light growth) to 4 (heavy growth). The experiment was validated using both a negative control (i.e., no sample) and positive control (i.e., cellulose). At 28 days, no growth was observed on the negative control specimens (rating of 0), and full growth coverage was observed on the positive control specimens (rating of 4).

Only traces of growth were observed on the samples without glycerol 28 days after inoculation. Therefore, it was concluded that agar and limestone did not serve as a primary carbon source for the growth of heterotrophic microorganisms, which is known to lead to biodeterioration. Agar is used as a microbiology culture media because it is non-nutritive for the vast majority of microorganisms making it a promising material candidate for construction application. However, glycerol provided an effective carbon source for fungal growth in samples containing it. Similar results were obtained by Nissa et al. [63], where 80% growth coverage (equivalent to ASTM G21 rating of 4) was observed on a starch-based bioplastic containing glycerol after just 10 days of incubation.

Microorganisms have evolved to use the hydrocarbons in petroleum as a source of carbon which can lead to biodegradation, which can worsen physical and chemical properties and reduced efficiency of petroleum products [64]. To combat potential biodeterioration, chemical biocides are often introduced. Notably, the chemical substances used to combat biodeterioration are often pollutive, mutagenic, and carcinogenic [65]. Comparatively, the fungal resistance properties of agar-based materials would limit the use and necessity of harmful biocides.

3.2.3. Moisture Sorption

The equilibrated moisture content of agar-based binders, calculated using Equation (1), is shown in Figure 9A. In general, agar-based binders absorbed a large percentage of moisture relative to the conditioned mass of samples (i.e., samples incubated at 30 °C until constant mass). This behavior was expected due to abundant hydroxyl functional groups (i.e., -OH) in polysaccharides like agar, and the hydrophilicity of hydrogels [66]. After sample conditioning, agar-based binders retained 5 to 23% of fresh-state mass and upon immersion in distilled water, agar-based binders swelled to 55 to 97% of fresh state mass.

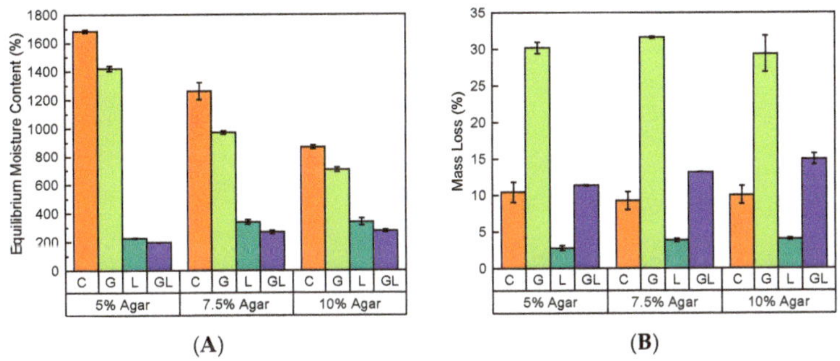

Figure 9. Equilibrium moisture content (**A**) and mass loss (**B**) of samples immersed in 20 ± 2 °C distilled water. Error bars represent one standard deviation of triplicate measurements.

The equilibrated moisture content for control agar-based samples (Figure 9A) ranged from 196% to 1682%, 268% to 1262%, and 275% to 868% for A5, A7.5, and A10 samples, respectively. The maximum moisture content of control and glycerol dosed agar-based binders decreased with an increase in agar-concentration. These experimental observations are consistent with previous studies on the water holding capacity of agar hydrogels. The swelling behavior of agar is highly dependent on the porous size between crosslinked bonds. An increase in agar content results in a dense polymer network (i.e., decreased porous space) which can hinder polymer mobility and decrease water holding capacity [52,67].

The use of glycerol decreased water sorption for samples herein. Consistently, samples with glycerol absorbed less than comparative samples without glycerol (i.e., G and GL compared to C and L, respectively). Previous work conducted on chitosan films have noted similar decreases in water sorption with the use of glycerol [68]. It is hypothesized that this behavior is due to the formation of bonds between biopolymers and glycerol, which increases the network density. Additionally, glycerol molecules take up intermolecular space which can decrease access to active sites for the biopolymers to bond with water.

Notably, the inclusion of limestone reduced the moisture sorption of agar-based materials. These findings align with literature on hydrogel-based composites. Rigid fillers can increase composite density and mechanical stability which can alter the swelling capacity of hydrogels [69,70]. Similarly to glycerol findings, the decrease in moisture sorption with the use of limestone filler may be due to a sort of barrier effect. The limestone filler molecules can occupy intermolecular space and oppose moisture transport through the material.

Sample deterioration (i.e., bath clouding, small sample fragments) was evident for the agar-based materials containing glycerol. This observation was validated by mass loss data obtained from re-conditioned sample masses using Equation (2) and shown in Figure 9B. Agar-based binders with glycerol lost 19–22% more mass than samples without glycerol. Similar mass loss due to glycerol leaching has been noted in previous work on polymers [71,72]. Interestingly, agar-based materials with both glycerol and limestone only lost 8–11% more mass than samples with only limestone. These results indicate that agar sample formulations with limestone may have a stronger and more dense gel network than formulations without limestone and may be more dimensionally stable and capable of withstanding the stresses due to drying and shrinkage more successfully [51]. Conversely, sample formulations with glycerol may be less apt at withstanding drying and shrinkage.

These findings substantiate a propensity for agar-based material to exhibit high moisture sensitivity (i.e., high equilibrium moisture contents) and propensities for dimensional change and deterioration due to wetting and drying (i.e., mass loss). The moisture sorption characteristics highlight an inherent weakness associated with biopolymers, namely hydrophilicity. Agar-based materials may be suitable for pavement applications if the hydrophilic nature of agar is improved through physical or chemical modification. Without treatment, agar-based materials are more appropriate for applications without significant moisture exposure.

3.3. Mechanical and Microstructural Characterization of Agar-Based Composites

3.3.1. X-Ray Tomography

An example of X-ray tomography reconstruction of the cross-section sampled from the center of the control section sample is illustrated in Figure 10. The figure provides an image of the reconstructed sample and a volume thickness map of the segmented interior pore network (i.e., top and bottom of figure, respectively). The volume thickness maps use a color spectrum ranging from a dark navy to red or white with the navy coloring representing the smallest measured distances between pores and a red or white color representing the largest distance between pores, measured in mm.

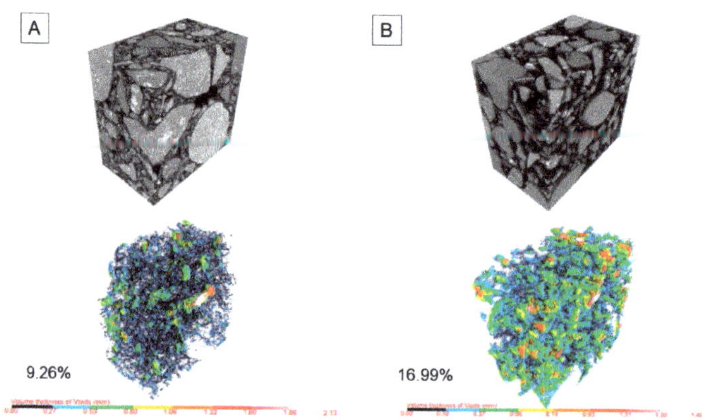

Figure 10. X-ray tomography reconstructions and volume thickness map of pore network for control composites with (**A**) 0.2 agar binder content and (**B**) 0.5 agar binder content. Sample porosity as % volume is shown to the bottom left.

The porosity of agar-based composites ranged from 5.79 to 16.99% for control samples, 6.25 to 15.21% for limestone-containing samples (L), 3.39 to 9.92% for sand-containing samples (S), and 5.60 to 16.55%, for limestone- and sand-containing (LS) samples. Comparatively, the measured porosity of agar-based composites is generally higher than the porosity of traditional concrete and asphalt (i.e., 5–15% and 2–7%, respectively). However, porosity measurements are within the same range as pervious concrete and porous asphalt (i.e., 15–25% and 10–15%, respectively) [73,74].

Porosity measurements obtained from segmented reconstructions of the entire tested composite (i.e., filled marker) and from a smaller internal cross-section sampled from the center of the tested composite (i.e., unfilled marker) are shown in Figure 11. In general, the porosity measured from the smaller segmented volume of the sample was slightly higher than the porosity calculated based on the entire sample volume. The LS-0.2 sample (i.e., limestone- and sand-reinforced, 0.2 binder content) was the only sample that presented an exception to this trend (i.e., porosity of the entire composite volume was approximately equivalent to the porosity of the segmented composite volume).

The volume thickness maps derived from this test (e.g., Figure 10) was used to identify the diameter of the largest sphere that can be bound within the segmented pore space of composite reconstructions. The maximum diameter of pores in samples ranged from 1.4 to 2.1 mm for all agar-based composites, with the exception of S-0.5 and LS-0.5 composites.

The maximum diameter of pores for these samples was larger and ranged from 3.1 to 3.6 mm. Detailed diameter data was processed in order to compute the 50th percentile (i.e., median) pore diameter for each sample and results are illustrated in Figure 12.

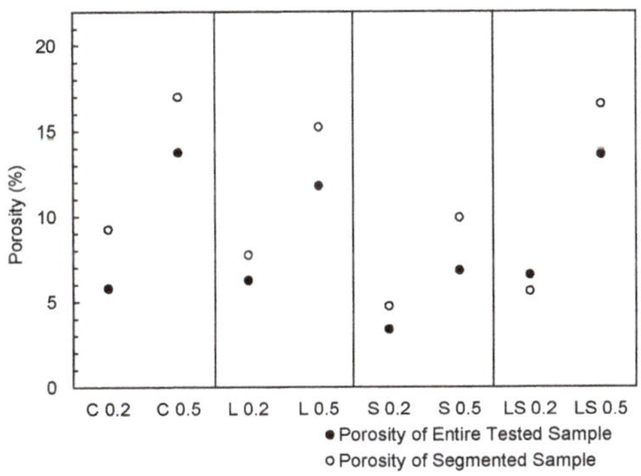

Figure 11. Porosity (% volume) of agar-based composites tested in X-ray tomography.

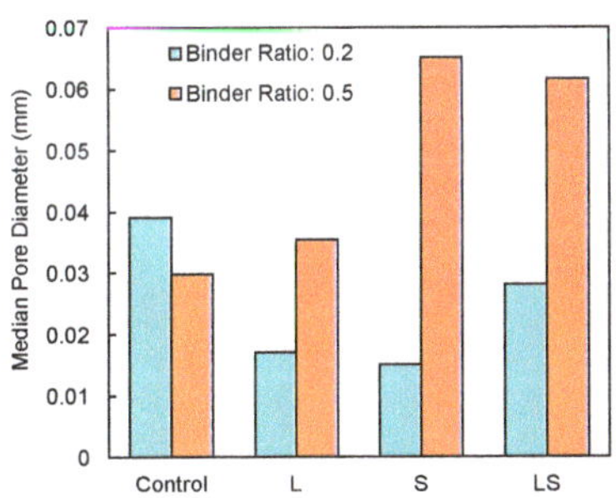

Figure 12. Median pore diameter (mm) of agar-based composites based on volume thickness map.

Several trends are visible in the porosity measurements obtained. Firstly, an increase in binder content led to an increase in the volume of pores (i.e., comparison of 0.2 and 0.5 binder content in Figure 11) and pore size (Figure 12). These results align with visual observations of composites. Agar-based composites with a 0.5 binder content had visible surface cracks concentrated between large aggregate particles (Figure 1). As biopolymers dehydrate, shrinkage stresses draw aggregate particles towards one another and compress the pore space of the composite [10]. The use of too much biopolymer in a composite can result in increased internal stresses during dehydration, which can lead to particle conglomeration and a sample structure with larger, more connected voids [75]. Porosity data suggests that a 0.5 binder content (i.e., by mass) surpasses the optimal agar proportion for the aggregate and filler compositions studied herein.

Second, porosity measurements indicate that the inclusion of sand led to a decrease in the porosity of samples, evident by comparison of sand reinforced samples and control samples. The porosity of sand reinforced samples with a 0.2 and 0.5 binder content was 2.4–4.5% and 6.9–7.1% lower than comparable control samples (i.e., C-0.2 and C-0.5), respectively. Conversely, the porosity of limestone- and sand-reinforced samples was greater than the porosity of samples with only sand (i.e., porosity of LS-0.2 and LS-0.5 samples was 1.8 to 2.2% and 6.6 to 6.8% higher than S-0.2 and S-0.5 samples, respectively).

2D cross-sections from X-ray tomography reconstructions (Figure 13) suggest that a film or coating forms on the surface of aggregate particles. In these cross-sections, the interparticle structure of samples is illustrated using a color map of segmented sample components. Dense particles (i.e., aggregate), less dense components (i.e., fine particles and biopolymer), and pores are shown in black, blue, and yellow, respectively. These results align with previous research, which has found that biopolymers like agar improve composite mechanical properties by forming a connected film around the surface of particles [8,12]. When biopolymer composites are dehydrated, this thin film remains and functions as a matrix for particles [8,75] improving interaggregate contact area [12,16], sample density, and porosity [11]. In a dehydrated state, mechanical improvements of biopolymer aggregate composites are primarily due to frictional forces in the sample rather than adhesion or charge interactions related to the gel strength of agar [8]. Our work found that sand reinforced samples demonstrated a condensed structure, evident by comparison of panels A and B with panels E and F in Figure 13 (i.e., Comparison of C-0.2 and C-0.5 with S-0.2 and S-0.5). Samples with limestone did not demonstrate the same trend. Although it appears the addition of limestone may have minimally improved the interparticle structure of control composites (i.e., comparison of panels A and B with panels C and D), the interparticle structure was largely similar to comparable control composites (i.e., large pore spaces are present between dense particles).

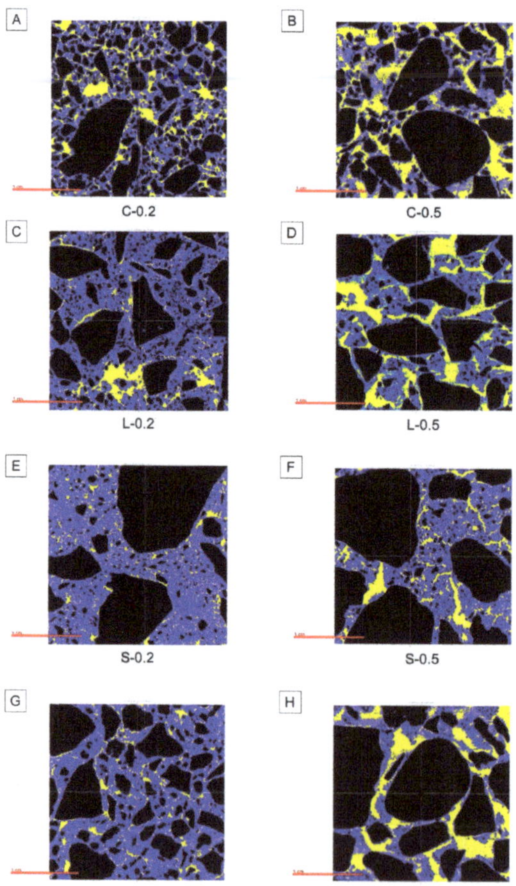

Figure 13. Representative cross-section of agar-based composites for (**A**) C-0.2; (**B**) C-0.5; (**C**) L-0.2; (**D**) L-0.5; (**E**) S-0.2; (**F**) S-0.5; (**G**) LS-0.2; and (**H**) LS-0.5 samples. Dense aggregate, binder and fines, and voids shown in black, blue, and yellow, respectively. Red scale bar represents 1 cm length.

3.3.2. Unconfined Compressive Strength

The results for unconfined compressive strength testing of agar-based composites are shown in Figure 14. Compressive strength results are within range of previous work conducted on biopolymers composites. Work conducted by Khatami and O'Kelly [9] on sand treated with 1, 2, and 4% agar found that the dehydrated (i.e., 30 days of acclimation) compressive strength of samples ranged from 158 to 487 kPa. Similarly, Fatehi et al. [10] reported unconfined compressive strengths as high as 225 kPa and 1800 kPa for agar composites composed of sand and clays, respectively, and Chang et al. [8] reported an unconfined compressive strength of 3190 kPa for agar and clay composites with thermal treatment. Notably, experimental results noted are for agar composites with fine particles (i.e., sand or clays) and not for composites with dense-graded, coarse aggregates.

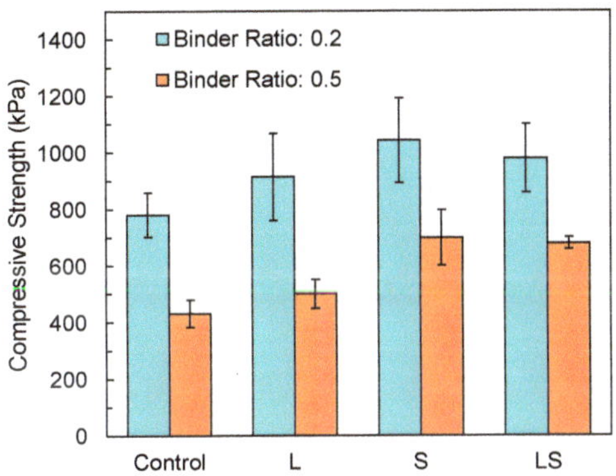

Figure 14. Compressive strength of agar-based composites. Error bars represent standard deviation of five samples.

Compressive strength data were further analyzed as a response variable in Scheirer Ray Hare (SRH) testing to determine the influence of binder content and aggregate and filler composition. Results of SRH statistical analysis (Table 6) indicate that both binder content ($p = 1.29 \times 10^{-6}$) and aggregate and filler composition ($p = 0.0401$) led to statistically significant differences in composite compressive strength. The interaction between factors was not statistically significant at a 95% confidence interval (i.e., p value > 0.05). A statistically significant interaction would indicate that the analyzed factors have a combined impact on the response variable that is not present in an analysis of each variable alone. These results indicate that the binder content had the same effect on compressive strength no matter the aggregate and filler composition used, and vice versa.

Table 6. Results for post hoc Dunn's analysis used to evaluate the impact of aggregate composition on the compressive strength of agar-based composites. An asterisk (*) represents statistically significant results.

Pairwise Sample Comparison	Z	p-Value
C-L	−0.9564	0.338
C-LS	−2.2570	0.024 *
L-LS	−1.3007	0.193
C-S	−2.5248	0.011 *
L-S	−1.5684	0.117
LS-S	−0.2679	0.789

Although the SRH test can indicate whether statistical differences exist in a group, these results do not explicitly indicate which specific groups differ from one another. In order to determine pairwise differences, a Dunn's test was applied to compare groups in detail. As there are only two binder contents (i.e., 0.2 vs. 0.5 binder content) and these sample types were determined to be significantly different in the SRH analysis, the Dunn's test was applied to analyze the impact of aggregate and filler composition on compressive strength.

Results in Table 6 show that, of the six pairwise comparisons analyzed, the compressive strength were statistically significant in two sample pairs (i.e., C-LS and C-S). This indicates that the inclusion of sand in the aggregate and filler composition of samples led to a statistically significant improvement in the compressive strength of agar-based composites as compared to sample equivalents without sand (i.e., control samples, or limestone reinforced). Statistical analysis indicates that unlike sand, the addition of limestone did not lead to a statistically significant impact on the compressive strength of composites.

3.3.3. Modulus of Resiliency

The moduli of resilience of agar-based composites are shown in Figure 15. The average modulus of resilience of samples ranged from 12,275 J/m^3 (i.e., sample C-0.5) to 18,934 J/m^3 (i.e., sample L-0.2).

These results align with findings in the literature. Namely, Cabalar et al. [76] reported an increase in the modulus of resilience (i.e., energy absorption capacity increased) to ~4000 J/m^3 for crushed rock stabilized with xanthan gum. Hamza et al. [75] found that high plastic clays stabilized with agar illustrated a modulus of resiliency over 30,000 J/m^3. Results in this work are within reasonable range considering that the particle size in this work is between that of the referenced literature.

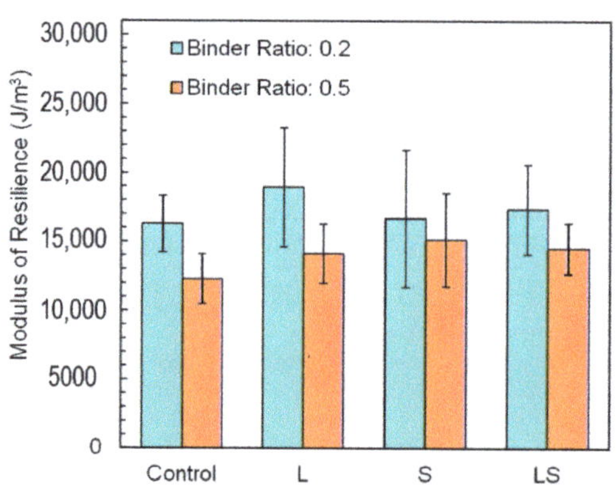

Figure 15. Modulus of resilience of agar-based composites. Error bars represent standard deviation of five samples.

The modulus of resilience results for agar-based composites were analyzed as a response variable in SRH testing using a 95% confidence interval in order to determine the influence of binder content and aggregate and filler composition. Results of SRH statistical analysis are shown in Table 7. SRH testing indicated that the aggregate and filler composition of composites did not lead to a statistically significant difference in the modulus of resilience of composites (p value = 0.70256). Similarly, the interaction between binder content and aggregate and filler composition was not statistically significant (i.e., p value = 0.51246).

Table 7. Results for Scheirer Ray Hare analysis used to evaluate modulus of resiliency results for agar-based composites. An asterisk (*) represents statistically significant results.

Source of Variation	df	SS	H	*p* Value
Binder Content	1	1060.9	7.76	0.00533 *
Aggregate and Filler Composition	3	193.0	1.41	0.70256
Interaction	3	314.3	2.30	0.51246
Residuals (Within Group)	32	3761.2		

Although the compressive strength of agar-based composites was significantly impacted by both experimental factors considered (i.e., binder content and the aggregate and filler composition produced a p value ≤ 0.025), the binder content was the only statistically significant factor for the modulus of resilience results (p-value = 0.00533). An increase in binder content of composites from 0.2 to 0.5 led to a 33%, 34%, 10% and 20% decrease in the average modulus of resilience for control, limestone reinforced, sand reinforced, and both limestone- and sand-reinforced samples, respectively.

3.3.4. Modulus of Elasticity

The modulus of elasticity (E) for samples is illustrated in Figure 16. Previous work conducted by Chang et al. [8] found the modulus of elasticity of 1% agar-treated clayey soils to range from 188 to 270 MPa depending on thermal treatment. The magnitude of difference in moduli results in this work was expected, as the agar-based composites investigated in this work utilize much larger particles (i.e., coarse aggregate instead of sand). A material with larger particles will often reflect a smaller modulus of elasticity due to increased porosity and reduced interparticle bonding and load transfer efficiency.

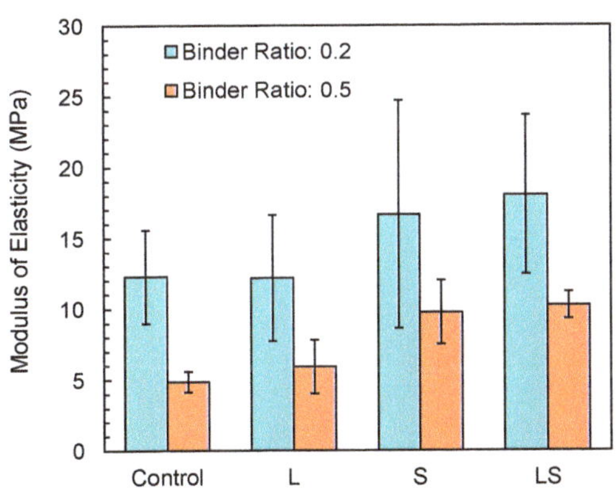

Figure 16. Modulus of elasticity of agar-based composites. Error bars represent standard deviation of five samples.

Results were analyzed as a response variable in SRH testing using a 95% confidence interval and results are shown in Table 8. Similarly to the modulus of resiliency, results from this analysis indicate that only binder content ($p = 3.0 \times 10^{-5}$) led to statistically significant differences in composite modulus of elasticity.

Table 8. Results for Scheirer Ray Hare analysis used to evaluate modulus of elasticity results for agar-based composites. An asterisk (*) represents statistically significant results.

Source of Variation	df	SS	H	p Value
Binder Content	1	2371.6	17.35	3.0×10^{-5} *
Aggregate and Filler Composition	3	965.8	7.07	0.0698
Interaction	3	166.6	1.22	0.7485
Residuals (Within Group)	32	1826.0		

4. Conclusions

This study evaluated agar-based binders as a direct alternative material for traditional petroleum-based asphalt binders. In particular, the study analyzed the rheological properties and durability of agar-based binders, as well as the mechanical and microstructural characterization of composites. Overall, this analysis is a step towards the practical adoption of more sustainable practices in the construction industry. The experimental data indicated the following conclusions:

- All 5% w/w agar-based binders passed the rotational viscosity (RV) threshold set by the performance-grade (PG) specification (<3 Pa·s), while all 7.5% and 10% w/w samples exceeded that same threshold. In general, agar-based binders exhibited more viscous behavior than traditional asphalt binders. However, RV testing of agar-based binders was completed at a lower temperature than stipulated by the PG specification (i.e., 80 °C vs. 135 °C). The lower temperature compatibility of agar-based materials indicates the possibility of lower mixing and compaction temperatures and an increase in energy efficiency of production as compared to petroleum-based binder production.
- Dynamic shear rheometry (DSR) revealed that all agar-based binders in this study exhibited adequate stiffness to resist early-age rutting at temperatures up to 80 °C. Agar-based materials generally illustrated a higher complex modulus and lower phase angle than traditional asphalt binders, which is advantageous for rutting. Relatedly, the high magnitude of the complex moduli at testing temperatures might indicate the possibility of low-temperature and fatigue-induced cracking.
- Penetration measurements performed on fully submerged, unaged agar-based binders ranged from 162 to greater than 350. While penetration readings were generally higher than those associated with traditional asphalt binders used in road applications, the results indicated that penetration consistency might be tailorable with the use of additives, such as ground limestone.
- Thermogravimetric analysis illustrated that all agar-based binders were thermally stable up to 210 °C. The onset of thermal decomposition occurs at a slightly lower temperature for agar-based binders as compared to petroleum-based materials (i.e., 210 °C vs. 250–300 °C).
- When subjected to ASTM G21 testing, agar-based binders without glycerol addition exhibited improved resistance to biodeterioration as compared to a positive control (i.e., cellulose). Samples with glycerol addition showed substantial growth, indicating that glycerol served as an effective carbon source for the growth of heterotrophic microorganisms.
- Moisture diffused rapidly in agar-based binders and the equilibrated moisture content for agar-based samples relative to conditioned mass ranged from 196% to 1682%. Relative to fresh-state sample mass, rehydrated agar-based binders swelled to between 55 and 97% moisture content. Limestone significantly reduced moisture sorption and improved mass loss during testing. However, glycerol plasticizer was found to leach from agar-based binder samples in isothermal sorption testing.

- As shown in X-ray tomography, agar-based material-coated aggregate particles, resulting in increased sample density and frictional contact between particles. Results indicate there is likely the existence of an optimum aggregate composition and binder content. Further, X-ray tomography illustrated the formation of large fractures due to shrinkage stresses in samples with a 0.5 agar binder content. Coupled with results from porosity measurements and mechanical data, this suggests that a 0.5 binder content (i.e., by mass) surpasses the optimal agar proportion for the aggregate and filler compositions studied herein.
- In microstructural evaluation, several trends are visible in porosity measurements. The porosity of agar-based composites ranged from 5.79 to 16.99% for control samples, 6.25 to 15.21% for limestone-containing samples (L), 3.39 to 9.92% for sand-containing samples (S), and 5.60 to 16.55% for limestone- and sand-containing (LS) samples. An increase in binder content led to an increase in the volume of pores and pore size. The inclusion of sand led to a decrease in the porosity of samples and porosity of sand reinforced samples with a 0.2 and 0.5 binder content was 2.4–4.5% and 6.9–7.1% lower than comparable control samples (i.e., C-0.2 and C-0.5), respectively. Conversely, the porosity of limestone- and sand-reinforced samples was greater than the porosity of samples with only sand.
- In mechanical characterization, the compressive strength of samples ranged from 431 to 780 kPa, 500 to 913 kPa, 699 to 1042 kPa, and 679 to 978 kPa for control, L, S, and LS samples, respectively. The average modulus of resilience of samples ranged from 12,275 J/m^3 (i.e., sample C-0.5) to 18,934 J/m^3 (i.e., sample L-0.2) and the average modulus of elasticity ranged from 4.87 MPa (i.e., sample C-0.5) to 18.03 MPa (i.e., sample LS-0.2). The use of a higher biopolymer content led to increased shrinking stresses and 0.5 binder content samples consistently illustrated lower mechanical properties.

Results herein illustrate that agar-based materials are more stiff than traditional asphalt binder materials, which may be problematic in regard to fatigue and low temperature stresses that accumulate in pavement materials. The rigid characteristics of agar-based materials indicate that these materials may perform adequately in regard to rutting, but they do not have the same ability as traditional petroleum-based asphalt binders to relax accumulated stresses.

Although this work highlights the benefits of agar-based material use (i.e., lower temperature requirements, advantageous rutting characteristics, strong biodeterioration resistance), portions of this work highlight inherent weaknesses associated with agar-based materials that need further research for its large-scale implementation. Namely, the moisture sorption properties of agar-based binders should be improved upon before implementation of agar-based binders. Possible solutions to address the high hydrophilicity and sensitivity to moisture of agar include physical (i.e., coatings, filler materials) and chemical (i.e., crosslinking) modification methods. These solutions, recommended for future research, could help expand the applicability of agar-based binders, which exhibited high penetration values aligned with applications in arctic climates.

Additional economic analyses are also recommended to better understand the broader implications of deploying agar-based binders. The literature includes large ranges of agar production costs that are mainly due to differences in the material quality of biopolymer. Utilizing the minimum unit cost found in the literature [77] (i.e., USD 2.21/kg) and the cost of purchase in this study (i.e., USD 397/kg), which is larger than the ones found in the literature, the cost of agar to prepare 5% w/w agar-based material could cost between USD 0.42 and USD 75.14 per gallon. These results do not consider the cost of the water source, long term maintenance, performance, and the environmental capital required to set up new forms of manufacturing/production. For reference, the cost of petroleum-based binder is approximately USD 2 per gallon [78]. This price estimate represents an exorbitant range and illustrates the need for future research on the economic costs associated with the deployment of this solution.

Application-specific results from this work indicate that agar-based materials are most appropriate for applications without significant moisture exposure. The moisture sensitivity, dimensional instability, and relative rigidity of agar-based materials may make it a less viable candidate for use as a surface pavement structure unless the hydrophilicity of agar is reduced through physical or chemical modification. The use of agar-based materials in the layers under a pavement surface course (i.e., base or subbase course) may be a more appropriate application as these layers are designed to be rigid and less exposed to surface moisture (i.e., they are protected by the surface course material).

Results from this work indicate that agar-based composites illustrate many properties in line with those of traditional engineering materials. Although the mechanical properties are lower than those of many typical engineering materials, results were in line with previous work on earthen material stabilization. Further, the porosity of agar-based composites was similar to that of pervious concrete or porous asphalt. This research is promising but highlights the need for additional work regarding optimization of binder and aggregate, the largescale environmental and economic impacts, the long-term impact of moisture on composite properties, and the evaluation of other durability mechanisms like UV.

Author Contributions: Conceptualization and methodology, all authors; data curation and formal analysis, M.R.F. and S.L.W.; resources, W.V.S.III and C.T.-M.; writing—original draft preparation and visualization, M.R.F.; writing—review and editing, S.L.W., W.V.S.III and C.T.-M.; supervision, project administration, and funding acquisition, W.V.S.III and C.T.-M. All authors have read and agreed to the published version of the manuscript.

Funding: This work was supported by the Colorado Department of Transportation under Grant No. R421.02; and the United States Department of Education Graduate Assistance in Areas of National Need Grant P200A180024. Publication of this article was funded by the University of Colorado Boulder Libraries Open Access Fund. Support for this work is gratefully acknowledged. This work represents the views of the authors and not necessarily those of the sponsors.

Institutional Review Board Statement: Not applicable.

Informed Consent Statement: Not applicable.

Data Availability Statement: Raw data were generated at the University of Colorado Boulder. Derived data supporting the findings of this study are available from the authors on request.

Conflicts of Interest: The authors declare no conflict of interest.

References

1. Asphalt Paving Association of Iowa. *Asphalt Paving Design Guide*; Asphalt Paving Association of Iowa: West Des Moines, IA, USA, 2020.
2. Penki, R.; Rout, S.K. Next-Generation Bitumen: A Review on Challenges and Recent Developments in Bio-Bitumen Preparation and Usage. *Biomass Conv. Bioref.* **2023**, *13*, 9583–9600. [CrossRef]
3. Fini, E.H.; Al-Qadi, I.L.; You, Z.; Zada, B.; Mills-Beale, J. Partial Replacement of Asphalt Binder with Bio-Binder: Characterisation and Modification. *Int. J. Pavement Eng.* **2012**, *13*, 515–522. [CrossRef]
4. Ahmed, R.B.; Hossain, K. Waste Cooking Oil as an Asphalt Rejuvenator: A State-of-the-Art Review. *Constr. Build. Mater.* **2020**, *230*, 116985. [CrossRef]
5. He, M.; Tu, C.; Cao, D.W.; Chen, Y.J. Comparative Analysis of Bio-Binder Properties Derived from Different Sources. *Int. J. Pavement Eng.* **2019**, *20*, 792–800. [CrossRef]
6. Armisen, R.; Galatas, F.; Spain, H.S.A. Agar. In *Handbook of Hydrocolloids*; Woodhead Publishing: Cambridge, UK, 2009.

7. Smitha, S.; Rangaswamy, K.; Keerthi, D.S. Triaxial Test Behaviour of Silty Sands Treated with Agar Biopolymer. *Int. J. Geotech. Eng.* **2021**, *15*, 484–495. [CrossRef]
8. Chang, I.; Prasidhi, A.K.; Im, J.; Cho, G.-C. Soil Strengthening Using Thermo-Gelation Biopolymers. *Constr. Build. Mater.* **2015**, *77*, 430–438. [CrossRef]
9. Khatami, H.R.; O'Kelly, B.C. Improving Mechanical Properties of Sand Using Biopolymers. *J. Geotech. Geoenviron. Eng.* **2013**, *139*, 1402–1406. [CrossRef]
10. Fatehi, H.; Ong, D.E.L.; Yu, J.; Chang, I. Biopolymers as Green Binders for Soil Improvement in Geotechnical Applications: A Review. *Geosciences* **2021**, *11*, 291. [CrossRef]
11. Verma, A.K.; Prasad, A.; Bonal, N.S. Investigation of the Long-Term Shear Strength Behavior of Municipal Solid Waste Fines Stabilized with Biopolymer: An Experimental Study. *J. Environ. Chem. Eng.* **2023**, *11*, 109805. [CrossRef]
12. Kantesaria, N.; Chandra, P.; Sachan, A. Stabilization of Expansive Soil Using Agar Biopolymer. In Proceedings of the IFCEE 2021: From Traditional to Emerging Geotechnics, Dallas, TX, USA, 10–14 May 2021; American Society of Civil Engineers: Reston, VA, USA, 2021; pp. 272–281.
13. Huang, J.; Kogbara, R.B.; Hariharan, N.; Masad, E.A.; Little, D.N. A State-of-the-Art Review of Polymers Used in Soil Stabilization. *Constr. Build. Mater.* **2021**, *305*, 124685. [CrossRef]
14. Losini, A.E.; Grillet, A.C.; Bellotto, M.; Woloszyn, M.; Dotelli, G. Natural Additives and Biopolymers for Raw Earth Construction Stabilization—A Review. *Constr. Build. Mater.* **2021**, *304*, 124507. [CrossRef]
15. *ASTM D75*; Standard Practice for Sampling Aggregates. American Society for Testing and Materials: West Conshohocken, PA, USA, 2019.
16. Smitha, S.; Sachan, A. Use of Agar Biopolymer to Improve the Shear Strength Behavior of Sabarmati Sand. *Int. J. Geotech. Eng.* **2016**, *10*, 387–400. [CrossRef]
17. Arham, R.; Salengke, S.; Metusalach, M.; Mulyati, M.T. Optimization of Agar and Glycerol Concentration in the Manufacture of Edible Film. *Int. Food Res. J.* **2018**, *25*, 1845–1851.
18. *ASTM D6373*; Standard Specification for Performance-Graded Asphalt Binder. American Society for Testing and Materials: West Conshohocken, PA, USA, 2021.
19. *ASTM D946*; Standard Specification for Penetration-Graded Asphalt Binder for Use in Pavement Construction. American Society for Testing and Materials: West Conshohocken, PA, USA, 2020.
20. *ASTM D4402*; Standard Test Method for Viscosity Determination of Asphalt at Elevated Temperatures Using a Rotational Viscometer. American Society for Testing and Materials: West Conshohocken, PA, USA, 2015.
21. Mao, B.; Bentaleb, A.; Louerat, F.; Divoux, T.; Snabre, P. Heat-Induced Aging of Agar Solutions: Impact on the Structural and Mechanical Properties of Agar Gels. *Food Hydrocoll.* **2017**, *64*, 59–69. [CrossRef]
22. Gudimettla, J.M.; Cooley, L.A.; Brown, E.R. Workability of Hot-Mix Asphalt. *Transp. Res. Rec.* **2004**, *1891*, 229–237. [CrossRef]
23. Dessouky, S.; Pothuganti, A.; Walubita, L.F.; Rand, D. Laboratory Evaluation of the Workability and Compactability of Asphaltic Materials Prior to Road Construction. *J. Mater. Civ. Eng.* **2013**, *25*, 810–818. [CrossRef]
24. *ASTM D7175*; Standard Test Method for Determining the Rheological Properties of Asphalt Binder Using a Dynamic Shear Rheometer. American Society for Testing and Materials: West Conshohocken, PA, USA, 2015.
25. Mallick, R.B.; El-Korchi, T. (Eds.) *Pavement Engineering: Principles and Practice*, 3rd ed.; CRC Press LLC: Boca Raton, FL, USA, 2018.
26. *ASTM D5*; Standard Test Method for Penetration of Bituminous Materials. American Society for Testing and Materials: West Conshohocken, PA, USA, 2020.
27. Williams, E.G. When to Use What Grade of Asphalt and Why. *Mater. Sci.* **1961**, 218–228. Available online: https://docs.lib.purdue.edu/cgi/viewcontent.cgi?article=3067&context=roadschool (accessed on 13 August 2025).
28. Billington, S.L.; Srubar, W.V.; Michel, A.T.; Miller, S.A. Renewable Biobased Composites for Civil Engineering Applications. *Sustain. Compos. Fibers Resins Appl.* **2014**, *2*, 313–348.
29. *ASTM G21*; Standard Practice for Determining Resistance of Synthetic Polymeric Materials to Fungi. American Society for Testing and Materials: West Conshohocken, PA, USA, 2021.
30. *ASTM D570*; Test Method for Water Absorption of Plastics. American Society for Testing and Materials: West Conshohocken, PA, USA, 2022.
31. Liu, J.; Wang, Y.; Wang, S.; Liu, Q.; Yu, B.; Wang, Q. Use of X-Ray Computed Tomography to Evaluate the Gradual Behaviour of Air Voids in Asphalt Mixtures during Permanent Deformation. *Int. J. Pavement Eng.* **2023**, *24*, 2134570. [CrossRef]
32. Miljković, M.; Griffa, M.; Münch, B.; Plamondon, M.; Lura, P. Mesostructural Evolution of Fine-Aggregate Bitumen Emulsion–Cement Composites by X-Ray Tomography. *Int. J. Pavement Eng.* **2024**, *25*, 2283610. [CrossRef]
33. *ASTM D7063*; Standard Test Method for Effective Porosity and Effective Air Voids of Compacted Bituminous Paving Mixture Samples. American Society for Testing and Materials: West Conshohocken, PA, USA, 2011.
34. *ASTM C642*; Standard Test Method for Density, Absorption, and Voids in Hardened Concrete. American Society for Testing and Materials: West Conshohocken, PA, USA, 2021.

35. du Plessis, A.; Boshoff, W.P. A Review of X-Ray Computed Tomography of Concrete and Asphalt Construction Materials. *Constr. Build. Mater.* **2019**, *199*, 637–651. [CrossRef]
36. Frazier, S.D.; Matar, M.G.; Osio-Norgaard, J.; Aday, A.N.; Delesky, E.A.; Srubar, W.V. Inhibiting Freeze-Thaw Damage in Cement Paste and Concrete by Mimicking Nature's Antifreeze. *Cell Rep. Phys. Sci.* **2020**, *1*, 100060. [CrossRef]
37. Koenig, A. Analysis of Air Voids in Cementitious Materials Using Micro X-Ray Computed Tomography (μXCT). *Constr. Build. Mater.* **2020**, *244*, 118313. [CrossRef]
38. Xu, X.; Xu, S.; Jin, L.; Song, E. Characteristic Analysis of Otsu Threshold and Its Applications. *Pattern Recognit. Lett.* **2011**, *32*, 956–961. [CrossRef]
39. ORS, Object Research System. *DragonFly Release Notes*; ORS: Montreal, QC, Canada, 2019; pp. 1–68.
40. Goodno, B.J.; Gere, J.M. *Mechanics of Materials*; Cengage Learning: Stamford, CT, USA, 2021.
41. Hossain, M.S.; Kim, W.S. *Estimation of Subgrade Resilient Modulus Using the Unconfined Compression Test*; Virginia Center for Transportation Innovation and Research: Charlottesville, VA, USA, 2014.
42. Hanz, A.J.; Bahia, H.U. Asphalt Binder Contribution to Mixture Workability and Application of Asphalt Lubricity Test to Estimate Compactability Temperatures for Warm-Mix Asphalt. *Transp. Res. Rec.* **2013**, *2371*, 87–95. [CrossRef]
43. Fernández, E.; López, D.; Mijangos, C.; Duskova-Smrckova, M.; Ilavsky, M.; Dusek, K. Rheological and Thermal Properties of Agarose Aqueous Solutions and Hydrogels. *J. Polym. Sci. Part B Polym. Phys.* **2008**, *46*, 322–328. [CrossRef]
44. Yu, Z.; Zhan, J.; Wang, H.; Zheng, H.; Xie, J.; Wang, X. Analysis of Influencing Factors on Viscosity of Agar Solution for Capsules. *J. Phys.Conf. Ser.* **2020**, *1653*, 012059. [CrossRef]
45. Vieira, M.G.A.; da Silva, M.A.; dos Santos, L.O.; Beppu, M.M. Natural-Based Plasticizers and Biopolymer Films: A Review. *Eur. Polym. J.* **2011**, *47*, 254–263. [CrossRef]
46. Yang, L.; Paulson, A.T. Mechanical and Water Vapour Barrier Properties of Edible Gellan Films. *Food Res. Int.* **2000**, *33*, 563–570. [CrossRef]
47. Kheradmand, B.; Muniandy, R.; Hua, L.T.; Yunus, R.B.; Solouki, A. An Overview of the Emerging Warm Mix Asphalt Technology. *Int. J. Pavement Eng.* **2014**, *15*, 79–94. [CrossRef]
48. Rubio, M.C.; Martínez, G.; Baena, L.; Moreno, F. Warm Mix Asphalt: An Overview. *J. Clean. Prod.* **2012**, *24*, 76–84. [CrossRef]
49. Norziah, M.H.; Foo, S.L.; Karim, A.A. Rheological Studies on Mixtures of Agar (*Gracilaria Changii*) and κ-Carrageenan. *Food Hydrocoll.* **2006**, *20*, 204–217. [CrossRef]
50. Somboon, N.; Karrila, T.T.; Kaewmanee, T.; Karrila, S.J. Properties of Gels from Mixed Agar and Fish Gelatin. *Int. Food Res. J.* **2014**, *21*, 485–492.
51. Labropoulos, K.C.; Rangarajan, S.; Niesz, D.E.; Danforth, S.C. Dynamic Rheology of Agar Gel Based Aqueous Binders. *J. Am. Ceram. Soc.* **2001**, *84*, 1217–1224. [CrossRef]
52. Ghebremedhin, M.; Seiffert, S.; Vilgis, T.A. Physics of Agarose Fluid Gels: Rheological Properties and Microstructure. *Curr. Res. Food Sci.* **2021**, *4*, 436–448. [CrossRef]
53. Zhang, L.; Xing, C.; Gao, F.; Li, T.; Tan, Y. Using DSR and MSCR Tests to Characterize High Temperature Performance of Different Rubber Modified Asphalt. *Constr. Build. Mater.* **2016**, *127*, 466–474. [CrossRef]
54. Gao, J.; Wang, H.; Liu, C.; Ge, D.; You, Z.; Yu, M. High-Temperature Rheological Behavior and Fatigue Performance of Lignin Modified Asphalt Binder. *Constr. Build. Mater.* **2020**, *230*, 117063. [CrossRef]
55. Naveed, H.; ur Rehman, Z.; Hassan Khan, A.; Qamar, S.; Akhtar, M.N. Effect of Mineral Fillers on the Performance, Rheological and Dynamic Viscosity Measurements of Asphalt Mastic. *Constr. Build. Mater.* **2019**, *222*, 390–399. [CrossRef]
56. Chen, Z.; Yi, J.; Chen, Z.; Feng, D. Properties of Asphalt Binder Modified by Corn Stalk Fiber. *Constr. Build. Mater.* **2019**, *212*, 225–235. [CrossRef]
57. Lv, S.; Xia, C.; Yang, Q.; Guo, S.; You, L.; Guo, Y.; Zheng, J. Improvements on High-Temperature Stability, Rheology, and Stiffness of Asphalt Binder Modified with Waste Crayfish Shell Powder. *J. Clean. Prod.* **2020**, *264*, 121745. [CrossRef]
58. Madera-Santana, T.J.; Robledo, D.; Azamar, J.A.; Ríos-Soberanis, C.R.; Freile-Pelegrín, Y. Preparation and Characterization of Low Density Polyethylene-Agar Biocomposites: Torque-Rheological, Mechanical, Thermal and Morphological Properties. *Polym. Eng. Sci.* **2010**, *50*, 585–591. [CrossRef]
59. Nishinari, K.; Fang, Y. Relation between Structure and Rheological/Thermal Properties of Agar. A Mini-Review on the Effect of Alkali Treatment and the Role of Agaropectin. *Food Struct.* **2017**, *13*, 24–34. [CrossRef]
60. Miandad, R.; Rehan, M.; Barakat, M.A.; Aburiazaiza, A.S.; Khan, H.; Ismail, I.M.I.; Dhavamani, J.; Gardy, J.; Hassanpour, A.; Nizami, A.-S. Catalytic Pyrolysis of Plastic Waste: Moving Toward Pyrolysis Based Biorefineries. *Front. Energy Res.* **2019**, *7*, 27. [CrossRef]
61. Bonati, A.; Merusi, F.; Polacco, G.; Filippi, S.; Giuliani, F. Ignitability and Thermal Stability of Asphalt Binders and Mastics for Flexible Pavements in Highway Tunnels. *Constr. Build. Mater.* **2012**, *37*, 660–668. [CrossRef]
62. Ragab, A.A.; Farag, R.K.; Kandil, U.F.; El-Shafie, M.; Saleh, A.M.M.; El-Kafrawy, A.F. Thermo-Mechanical Properties Improvement of Asphalt Binder by Using Methylmethacrylate/Ethylene Glycol Dimethacrylate. *Egypt. J. Pet.* **2016**, *25*, 397–407. [CrossRef]

63. Nissa, R.C.; Fikriyyah, A.K.; Abdullah, A.H.D.; Pudjiraharti, S. Preliminary Study of Biodegradability of Starch-Based Bioplastics Using ASTM G21-70, Dip-Hanging, and Soil Burial Test Methods. *IOP Conf. Ser. Earth Environ. Sci.* **2019**, *277*, 012007. [CrossRef]
64. Yemashova, N.A.; Murygina, V.P.; Zhukov, D.V.; Zakharyantz, A.A.; Gladchenko, M.A.; Appanna, V.; Kalyuzhnyi, S.V. Biodeterioration of Crude Oil and Oil Derived Products: A Review. *Rev. Environ. Sci. Biotechnol.* **2007**, *6*, 315–337. [CrossRef]
65. Obahiagbon, K.O.; Amenaghawon, A.N.; Onyia, C. Prevention of Biodeterioration of Crude Oil in Tanks Using Anti-Microbial Agents. *Int. J. Sci. Res. Environ. Sci.* **2014**, *2*, 56–62. [CrossRef]
66. Liu, Y.; Liu, X.; Duan, B.; Yu, Z.; Cheng, T.; Yu, L.; Liu, L.; Liu, K. Polymer–Water Interaction Enabled Intelligent Moisture Regulation in Hydrogels. *J. Phys. Chem. Lett.* **2021**, *12*, 2587–2592. [CrossRef]
67. Rani, G.U.; Konreddy, A.K.; Mishra, S. Novel Hybrid Biosorbents of Agar: Swelling Behaviour, Heavy Metal Ions and Dye Removal Efficacies. *Int. J. Biol. Macromol.* **2018**, *117*, 902–910. [CrossRef]
68. Kusmono; Abdurrahim, I. Water Sorption, Antimicrobial Activity, and Thermal and Mechanical Properties of Chitosan/Clay/Glycerol Nanocomposite Films. *Heliyon* **2019**, *5*, e02342. [CrossRef] [PubMed]
69. Dave, P.N.; Gor, A. Chapter 3—Natural Polysaccharide-Based Hydrogels and Nanomaterials: Recent Trends and Their Applications. In *Handbook of Nanomaterials for Industrial Applications*; Mustansar Hussain, C., Ed.; Micro and Nano Technologies; Elsevier: Amsterdam, The Netherlands, 2018; pp. 36–66. ISBN 978-0-12-813351-4.
70. Utech, S.; Boccaccini, A.R. A Review of Hydrogel-Based Composites for Biomedical Applications: Enhancement of Hydrogel Properties by Addition of Rigid Inorganic Fillers. *J. Mater. Sci.* **2016**, *51*, 271–310. [CrossRef]
71. Jones, A.; Zeller, M.A.; Sharma, S. Thermal, Mechanical, and Moisture Absorption Properties of Egg White Protein Bioplastics with Natural Rubber and Glycerol. *Prog. Biomater.* **2013**, *2*, 12. [CrossRef] [PubMed]
72. Wei, X.-F.; Linde, E.; Hedenqvist, M.S. Plasticiser Loss from Plastic or Rubber Products through Diffusion and Evaporation. *Npj Mater. Degrad.* **2019**, *3*, 18. [CrossRef]
73. Chen, J.-S.; Yang, C.H. Porous Asphalt Concrete: A Review of Design, Construction, Performance and Maintenance. *Int. J. Pavement Res. Technol.* **2020**, *13*, 601–612. [CrossRef]
74. Moretti, L.; Di Mascio, P.; Fusco, C. Porous Concrete for Pedestrian Pavements. *Water* **2019**, *11*, 2105. [CrossRef]
75. Hamza, M.; Nie, Z.; Aziz, M.; Ijaz, N.; Akram, O.; Fang, C.; Ghani, M.U.; Ijaz, Z.; Noshin, S.; Madni, M.F. Geotechnical Behavior of High-Plastic Clays Treated with Biopolymer: Macro–Micro-Study. *Environ. Earth Sci.* **2023**, *82*, 91. [CrossRef]
76. Cabalar, A.F.; Akbulut, N.; Demir, S.; Yildiz, O. Use of a Biopolymer for Road Pavement Subgrade. *Sustainability* **2023**, *15*, 8231. [CrossRef]
77. Jang, J. A Review of the Application of Biopolymers on Geotechnical Engineering and the Strengthening Mechanisms between Typical Biopolymers and Soils. *Adv. Mater. Sci. Eng.* **2020**, *2020*, 1465709. [CrossRef]
78. Fini, E.H.; Kalberer, E.W.; Shahbazi, A.; Basti, M.; You, Z.; Ozer, H.; Aurangzeb, Q. Chemical Characterization of Biobinder from Swine Manure: Sustainable Modifier for Asphalt Binder. *J. Mater. Civ. Eng.* **2011**, *23*, 1506–1513. [CrossRef]

infrastructures

MDPI

Article

Towards Sustainable Road Pavements: Sound Absorption in Rubber-Modified Asphalt Mixtures

Freddy Richard Apaza [1,*], Víctoriano Fernández Vázquez [2], Santiago Expósito Paje [2], Federico Gulisano [1], Valerio Gagliardi [3], Leticia Saiz Rodríguez [4] and Juan Gallego Medina [1]

1 Departamento de Ingeniería del Transporte, Territorio y Urbanismo, Universidad Politécnica de Madrid, C/Profesor Aranguren 3, 28040 Madrid, Spain; federico.gulisano@upm.es (F.G.); juan.gallego@upm.es (J.G.M.)
2 Laboratory of Acoustics Applied to Civil Engineering, University of Castilla—La Mancha, Avda. Camilo José Cela s/n, 13071 Ciudad Real, Spain; victoriano.fernandez@uclm.es (V.F.V.); santiago.exposito@uclm.es (S.E.P.)
3 Department of Civil, Computer Science and Aeronautical Technologies Engineering, Roma Tre University, Via Vito Volterra 62, 00146 Rome, Italy; valerio.gagliardi@uniroma3.it (V.G.)
4 Signus Ecovalor S.L. Calle Caleruega, 102, 5°, 28033 Madrid, Spain; lsaiz@signus.es (L.S.R.)
* Correspondence: fr.aapaza@upm.es

Abstract: In the last decade, various asphalt paving materials have undergone investigation for sound attenuation purposes. This research aims to delve into the innovative design of sustainable road pavements by examining sound absorption in rubber-modified asphalt mixtures. More specifically, the impact of alternative sustainable materials on the sound absorption of asphalt mixtures across different temperatures, precisely crumb rubber (CR) derived from recycling of end-of-life tires, was investigated. The acoustic coefficient and its Gaussian fit parameters (Peak, BandWidth, and Area Under the Curve) were evaluated. Five different types of asphalt mixtures were studied, encompassing dense, discontinuous, and open mixtures with 0%, 0.75%, and 1.50% CR incorporated through the dry process (DP). The results of sound absorption indicated a slight influence of crumb rubber at temperatures ranging from 10 °C to 60 °C, particularly in mixtures with high void content. On the other hand, as expected, the void content proved to be highly correlated with sound absorption. These findings facilitated the establishment of predictive models that correlate acoustic absorption spectra with the characteristics of asphalt mixtures. As a result, these models will be valuable in the design of the next generation of sound-absorbing pavements.

Keywords: asphalt pavements; crumb rubber; sound absorption; impedance tube; sustainable road pavements; acoustic absorption spectra; rubber-modified asphalt mixtures

1. Introduction

Noise pollution is one of the major environmental problems affecting populations around the world. In the last decade, the increase in traffic has led to rising levels of noise pollution [1–3]. This issue poses significant health problems for people living near roads.

To reduce traffic noise to acceptable thresholds, "sound-absorbing pavements" could become one of the most effective alternatives. The majority of transportation modes contribute to increasing noise pollution. This is particularly true for vehicle traffic on asphalt pavements which predominantly accounts for elevated noise levels in urban environments. The mechanisms that dominate noise generation and propagation are influenced by vibrations of moving vehicles [4]. Researchers have investigated the interaction between tires and pavement at speeds exceeding 40 km/h [5], demonstrating that it is the main source of noise as opposed to aerodynamic and mechanical noise.

More specifically, the physical properties of tires and pavements are factors that influence the generation of rolling noise. In recent decades, an extensive literature has been focused on tire characteristics such as tread shape, carcass, and stiffness patterns and their effects on noise generation [6–8]. On the other hand, the impact of the physico-mechanical properties of pavements on traffic noise reduction has been investigated, as well as the effects related to surface texture, maximum aggregate size, layer thickness, drainage capacity of the asphalt pavement [9], and other material properties, such as stiffness [10,11] and the shape of the aggregates [12].

In addition, environmental effects, including climate change and changes in air temperature, can significantly affect the stiffness of the asphalt mixture. This effect was measured by [10], demonstrating that with an increase in temperature, the stiffness of the asphalt mixture and its modulus decrease [10]. This phenomenon explains that temperature can significantly influence the rolling noise absorption of a pavement. In the literature, it has been reported that temperature changes can impact the noise generation in asphalt pavements [13,14]. However, the effect of temperature on the sound absorption of asphalt mixtures was unclear, and therefore the study of the sound absorption levels of specimens at various temperatures is fundamental.

The increase in temperature at the pavement surface can rapidly transfer heat to the tire tread, making the two stiffnesses similar to softening. Additionally, this aspect influences the reduction of noise generated by tire vibrations at frequencies between 1000 Hz and 2000 Hz [15]. Some authors have tried to show that the generated traffic noise and its propagation depend on the tire/pavement temperatures [16]. They recommend that noise measurements should not be made if the air temperature is below 15 °C or above 35 °C and propose a temperature correction of ±0.09 dB (A)/°C considering a reference air temperature of 20 °C [17–20] with noise being higher at 15 °C and lower at 35 °C. However, given this variability, the level of noise absorption at different temperatures is still unclear and further research is needed.

Other researchers have concentrated on the development of noise-reducing pavements utilizing crumb rubber (CR) obtained from end-of-life tires (ELTs) as an alternative and sustainable material, aiming to establish a quiet pavement variant using porous asphalt (PA). The addition of rubber in the form of aggregates in the porous–elastic pavement structure (PERS) achieves an attenuation of up to 8 dB (A) in the PA and 11 dB (A) in the PERS, according to the proximity methodology test (CPX) [17]. The attenuation achieved is relative to a dense pavement with a maximum aggregate size of 11 mm [21–23].

According to the literature, it is evident that the study of the effect of pavement temperature on noise pollution is of great interest to researchers [17,24,25]. This holds especially true for asphalt mixtures with rubber, as this material has been proposed as an alternative sustainable noise-attenuating additive in asphalt pavements.

Therefore, this research aims to evaluate the effect of recycled tire rubber at the end of its useful life as an alternative, sustainable, potential acoustic attenuator in asphalt mixtures, as well as the influence of temperature on this attenuation. For this purpose, the acoustic attenuation capacity has been evaluated by means of the absorption coefficient (α). In addition, a fitting curve of Gaussian treatment has been made for the sound absorption measurement. To this purpose, several experimental spectra were investigated. The proposed approach allows the establishment of models relating the characteristics of asphalt mixtures to their sound absorption. To achieve these objectives, several asphalt mixtures were produced with different aggregate gradations and in various percentages of crumb rubber (CR) (0.0, 0.75, and 1.50%) by the dry method according to the total weight of the asphalt mixture. The acoustic performance of these mixtures was evaluated by conducting an experimental laboratory activity by means of an impedance tube, which made it possible to evaluate the sound absorption in different temperature conditions (10 °C to 60 °C). Finally, predictive models of the sound absorption spectra of the materials studied have been proposed.

2. Methodology and Materials

The methodology of this research is based on the comparison of the sound absorption of different categories of asphalt mixtures with and without the addition of crumb rubber (CR) and the proposal of predictive models of this absorption through the characteristics of the asphalt mixture and the temperature at which the absorption is measured. The CR was added by dry process (DP) technology as a fraction of the fine aggregate during the manufacture of the asphalt mixture. The CR has been incorporated by the dry process in proportions of 0.0% (reference mixtures), 0.75%, and 1.50% by weight of the total asphalt mixture. The asphalt mixtures are asphalt concrete (AC), stone mastic asphalt (SMA), and béton bitumineux mince (BBTM).

An experimental investigation was carried out as reported in Figure 1. Firstly, the asphalt mixture samples were manufactured in the laboratory and a characterization of their volumetric properties (air void contents, maximum and apparent densities) was carried out. Furthermore, non-destructive sound absorption tests were carried out using an impedance tube. A thermostatic chamber was used for temperature control. Lastly, the obtained acoustic results were treated by means of Gaussian fitting (GF) for each sound absorption spectrum, evaluating several characteristics such as Peak (P) shape, Area Under the Curve (AUC), and BandWidth (BW). This procedure was performed for five categories of asphalt mixtures, proposed in this study. The results obtained have been statistically treated, establishing models correlating the characteristics of asphalt mixtures and sound absorption. According to these models, the air void content of the mixture is the most influential factor affecting the absorption Peak, while the BandWidth and the Area Under the Curve of sound absorption depend on both the voids in the mixture and the CR content.

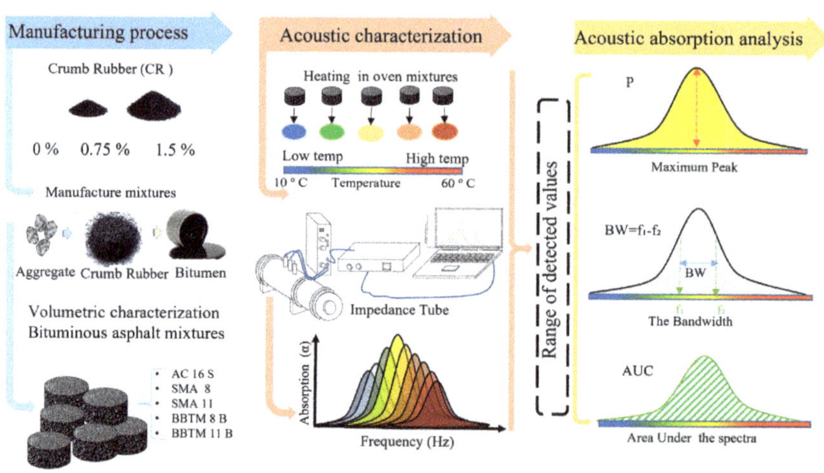

Figure 1. Methodology, including manufacturing process, evaluation, and analysis.

2.1. Sound Absorption

In order to evaluate the sound absorption and assess the contribution of CR as an attenuating material at different temperatures of asphalt mixture samples, a 4206 Bruel & Kjaer (B&K) impedance tube with frequency range of 100–1600 Hz was used. In this study, standard normal incidence sound absorption spectra were measured on asphalt mix samples with different crumb rubber contents and at different temperatures. Prior to the acoustic evaluation, the samples were laterally coated with a thin Teflon film (polytetrafluoroethylene—PTFE), as shown in Figure 2, to eliminate the clearance between the sample contour and the tube walls [26]. The EN ISO 10534-1 [27] and EN ISO 10534-2 [28] standards describe the impedance tube method and the transfer function technique (FFT). The signal emission equipment consists of the PULSE multianalyzer system type 3560 B-T06, and a B&K 2716C amplifier. The two 4187 microphones [29] are placed along the length of the tube, which allows the sound wave signals to be received. By means of the transfer functions (FFT), the sound absorption coefficient (α) is determined with a frequency resolution of 1 Hz from a total of 100 averages per sound absorption spectrum at the microphone positions [29].

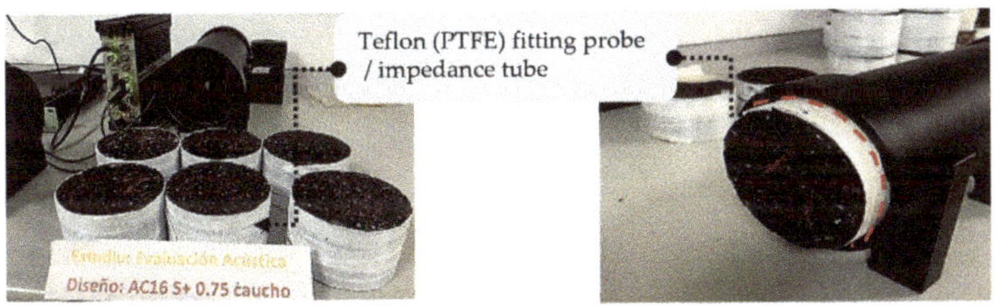

Figure 2. Impedance tube and specimens of asphalt mixtures with PTFE on their lateral surface.

2.2. *Sound Absorption Analysis by Gaussian Adjustment*

Using the Gaussian fit (GF), three parameters were determined from the acoustic response spectrum: Peak (P), BandWidth (BW), and Area Under the Curve (AUC). Based on the parameters and the response spectrum analysis techniques, it is possible to characterize the absorption by GF [30]. To explore the characteristics of the acoustic absorption spectrum, Gaussian fits have been successfully used in the analysis of the acoustic response of various materials [31,32]. All the fittings described in this subsection were carried out using the software Origin 8.0. The values of these parameters (P, BW, and AUC) are based on the general shape of the bell-shaped Gaussian function (Figure 3).

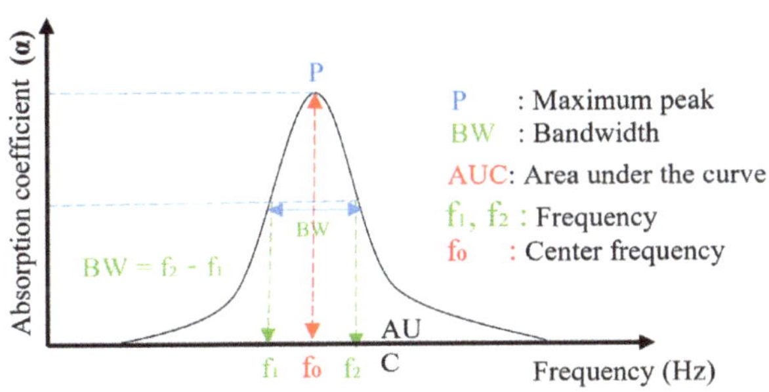

Figure 3. Geometric parameters for the characterization of the Gaussian fit.

In Figure 4, graph (a) shows a plot of the measured acoustic spectra for three types of asphalt mixtures (AC, SMA, and BBTM) and graph (b) shows the corresponding schemes of the parameters of the Gaussian fit: the fitted maximum Peak (P), the fitted BW, and the fitted AUC for each type of asphalt mixture. The values characterized by the parameters of the Gaussian fit were used to build correlation models with different volumetric characteristics of the asphalt mixtures, such as air void content, densities, permeability, and average depth of the surface texture. This analysis was established from consecutive and statistically representative measurements for the different pavements built with the asphalt mixtures studied.

Higher Gaussian adjustment parameters (P, BW, and AUC) are favorable for sound absorption. The P is related to the maximum sound absorption. The BW parameter could be related to sound absorption frequencies, as wider BandWidth allows a wider frequency range of sound absorption. The AUC parameter seems to most likely relate to the volumetric parameters, as its larger area would denote higher efficiency of the sound wave absorption process through the interconnected voids in the asphalt mixture.

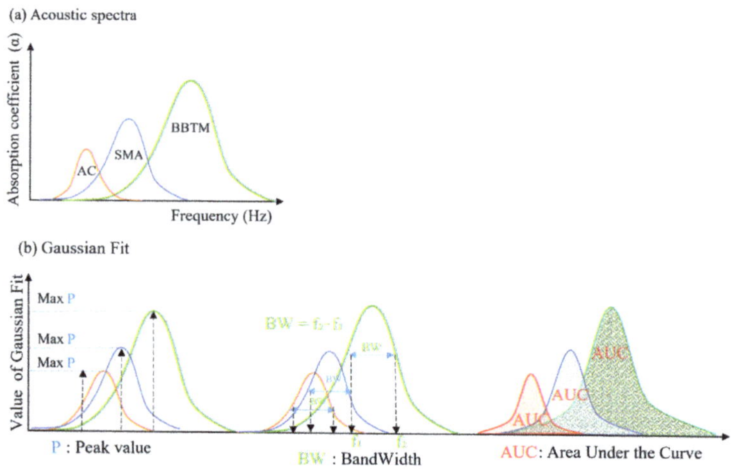

Figure 4. (**a**) Schematic of sound absorption spectra (ABS) for three types of asphalt mixtures, (**b**) Schematic of the Gaussian fits describing the ABS for three types of mixtures.

2.3. Determination of Volumetric and Permeability Characteristics

Volumetric characteristics are properties measured after the compaction procedure of asphalt mixtures. Density comprises two variants, (a) apparent density (G_{mb}) (g/cm^3) (EN 12697-6) [33] and (b) maximum density G_{mm} (g/cm^3) (EN-12697-5) [34]. The G_{mb} is the compacted density, including air voids obtained according to EN 12697-8 [33]. The G_{mm} is a maximum value of the density that the mixture would have if it did not have air voids, and it is obtained by the pycnometer method and is calculated according to EN 12697-5 [34]. Together, G_{mb} and G_{mm} allow the determination of the air void content V_a in (%) of a compacted asphalt mixture by the following expression:

$$V_a = \frac{G_{mm} - G_{mb}}{G_{mm}} * 100 \quad (1)$$

where: G_{mm} is the maximum density in g/cm^3, G_{mb} is the compacted apparent density in g/cm^3.

Macrotexture is measured by the mean texture depth (MTD) through the volumetric method of the sand patch test in mm. The procedure used for this work is based on EN 13036-1 [35] adapted to laboratory specimens. This method explains that the coarser the texture, the smaller the circle that can be covered by spreading the standard amount of sand. The calculation of the MTD value is expressed in Equation (2).

$$\text{MTD} = \frac{4\text{V}}{\pi \cdot \text{D}^2 avg} \quad (2)$$

where: V is the volume of the glass spheres in mL, $Davg$ is the mean diameter of the sand stain in mm.

The vertical permeability of asphalt mixture specimens was determined using the standard EN 12697-19 [36]. This method measures the permeability in the vertical direction by means of the time of drainage of a known volume of water through an asphalt sample. It determines the interconnection of internal voids in draining asphalt mixtures using Equations (3) and (4).

$$Q_v = \frac{(m_2 - m_1)}{t} \times 10^{-6} \quad (3)$$

where: Q_v is the flow rate through the test tube (m^3/s); m_1 is the initial mass of the water on the sample (g); m_2 is the final mass of water (g); t is the time of water drainage (s).

$$K_v = \frac{4 \cdot Qv \cdot L}{(h \cdot \pi \ D^2)} \tag{4}$$

where: K_v is the vertical permeability (m/s), Q_v is the vertical flow through the test sample (m^3/s), L is the thickness of the sample (m), h is the height of the water (m), and D is the diameter of the sample (m).

2.4. Statistical Analysis

A statistical analysis was performed to evaluate the effect of temperature, CR content, and volumetric properties of the asphalt mixtures (independent variables) on the Gaussian fitting parameters of sound absorption, i.e., P, BW, and AUC.

Prior to this, a correlation analysis was carried out to check for possible links between the independent variables, which could give rise to multicollinearity problems.

The Pearson correlation coefficient (r) has a range between [−1 and 1] [37]. The higher the absolute value of the coefficient r, the stronger the relationship between the two variables studied will be. Conversely, with values close to 0, the weaker the association between the two variables will be. The Pearson correlation coefficient is calculated using Equation (5).

$$r = \frac{\sum_{i=1}^{n} (x_i - \overline{x})(y_i - \overline{y})}{\sqrt{\sum_{i=1}^{n} (x_i - \overline{x})^2} \sqrt{\sum_{i=1}^{n} (y_i - \overline{y})^2}} \tag{5}$$

where: n is the sample size, x_i and y_i are the ith sample points, and $\overline{x}$ and $\overline{y}$ are the sample means.

The degree of correlation between volumetric characteristics, CR content, and sound absorption measurement temperature was established according to Pearson's correlation grading rules (r) [37] which are shown in Table 1. The sign of the coefficient indicates the direction of the relationship (positive or negative). A positive value (r) indicates that as one variable grows so does the other, while a negative value (r) means the opposite trend.

Table 1. Degree of correlation by Pearson's coefficient.

Classification Rules	Degree of Correlation
$r = 0$	No correlation
$0 < \lvert r \rvert \leq 0.19$	Very weak
$0.20 < \lvert r \rvert \leq 0.39$	Weak
$0.40 < \lvert r \rvert \leq 0.59$	Moderate
$0.60 < \lvert r \rvert \leq 0.79$	Strong
$0.80 < \lvert r \rvert \leq 1.00$	Very strong
$\lvert r \rvert = 1$	Monotonic correlation

On the other hand, the effects of volumetric characteristics, CR content, and temperature on the peak Gaussian fit parameters P, BW, and AUC obtained from the acoustic spectra have been evaluated by means of a multiple linear regression (MLR) analysis. Linear models have been estimated to predict the coefficients of the Gaussian fit P, BW, and AUC. The multiple linear regression model can be expressed as (Equation (6)):

$$y = \beta_0 + \beta_1 x_1 + \beta_2 x_2 + \cdots + \beta_n x_n \tag{6}$$

where: y is the dependent variable (in this case P, BW, or AUC), x_1 to x_n are the independent or predictor variables (in this case CR content, temperature, void content, average texture depth, permeability, and apparent density), β_0 is the y-intercept (constant term), and β_1 to β_n are the regression coefficients estimated to improve the model fit.

Air void content V_a, CR content (%), and temperature (°C) were considered as potential predictors of Peak (P), BandWidth (BW), or Area Under the Curve (AUC) models, and the ordinary least squares (OLS) regression technique was used to estimate the regression coefficients. It is noteworthy that the type of mixture was not considered as a potential predictor, since it is believed that the main differences among the mixtures can be captured by variations in air void contents. The model estimation was carried out using "Statmodels", a Python library built specifically for statistical computing. A backward stepwise regression approach was used, with a significance level of $\alpha = 0.01$. First, all potential predictor variables were included in the models excluding those that showed collinearity. Next, non-significant predictors were excluded from the models and the regression analysis was repeated. The procedure was stopped when the optimal model was identified after selecting the most appropriate and significant variables. Furthermore, the goodness of fit of the models was assessed by the coefficient of determination (R^2).

2.5. Aggregates

The aggregates used in this research are porphyry (of magmatic origin), generally used as a paving material for wearing courses in Spain. Due to the crushing process, the aggregate has an angular shape, which allows for a good mineral skeleton. The gradation of the aggregates is based on particle size curves according to articles 542, 543, and 544 of PG-3 [38]. The grading curves of the crushed aggregate of the five different mixtures (AC, SMA, and BBTM) are shown in Figure 5.

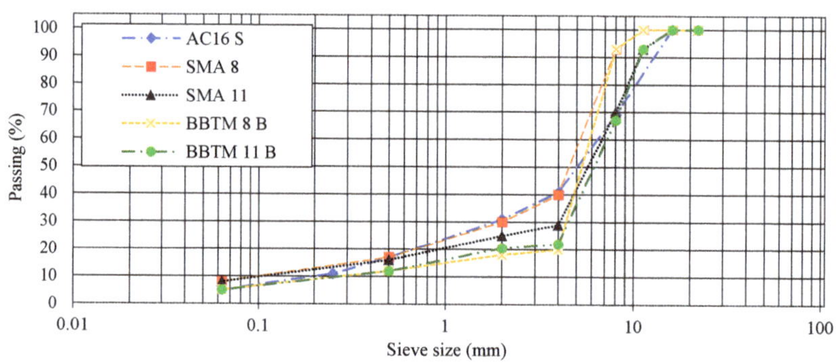

Figure 5. Aggregate gradation curves used in asphalt mixture design.

2.6. Crumb Rubber (CR)

Crumb rubber (CR) has been widely used in asphalt mixtures in several countries all over the world. For this research, a maximum particle size of 0.6 mm has been used. Table 2 presents the gradation curve of CR (EN 933-1, 2013). In this study, CR contents of 0% (reference), 0.75%, and 1.50% of the total weight of the asphalt mixtures were used.

Table 2. Gradation curve of Crumb Rubber (CR).

Sieve (mm) EN 933-1-13	2	1.5	1	0.5	0.25	0.125	0.063
Pass (%)	100	100	100	94.1	23.7	3.7	0.4

2.7. Asphalt Mixture Investigation: Laboratory Experimental Activity

In this research, five asphalt mixtures, and more specifically AC 16 S, SMA 8, SMA 11, BBTM and BBTM 11 B, were manufactured with three different CR contents: 0%, 0.75%, and 1.50%. All mixtures were initially defined on the basis of a previous research work that demonstrated adequate mechanical behavior. For each mixture, six compacted specimens were manufactured, according to the Marshall EN 12697-34 method [39]. A total of 90 specimens (5 mixtures for each of the 3 rubber contents for 6 specimens) were implemented and tested for the study.

Table 3 shows the manufacturing characteristics of the different asphalt mixtures studied in this research. The type of bitumen used is 50/70 asphalt binder according to EN 1426-07 [40]. This binder has been used because it has a good affinity with rubber, as widely demonstrated by previous works [26,41,42]. The samples were manufactured in the laboratory in a cylindrical shape of 60 mm in height and 99.5 mm in diameter to facilitate their accommodation in the impedance tube, which has an inner diameter of 100 mm.

Table 3. Manufacturing parameters.

Mixture	Max. Size (mm)	Type of Mixture	CR (%)	Binder Content (%)	MT (°C)	Digestion Temperature (°C)
AC 16 S	16	AC 16 S—REF AC 16 S—0.75% AC 16 S—1.50%	0.00 0.75 1.50	5.1 5.3 5.5	170	160
SMA 8	8	SMA 8—REF SMA 8—0.75% SMA 8—1.50%	0.00 0.75 1.50	6.0 6.1 6.2	170	160
SMA 11	11.2	SMA 11—REF SMA 11—0.75% SMA 11—1.50%	0.00 0.75 1.50	6.0 6.1 6.2	170	160
BBTM 8 B	8	BBTM 8 B—REF BBTM 8 B—0.75% BBTM 8 B—1.50%	0.00 0.75 1.50	5.0 5.1 5.2	170	160
BBTM 11 B	11.2	BBTM 11 B—REF BBTM 11 B—0.75% BBTM 11 B—1.50%	0.00 0.75 1.50	5.0 5.1 5.2	170	160

Note: CR: Crumb Rubber. MT: Manufacturing Temperature.

The increase in binder viscosity with the addition of CR is a well-known effect [43]. The increase in viscosity is caused by the integration of the rubber into the mixture once it comes into contact with the bitumen at high temperatures, resulting in an absorption of the lighter fractions of the bitumen (digestion process). It is also due to the fact that the (EN 12697-30, 2007) rubber increases in volume. This effect leads to a reduction of the distance between the particles and an increase in viscosity of the binder. For this reason, the manufacturing temperature of the mixtures with rubber has been set at 170 °C to facilitate the digestion process and the mixing of aggregates, binder, and CR.

Regarding the manufacture temperature (MT) of the test specimens, the aggregates were heated to 170 °C, and mixed with the crumb rubber for 30 s. The bitumen was then added at 170 °C and mixed with the aggregates and CR for approximately 30 s. The filler was then incorporated into the mixture and mixed for 120 s. The samples are then thermally conditioned to start the oven digestion temperature (DT) process at 160 °C, prior to the compaction process. The oven digestion time allows the asphalt mixtures to be kept at a constant temperature for 90 min so that the crumb rubber can better integrate with the bitumen and ensure proper digestion. Finally, all samples are compacted at 160 °C in the Marshall impact compactor in accordance with EN 12697-30 with application of 75 blows/surface for AC mixtures and 50 blows/surface for SMA and BBTM mixtures. Table 3 details the manufacturing parameters.

2.8. Thermal Conditioning for Absorption Measurement

For the evaluation of the influence of temperature on the sound absorption coefficients, the asphalt mixture samples were thermally conditioned, and their temperature was monitored using a FLIR C2 thermal imaging camera. The average surface temperature of the sample was obtained with the FLIR tools analysis software. All samples were kept at an initial temperature of 10 °C, then conditioned per sample group until the desired temperatures of 20, 30, 40, 50, and 60 °C were reached in the oven (Figure 6a). The thermal conditioning is prolonged by 8 h for each temperature step. The effects of different temperatures are controlled by ensuring that the specimens are thermally stabilized before starting the test and confirming that there are no substantial variations during the measurement. Figure 6b shows the temperature check of the specimens with a thermographic camera.

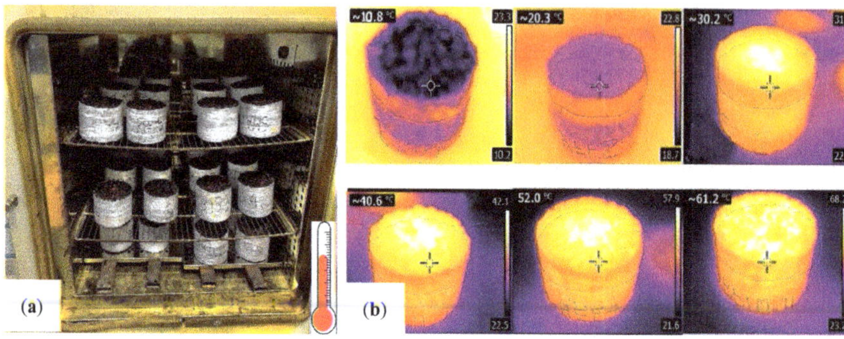

Figure 6. (**a**) Sample conditioning oven, (**b**) Thermographic imaging (10 °C to 60 °C).

3. Results and Discussion

3.1. Volumetric Characterization, Permeability, and Macrotexture

The volumetric parameters of the reference asphalt samples without CR and the mixtures with CR were determined in the laboratory. The maximum density (G_{mm}), apparent density (G_{mb}), air void content (V_a), hydraulic permeability (K_v), and mean texture depth (MTD) were obtained from the asphalt samples after compaction.

The volumetric characterization results for each type of asphalt mixture are shown in Table 4. From the results obtained, it can be seen that the reference mixtures (without CR) have a higher maximum density (G_{mm}) verified for all five types of asphalt mixtures, compared to the mixtures containing 0.75 and 1.50% CR. This is because the reference sample does not contain crumb rubber and its maximum density is related to the specific density of the aggregates and bitumen. However, the CR additions have a lower density than the aggregates. Therefore, the reduction in maximum density in the mixtures with CR is closely related to the presence of CR. This behavior of CR in asphalt mixtures was extensively studied by Gallego et al. [44].

The presence of CR slightly influences the air void content in AC asphalt mixtures. This could be due to its continuous particle size where the rubber can be better integrated into the mixture mass. However, rubber appears to increase voids in the SMA 8 and BBMT 8 B mixtures. This is probably due to the strong mineral skeleton of these two mixture types. It can be observed that the maximum aggregate size of 8 mm would be more affected by rubber particles in the compaction stage. The percentage increase in the voids with the addition of 1.50% CR was determined as follows: in AC16 it increases by 1.5% V_a, in SMA 8 by 38.4% V_a, SMA 11 by 1.6% V_a, in BBTM 8 by 10.40% V_a, and BBTM 11 by 0.79% V_a. The asphalt mixture exhibiting a significant increase in air void content is SMA 8 (38.4%). This is attributed to the expansion of the CR during the dry process of digestion. The lowest effect of CR is seen in BBTM 11 (0.79%). The presence of CR mainly impacted semidense mixtures with dimensions of the aggregate of 8 mm, while in the larger size of the aggregate, there was no significant effect of CR. This is due to internal porosity of open mixtures which allows the CR to swell and cover the internal pores without affecting it significantly. The presence of rubber in the SMA 11 and BBTM 11 B mixtures does not seem to generate as strong increases as in the SMA 8 and BBTM 8 B mixtures. This could be due to the size of the aggregate and the mineral skeleton it forms, which is smaller in the case of the 8 mixtures and more vulnerable to distortion by the presence of CR particles.

Table 4. Manufacturing detail, volumetric characteristic, permeability, and texture.

Mixtures	CR (%)	MT (°C)	DT (°C)	G_{mb} (g/cm^3)	G_{mm} (g/cm^3)	V_a (%)	K_v (m/s)	MTD (mm)
AC 16 S	0.00	160	170	2.406	2.540	5.27	*-	0.985
	0.75	160	170	2.358	2.487	5.24	*-	0.955
	1.50	160	170	2.295	2.424	5.35	*-	0.846
SMA 8	0.00	160	170	2.385	2.507	4.86	4.41×10^{-4}	0.990
	0.75	160	170	2.342	2.483	5.67	7.53×10^{-5}	0.984
	1.50	160	170	2.296	2.461	6.73	7.18×10^{-4}	0.932
SMA 11	0.00	160	170	2.349	2.517	6.66	1.97×10^{-4}	1.607
	0.75	160	170	2.316	2.483	6.73	9.26×10^{-4}	1.320
	1.50	160	170	2.252	2.442	7.78	6.34×10^{-4}	1.221
BBTM 8 B	0.00	160	170	2.210	2.650	16.61	2.98×10^{-4}	2.975
	0.75	160	170	2.136	2.591	17.56	1.98×10^{-4}	2.573
	1.50	160	170	2.117	2.592	18.34	1.93×10^{-4}	1.934
BBTM 11 B	0.00	160	170	2.148	2.643	18.72	2.79×10^{-4}	3.054
	0.75	160	170	2.140	2.634	18.76	3.05×10^{-4}	2.873
	1.50	160	170	2.135	2.632	18.87	3.03×10^{-4}	2.711

Note: CR: Crumb rubber, MT: Manufacturing temperature, DT: Digestion temperature, G_{mb}: Apparent density, G_{mm}: Maximum density, V_a: Air voids, K_v: Permeability, MTD: Mean texture depth, *-: Impermeable samples.

3.2. *Sound Absorption Coefficient: Influence of Mixture Type, Temperature, and CR Content*

In order to evaluate the influence of temperature on the sound absorption coefficient, several measurements were made at the following temperatures: 10 °C, 20 °C, 30 °C, 40 °C, 50 °C, and 60 °C (±2 °C). The sound absorption evaluations were quantified through the spectrum of the sound absorption coefficient (α) in the samples previously conditioned at different temperatures. Figure 7 shows the sound absorption spectra for normal incidence at 20 °C for all AC, SMA, and BBTM asphalt mixtures with 0%, 0.75%, and 1.50% CR, at frequencies between 100 Hz and 1600 Hz. As shown in Figure 7, there is a strong difference in sound absorption at 20 °C between the different asphalt mixtures, with the BBTM mixtures having the highest sound absorption coefficient.

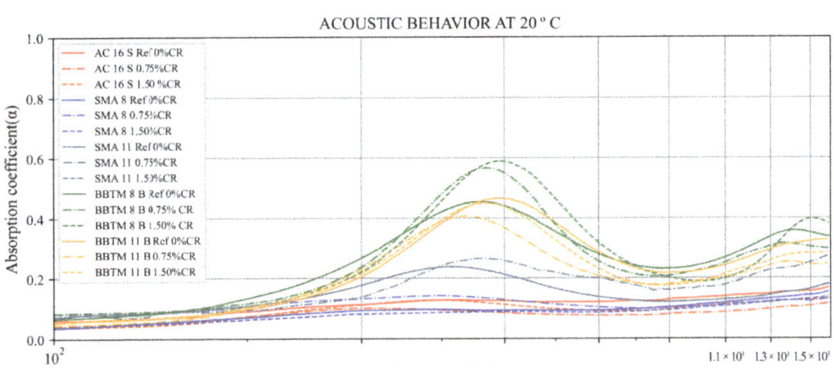

Figure 7. Sound absorption coefficient of the asphalt mixtures studied at 20 °C.

The behavior of the different sound absorption peaks might indicate that some asphalt samples show better sound absorption within a specific low-frequency range from 400 Hz to 700 Hz. Experimental studies from different investigations on asphalt mixtures showed similar sound absorption behavior at different frequencies [45,46], confirming that the peaks of maximum sound absorption are associated with the low-frequency range.

In addition, as shown in Figure 8, as the temperature increases, the levels of the maximum sound absorption coefficient also increase in the discontinuous BBTM mixtures. However, the changes in the maximum absorption coefficient in the dense AC and SMA mixtures due to the increasing effect of temperature are not evident.

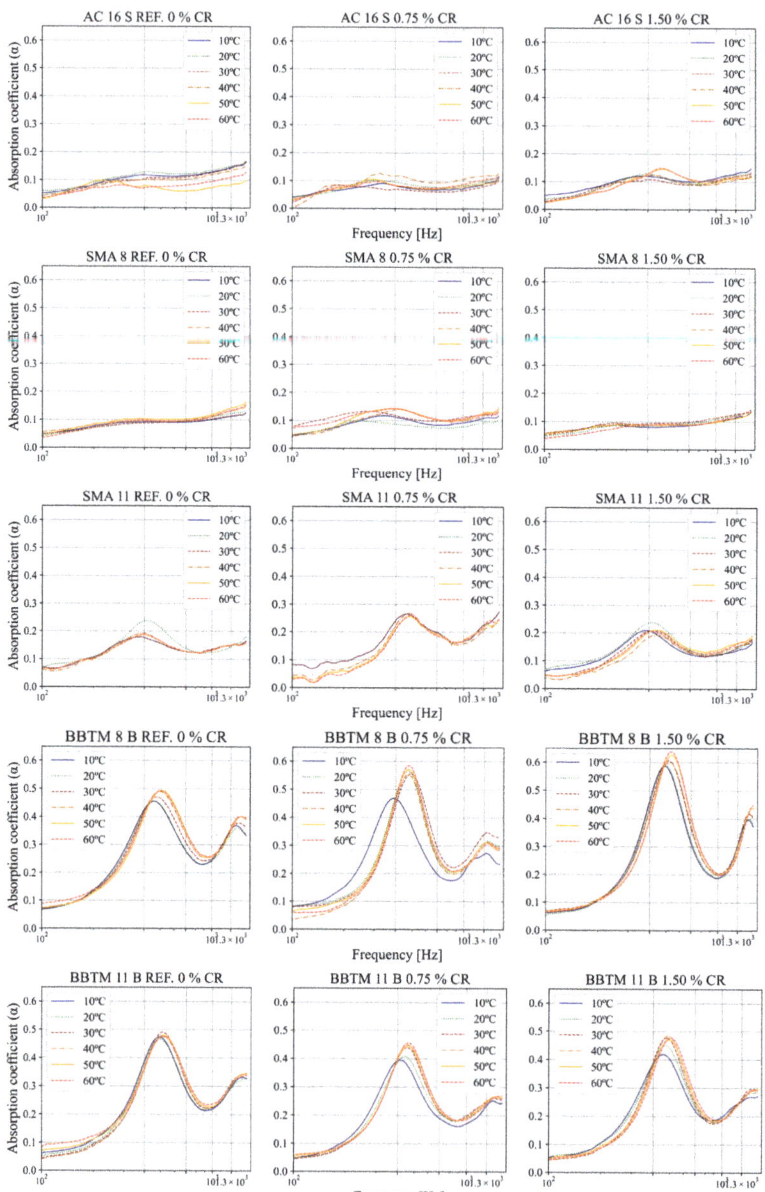

Figure 8. Sound absorption coefficient of the asphalt mixtures studied at different temperatures.

On the other hand, it is interesting to mention the behavior of the maximum sound absorption recorded in the BBTM mixtures, which show a positive behavior of the spectra and peaks of maximum sound absorption with slight shifts at higher frequencies. However, it is necessary to conclude that for practical purposes the measurement temperature has little influence, except for the BBTM mixture, where at 60 °C the sound absorption seems to be approximately 10% higher than at 10 °C. As for the CR content, it seems to have little influence on the sound absorption. The statistical significance of these variables will be analyzed in the following section.

3.3. Gaussian Goodness of Fit

Gaussian goodness of fit was assessed using the R^2 statistic (coefficient of determination) to check how well the normal distribution fits the observed data (sound absorption spectra). Figure 9 presents an example of this procedure applied for the BBTM 11 B with 1.50% CR mixtures, showing excellent goodness of fit results.

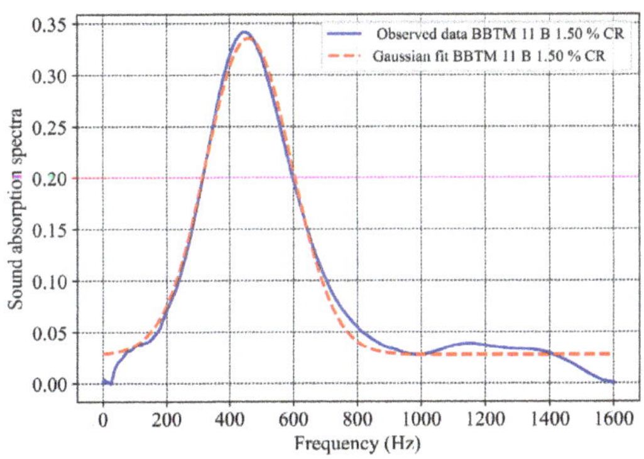

Figure 9. Gaussian fitting applicated for sound absorption spectra.

Table 5 illustrates the goodness of fit performance of all the mixtures under study, with values of R^2 always greater than 0.92.

Table 5. Result for the goodness of Gaussian fitting.

Mixtures	CR (%)	No. Observations	R^2	Adjusted R^2	Result
AC 16 S	0.00	801	0.923	0.923	Accepted
	0.75	801	0.966	0.966	Accepted
	1.50	801	0.943	0.943	Accepted
SMA 8	0.00	801	0.986	0.984	Accepted
	0.75	801	0.922	0.922	Accepted
	1.50	801	0.984	0.985	Accepted
SMA 11	0.00	801	0.924	0.924	Accepted
	0.75	801	0.941	0.940	Accepted
	1.50	801	0.963	0.962	Accepted
BBTM 8 B	0.00	801	0.969	0.968	Accepted
	0.75	801	0.958	0.958	Accepted
	1.50	801	0.967	0.966	Accepted
BBTM 11 B	0.00	801	0.993	0.931	Accepted
	0.75	801	0.989	0.989	Accepted
	1.50	801	0.981	0.980	Accepted

3.4. Regression Models of Asphalt Mixture Characteristics as a Function of Gaussian Parameters

The obtained results presented in the previous section have been analyzed through a statistical analysis. The first step consists in a correlation analysis, which is carried out to investigate and assess potential correlations between the independent variables to prevent possible multicollinearity problems. To this purpose, the correlation between all potential predictor variables has been examined in order to determine the relations and to discard those that are correlated, thus avoiding multicollinearity problems in the predictive models of this research. Figure 10 shows the correlation matrix in which the relations between volumetric characteristics, temperature, and CR content, according to the Pearson coefficient, are shown.

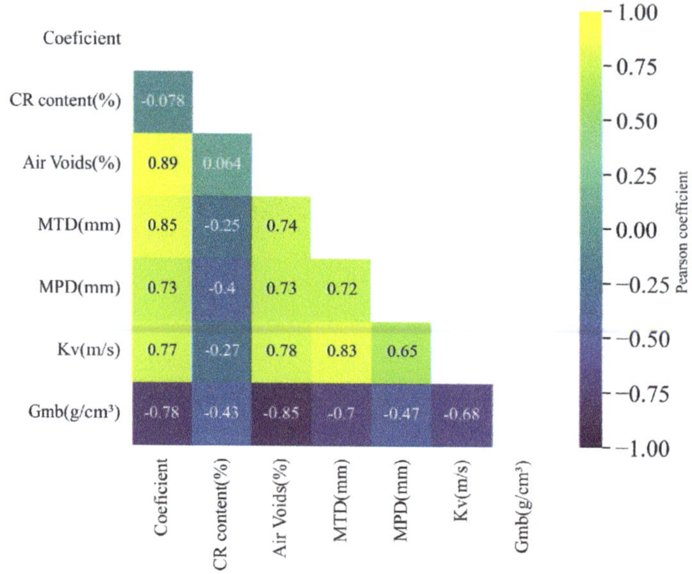

Figure 10. Pearson correlation analysis of temperature, rubber, and intrinsic parameters of asphalt mixtures.

The volumetric characteristics of the asphalt mixtures (AC16, SMA 8, SMA 11, BBTM 8 b, and BBTM 11 B) taken into consideration are: air void content (V_a), hydraulic permeability (K_v), mean texture depth (MTD), and the apparent density of the mixtures (G_{mb}). The maximum density G_{mm} has not been considered, as it depends on the void content (V_a) and the apparent density (G_{mb}) by Equation (1). The volumetric characteristics of the mixtures have been found to be related to each other. However, the rubber content is only moderately related to G_{mb}.

A high degree of correlation between the volumetric properties of the mixtures has thus been found (Figure 10). More specifically, strong correlations have been found between the V_a and the MTD, the K_v, and the apparent density G_{mb}. It is believed that the phenomenon of sound absorption is mainly due to the interconnected air voids within the mixture structure. For this reason, it has been decided to keep the air void variable V_a in the predictive models and to remove MTD, K_v, and apparent density G_{mb}, in order to avoid collinearity and thus facilitate the interpretation of the predictive models presented below.

Furthermore, it has been observed that CR content and temperature did not show any strong correlation with the other variables, therefore they have been included in the predictive models for P, BW, and AUC.

The regression models of the Gaussian parameters, i.e., Peak (P), BandWidth (BW), and Area Under the Curve (AUC), are presented as a function of the air void content (V_a), the rubber content (% CR), and the temperature (°C) at which the measurement is made. It should be noted that the other volumetric properties (maximum density (G_{mm}), apparent density (G_{mb}), hydraulic permeability (K_v), and mean texture depth (MTD)) were not considered as potential predictors, as they exhibited a strong correlation with void content as observed in Figure 10.

The results obtained were pre-treated for all the asphalt mixtures in Table 5 with the Gaussian curve fitting detailed in Section 3.3. Several attempts were made for each model, including the potential predictors (void content (V_a), absorption (α) measurement temperature, and CR content), Furthermore, the process was iterated to obtain optimized models with statistically significant predictors. Tables 6–8 show the degrees of significance of all predictor variables used for the prediction models. In the Peak (P) model it has been observed that the CR content and temperature variables do not show high significance, while, for the BandWidth (BW) and Area Under the Curve (AUC) models, only temperature does not show significance.

Table 6. Peak (P) model.

	Coefficients		Statistics		95% Confidence Interval for β	
	β	Standard Error	t	p-Value	Lower Bound	Upper Bound
Intercept	−0.0301	0.028	1.055	0.292	−0.026	0.086
Air voids (%)	0.0708	0.002	44.611	<0.001	0.068	0.074
CR content (%)	−0.0061	0.015	−0.396	0.693	−0.036	0.024
Temperature (°C)	0.0005	0.001	0.919	0.358	−0.001	0.002
No. observations	540					
R^2	0.788					
Adjusted R^2	0.787					

R^2: Correlation coefficient, β: Estimated correlations coefficient, t: *t*-Student, *p*-value: Probability value.

Table 7. BandWidth (BW) model.

	Coefficients		Statistics		95% Confidence Interval for β	
	β	Standard Error	t	p-Value	Lower Bound	Upper Bound
Intercept	416.770	11.341	36.748	<0.001	394.495	439.053
Air voids (%)	−10.650	0.811	−13.126	<0.001	−12.242	−9.055
CR content (%)	79.450	7.864	10.103	<0.001	64.004	94.900
Temperature (°C)	−0.352		−1.180	0.239	−0.941	0.235
No. observations	540					
R^2	0.327					
Adjusted R^2	0.323					

R^2: Correlation coefficient, β: Estimated correlations coefficient, t: *t*-Student, *p*-value: Probability value.

Table 8. Area Under the Curve (AUC) model.

	Coefficients		Statistics		95% Confidence Interval for β	
	β	Standard Error	t	p-Value	Lower Bound	Upper Bound
Intercept	18.504	2.931	6.314	<0.001	12.747	24.261
Air voids (%)	8.157	0.210	38.907	<0.001	7.745	8.569
CR content (%)	−5.921	2.032	−2.914	0.004	−9.913	−1.929
Temperature (°C)	0.1338	0.077	1.734	0.084	−0.018	0.285
No. observations	540					
R^2	0.740					
Adjusted R^2	0.738					

R^2: Correlation coefficient, β: Estimated correlations coefficient, t: *t*-Student, *p*-value: Probability value.

Once the significant predictor variables have been determined (see Section 2.4), the results of the optimized models for Peak (P), BandWidth (BW), and Area Under the Curve (AUC) based on the 540 data set (5 mixtures × 3 CR contents × 6 measurement temperatures × 6 specimens) were obtained and are shown in Tables 9–11, respectively.

Table 9. Statistical parameters of the Peak (P) model.

	Coefficients		Statistics		95% Confidence Interval for β	
	β	Standard Error	t	*p*-Value	Lower Bound	Upper Bound
Intercept	0.0424	0.020	2.164	0.031	−0.004	0.081
Air voids (%)	0.0708	0.002	44.720	<0.001	0.068	0.074
No. observations	540					
R^2	0.788					
Adjusted R^2	0.788					

R^2: Correlation coefficient, β: Estimated correlations coefficient, t: *t*-Student, *p*-value: Probability value.

Table 10. Statistical parameters of the BandWidth (BW) model.

	Coefficients		Statistics		95% Confidence Interval for β	
	β	Standard Error	t	*p*-Value	Lower Bound	Upper Bound
Intercept	416.77	11.341	36.748	<0.001	394.495	439.053
Air voids (%)	−10.65	0.811	−13.126	<0.001	−12.242	−9.055
CR content (%)	79.45	7.864	10.103	<0.001	64.004	94.900
No. observations	540					
R^2	0.325					
Adjusted R^2	0.322					

R^2: Correlation coefficient, β: Estimated correlations coefficient, t: *t*-Student, *p*-value: Probability value.

Table 11. Statistical parameters of the Area Under the Curve (AUC) model.

	Coefficients		Statistics		95% Confidence Interval for β	
	β	Standard Error	t	*p*-Value	Lower Bound	Upper Bound
Intercept	18.504	2.931	6.314	<0.001	12.747	24.261
Air voids (%)	8.157	0.210	38.907	<0.001	7.745	8.569
CR content (%)	−5.921	2.032	−2.914	0.004	−9.913	−1.929
No. observations	540					
R^2	0.738					
Adjusted R^2	0.737					

R^2: Correlation coefficient, β: Estimated correlations coefficient, t: *t*-Student, *p*-value: Probability value.

In the Peak (P) model in Table 6, the variables temperature and rubber content (CR) were found to be non-significant and were therefore excluded from the model construction (Table 9). The model has a coefficient of determination R^2 = 0.788. The only significant variable was air void content (V_a), and the coefficient is positive, meaning that an increase in air voids in the asphalt mixtures leads to a higher Peak (P) sound absorption. More specifically, each one unit change in the air void content of the mixtures corresponds to an increase in Peak (P) by a factor of 0.0708.

The fitted line and the relationship between air void content (V_a) and Peak (P) are shown in Figure 11, and the model can be expressed as Equation (7):

$$P = 0.0424 + 0.0708 \cdot (V_a) \quad (7)$$

It should be noted that the model obtained is valid in the range of air void content considered in this study (4% < V_a < 19%).

In the BandWidth (BW) regression model, two significant variables were identified, V_a and CR content (%). On the other hand, the temperature of the mixtures was found to be non-significant as seen in Table 6 and was therefore excluded from the model in Table 10. The coefficient of determination R^2 of the model was quite low, equal to 0.325, which means that the model can only explain 32.5% of the BW variability. The sign of the air void content (V_a) was negative (β = −10.648), which means that an increase in air voids produces a decrease in the BandWidth (BW). In other words, a narrower BW concentrates more sound energy absorption in the frequencies that are most easily attenuated in the network of tunnels formed by the interconnected voids.

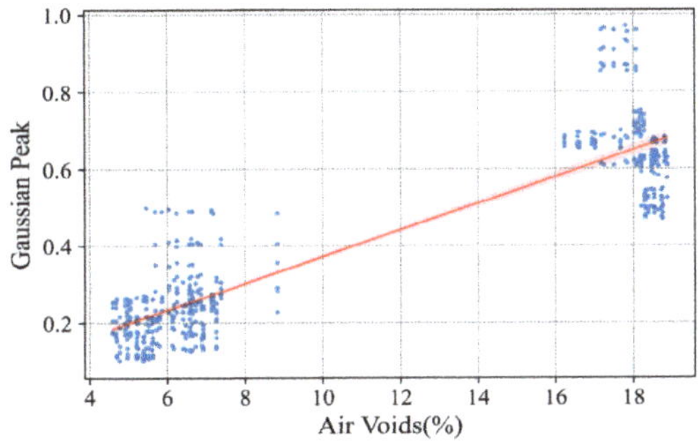

Figure 11. Linear correlation analysis of void content (V_a) and Peak (P).

On the other hand, the rubber content (CR) showed a positive coefficient, which means that the addition of CR tends to increase the BW. That is, it widens the range of frequencies of sound waves that are absorbed. Since the coefficient is high (β = 79.452) the effect is noticeable despite the small range of variation in crumb rubber content.

Figure 12 shows a 3D visualization of the multiple regression model for a better interpretation of the relationship between air voids, CR content, and BW. This model is within the limits of the void contents studied in this project ($V_a > 5\%$ and $V_a < 19\%$). This relationship can be expressed as:

$$BW = 416.77 - 10.648 \cdot V_a\ (\%) + 79.452 \cdot CR\ content\ (\%) \tag{8}$$

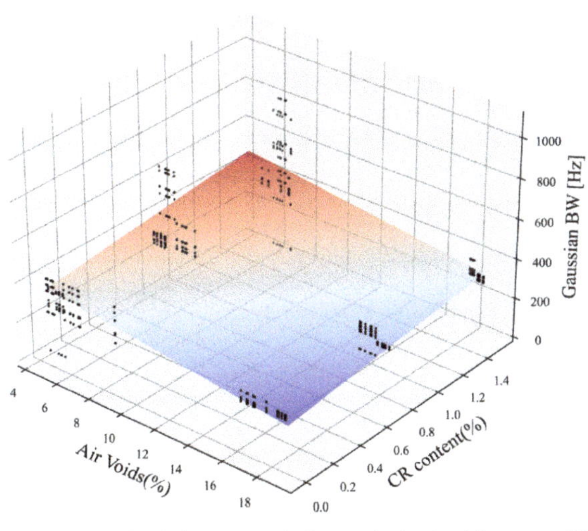

Figure 12. Multiple linear correlation analysis of void content, CR content, and BandWidth (BW).

Finally, in the Area Under the Curve (AUC) model, two significant variables, V_a and CR content, were found. Again, temperature appears not to affect the acoustic response of the asphalt mixture (Table 8), is not significant, and is excluded in Table 11 of the optimized model. The coefficient of determination R^2 of the model was equal to 0.74, representing a strong relationship between the predictors and the dependent variable AUC. The coefficient of the air void content (V_a) was positive, meaning that an increase in air voids in the mixture results in an increase in the AUC. Each one-unit change corresponds to an increase in AUC by a factor of 8.157, when the other predictor variable (CR content) is held constant [47–52].

On the other hand, CR content exhibited a negative coefficient, meaning that its effect is to reduce AUC.

Figure 13 shows the combined effect of air void content V_a and CR content on the AUC. The model can be expressed as:

$$AUC = 18.504 + 8.157 \cdot V_a(\%) - 5.921 \cdot CR\ content(\%) \quad (9)$$

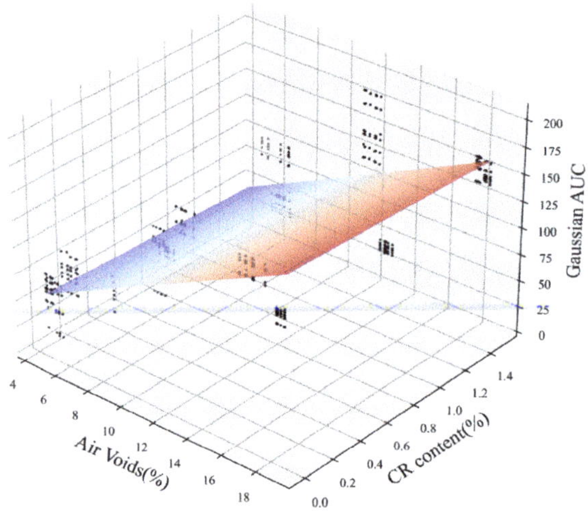

Figure 13. Multiple linear correlation between void content, CR content, and Area Under the Curve (AUC).

The results revealed that the Peak (P) of the Gaussian curve associated with the sound absorption is highly correlated and depends on the air voids (V_a) of the asphalt mixture. In other words, the results demonstrate that an increase in the porosity significantly improves the acoustic absorption performance. This finding aligns with the results obtained by other researchers [51]. More specifically, ref. [52] identified a linear relationship between the air voids of the mixture and the sound absorption coefficient with $R^2 = 0.94$.

Nevertheless, the Peak (P) does not stand as the unique variable delineating the acoustic response of asphalt mixtures. In addition, it was found that the CR had no significant effect on the Peak (P) sound absorption of asphalt mixtures. Experimental laboratory observations reveal that both air voids (V_a) and crumb rubber (CR) content exert discernible influences on the Area Under the Curve (AUC). The AUC serves as a comprehensive metric characterizing the acoustic response of asphalt mixtures across a spectrum of frequencies. Despite the proclivity of V_a to positively correlate with AUC, indicating an augmentation in sound absorption properties, it is discernible that the introduction of CR marginally attenuates AUC. Corroborating this, Shatanawi et al. [47] have asserted that CR does not directly elicit an enhancement in sound absorption. Furthermore, the findings manifest that CR particles induce alterations in the volumetric properties of the mixtures, thereby influencing sound absorption characteristics.

The BandWidth (BW) provides information about the frequency-dependent variability of sound absorption properties. Mixtures without CR exhibit a tendency to concentrate absorption characteristics predominantly around the Peak, whereas with CR mixtures, the acoustic response manifests heightened stability surrounding the Peak. Furthermore, it is noteworthy that V_a exhibits a reducing effect on the BW. Nonetheless, it is imperative to underscore the relatively low coefficient of determination for the model ($R^2 = 0.325$), indicative of its limited reliability. Subsequent investigations are warranted to advance the comprehension of sound absorption properties inherent in sustainable road materials integrated into asphalt mixtures, with a specific focus on rubber. This entails the conduction of additional laboratory experiments and the exploration of materials aimed at amplifying the sustainability quotient of road construction materials.

4. Conclusions

The current research entails the assessment of sound absorption implementing an impedance acoustic tube for asphalt mixtures, both with and without crumb rubber (CR), across varied temperatures. The study encompasses dense asphalt concrete (AC), discontinuous stone matrix asphalt (SMA), and open-graded béton bitumineux mince (BBTM) formulations, with CR content of 0%, 0.75%, and 1.50% incorporated through the dry process (DP).

In-depth analyses of sound absorption spectra were conducted at diverse measurement temperatures ranging from 10 °C to 60 °C. Gaussian curve fitting was employed to extract parameters such as Peak (P), BandWidth (BW), and Area Under the Curve (AUC). The primary objective of this research is to explore potential correlations between the intrinsic characteristics of the asphalt mixtures and the Gaussian fit parameters derived from sound absorption results.

The establishment of prediction models for Gaussian absorption parameters, grounded in the distinctive features of the mixtures, allows the following conclusions:

- The Peak (P) absorption improves with the air void content (V_a) in asphalt mixtures. However, the rubber content (CR) and the temperature at which the measurement is made do not seem to influence the Peak parameter. The model fit is $R^2 = 0.788$.
- The air void content (V_a) tends to reduce the BandWidth (BW), while the crumb rubber content increases it. However, the model has limitations in terms of low goodness of fit, with $R^2 = 0.325$.
- Void content positively affects the Area Under the Curve (AUC). On the other hand, rubber content slightly reduces the AUC. This model has a good fit with a coefficient of $R^2 = 0.738$.

Based upon the finding of this research and in the context of practical applications, it can be asserted that:

- The main factor affecting the sound absorption is clearly the void content of the asphalt mixture. The maximum sound absorption coefficient was obtained for asphalt mixtures with high air void content, in this research the discontinuous BBTM open mixtures.
- Crumb rubber has limited influence on the sound absorption of AC and SMA mixtures. In open-mixture-type BBTM it seems to slightly improve the sound absorption.
- The temperature (10 °C to 60 °C) has a limited influence on the results.

The novelty of this research is the formulation of models describing the acoustic response of asphalt mixtures as a function of their volumetric characteristics and crumb rubber content. The results of this work constitute a useful starting point for the development and design of innovative sustainable sound-reducing pavements. However, there are other variables that need to be analyzed to improve the accuracy of the presented models and to gain a deeper understanding of this complex phenomenon. In this context, the effect of pavement aging, the nature of the aggregates, and the incorporation of different additives could be analyzed in future research. In addition, additional mixtures with a wider range of air voids should be studied. Future further analyses, including the realization of full-scale projects, would be crucial for the assessment of these technologies, which would allow for the mitigation of road noise pollution.

Author Contributions: Conceptualization, J.G.M., F.R.A. and V.F.V.; methodology, F.R.A., V.F.V. and J.G.M.; software, S.E.P.; validation, V.F.V., S.E.P. and V.G.; formal analysis, F.R.A., F.G. and V.G.; investigation, F.R.A.; resources, L.S.R.; data curation, V.F.V., F.G. and V.G.; writing—original draft preparation, F.R.A.; writing—review and editing, J.G.M., V.F.V., F.G. and V.G.; visualization, F.R.A., F.G. and V.G.; supervision, J.G.M.; project administration, S.E.P.; funding acquisition, L.S.R. All authors have read and agreed to the published version of the manuscript.

Funding: This work has been partially supported by the Ministry of Economy and Competitiveness in the project PID2020-118831RB-I00/AEI/10.13039/501100011033 in the framework of the National Plan for Scientific Research and by the Junta de Comunidades de Castilla-La Mancha (European Regional Development Funds—ERDF) in the project SBPLY/19/180501/000313. In addition, support and technical collaboration has been provided by Signus Ecovalor, a non-profit organization dedicated to the management of end-of-life tires in Spain.

Data Availability Statement: Data are contained within the article.

Conflicts of Interest: The authors declare no conflicts of interest.

References

1. European Environment Agency. *Environmental Noise in Europe—2020*; European Environment Agency: Copenhagen, Denmark, 2020; p. 104.
2. Klompmaker, J.O.; Hoek, G.; Bloemsma, L.D.; Wijga, A.H.; van den Brink, C.; Brunekreef, B.; Lebret, E.; Gehring, U.; Janssen, N.A.H. Associations of Combined Exposures to Surrounding Green, Air Pollution and Traffic Noise on Mental Health. *Environ. Int.* **2019**, *129*, 525–537. [CrossRef]
3. EEA. *EEA Report No 10/2014—Noise in Europe 2014*; EEA: Copenhagen, Denmark, 2014; ISBN 978-92-9213-505-8.
4. Sandberg, U. Tyre/Road Noise—Myths and Realities. In Proceedings of the 2001 International Congress and Exhibition on Noise Control Engineering, The Hague, The Netherlands, 27–30 August 2001; p. 22.
5. Greer, G. Stone Mastic Asphalt—A Review of Its Noise Reducing and Early Life Skid Resistance Properties. In Proceedings of the 1st Australasian Acoustical Societies' Conference 2006, ACOUSTICS 2006: Noise of Progress, Christchurch, New Zealand, 20–22 November 2006; pp. 319–323.
6. Li, T. A Review on Physical Mechanisms of Tire-Pavement Interaction Noise. *SAE Int. J. Veh. Dyn. Stab. NVH* **2019**, *3*, 87–112. [CrossRef]
7. Bennert, T.; Hanson, D.; Maher, A.; Vitillo, N. Influence of Pavement Surface Type on Tire/Pavement Generated Noise. *J. Test. Eval.* **2005**, *33*, 94–100. [CrossRef]
8. Ling, S.; Yu, F.; Sun, D.; Sun, G.; Xu, L. A Comprehensive Review of Tire-Pavement Noise: Generation Mechanism, Measurement Methods, and Quiet Asphalt Pavement. *J. Clean. Prod.* **2021**, *287*, 125056. [CrossRef]
9. Meiarashi, S.; Ishida, M.; Nakashiba, F.; Niimi, H.; Hasebe, M.; Nakatsuji, T. Improvement in the Effect of Drainage Asphalt Road Surface on Noise Reduction. *Appl. Acoust.* **1996**, *47*, 189–204. [CrossRef]
10. Liao, G.; Heitzman, M.; West, R.; Wang, S.; Ai, C. Temperature Effects on the Correlations between Tire-Pavement Noises and Pavement Surface Characteristics. In Proceedings of the New Frontiers in Road and Airport Engineering—Selected Papers from the 2015 International Symposium on Frontiers of Road and Airport Engineering, IFRAE, Shanghai, China, 26–28 October 2015; pp. 219–232. [CrossRef]
11. Vázquez, V.F.; Terán, F.; Paje, S.E. Dynamic Stiffness of Road Pavements: Construction Characteristics-Based Model and Influence on Tire/Road Noise. *Sci. Total Environ.* **2020**, *736*, 139597. [CrossRef]
12. Cong, L.; Swiertz, D.; Bahia, H. Mix Design Factors to Reduce Noise in Hot-Mix Asphalt. *Transp. Res. Rec.* **2013**, *2372*, 17–24. [CrossRef]
13. Bueno, M.; Luong, J.; Viñuela, U.; Terán, F.; Paje, S.E. Pavement Temperature Influence on Close Proximity Tire/Road Noise. *Appl. Acoust.* **2011**, *72*, 829–835. [CrossRef]
14. Bühlmann, E.; Van Blokland, G. Temperature Effects on Tyre/Road-Noise-A Review of Empirical Research. In Proceedings of the Forum Acousticum, Krakow, Poland, 7–12 September 2014.
15. Bendtsen, H.; Lu, Q.; Kohler, E. *Temperature Influence on Road Traffic Noise: California OBSI Measurement Study*; University of California Pavement Research Center: Davis, CA, USA, 2010; ISBN 9788792094476.
16. Haider, M.; Conter, M.; Wehr, R.; Sandberg, U.; Anfosso, F. Project ROSANNE: Rolling Resistance, Skid Resistance, and Noise Emission Measurement Standards for Road Surfaces. In Proceedings of the INTERNOISE 2014—43rd International Congress on Noise Control Engineering: Improving the World through Noise Control, Melbourne, Australia, 16–19 November 2014.
17. Ejsmont, J.; Sandberg, U.; Wieczko-Zurek, B.; Mioduszewski, P. Tyre/Road Noise Reduction by a Poroelastic Road Surface. In Proceedings of the INTERNOISE 2014—43rd International Congress on Noise Control Engineering: Improving the World through Noise Control, Melbourne, Austrilia, 16–19 November 2014; pp. 1–12.

18. Bühlmann, E.; Schlatter, F.; Sandberg, U. Temperature Influence on Tire/Road Noise Measurements: Recently Collected Data and Discussion of Various Issues Related to Standard Testing Procedures. In Proceedings of the INTER-NOISE 2021—2021 International Congress and Exposition of Noise Control Engineering, Washington, DC, USA, 1 August 2021. [CrossRef]
19. Del Pizzo, A.; Teti, L.; Moro, A.; Bianco, F.; Fredianelli, L.; Licitra, G. Influence of Texture on Tyre Road Noise Spectra in Rubberized Pavements. *Appl. Acoust.* **2020**, *159*, 107080. [CrossRef]
20. Mioduszewski, P.; Taryma, S.; Woźniak, R. Temperature Influence on Tyre / Road Noise on Poroelastic Road Surface Based on Laboratory Measurements. *Euronoise* **2018**, *2012*, 2711–2718.
21. Taryma, S.; Woźniak, R.; Ejsmont, J.; Mioduszewski, P.; Ronowski, G. Tire/Road Noise and Tire Rolling Resistance on the Prototype PERS Surface. *IOP Conf. Ser. Mater. Sci. Eng.* **2018**, *421*, 022035. [CrossRef]
22. Sandberg, B.U. The Global Experience in Using Low-Noise Road Surfaces: A Benchmark Report. Hong Kong Environmental Protection Department Project No. AN 06-004 "Reviewing the Trial of Low Noise Road Surface (LNRS) in Hong Kong and Benchmarking the Experience on the Use of LNRS in Overseas Countries"; Sweden. 2009. Available online: https://www.yumpu.com/en/document/view/11548752/the-global-experience-in-using-low-noise-road (accessed on 15 February 2024).
23. Vieira, T.; Sandberg, U.; Erlingsson, S. Negative Texture, Positive for the Environment: Effects of Horizontal Grinding of Asphalt Pavements. *Road Mater. Pavement Des.* **2019**, *22*, 1–22. [CrossRef]
24. Sandberg, U. *Road Traffic Noise the Influence of the Road Surface and Its Characterization*; Elsevier: Amsterdam, The Netherlands, 1987; Volume 21.
25. Anfosso-Lédée, F.; Pichaud, Y. Temperature Effect on Tyre-Road Noise. *Appl. Acoust.* **2007**, *68*, 1–16. [CrossRef]
26. Luong, J.; Bueno, M.; Vázquez, V.F.; Paje, S.E. Ultrathin Porous Pavement Made with High Viscosity Asphalt Rubber Binder: A Better Acoustic Absorption? *Appl. Acoust.* **2014**, *79*, 117–123. [CrossRef]
27. *EN ISO 10534-1*; Determination of Sound Absorption Coefficient and Acoustic Impedance in Impedance Tubes Part 1: Standing Wave Range Method. ISO: Geneva, Switzerland, 2002; pp. 32–34.
28. *EN 10534-2*; Determination of Sound Absorption Coefficient and Acoustic Impedance in Impedance Tubes Part 2: Transfer Function Method. ISO: Geneva, Switzerland, 2002.
29. Paje, S.E.; Bueno, M.; Terán, F.; Viñuela, U.; Luong, J. Assessment of Asphalt Concrete Acoustic Performance in Urban Streets. *J. Acoust. Soc. Am.* **2008**, *123*, 1439–1445. [CrossRef]
30. Lalanne, C. *Mechanical Vibration and Shock Measurements*, 2nd ed.; Bruel & Kjaer: Nærum, Denmark, 1984; Volume 4, pp. 1–436.
31. Iannace, G.; Ciaburro, G. Modelling Sound Absorption Properties for Recycled Polyethylene Terephthalate-Based Material Using Gaussian Regression. *Build. Acoust.* **2021**, *28*, 185–196. [CrossRef]
32. Puyana-Romero, V.; Iannace, G.; Cajas-Camacho, L.G.; Garzón-Pico, C.; Ciaburro, G. Acoustic Characterization and Modeling of Silicone-Bonded Cocoa Crop Waste Using a Model Based on the Gaussian Support Vector Machine. *Fibers* **2022**, *10*, 25. [CrossRef]
33. *EN 12697-6*; Test Methods for Hot Bituminous Mixtures Part 6: Determination of the Bulk Density of Bituminous Test Specimens by the Hydrostatic Method. Slovenian Institute for Standardization: Ljubljana, Slovenia, 2003.
34. *UNE-EN 12697-5:2003*; Métodos de Ensayo Para Mezclas Bituminosas En Caliente Parte 5: Determinación de La Densidad Máxima. Asociación Española de Normalización y Certificación: Madrid, Spain, 2007.
35. *EN 13036-1*; Road Surface Characteristics of Roads and Surfaces Test Methods Part 1: Measurement of the Depth of Pavement Surface Macrotexture by the Sand-Circle Method. Slovenian Institute for Standardization: Ljubljana, Slovenia, 2002.
36. *UNE-EN 12697-19*; Mezclas Bituminosas Metodo de Ensayo Para Mezclas Bituminosas En Caliente Parte19: Permeabilidad de Las Probetas. Swiss Standards: Winterthur, Switzerland, 2013.
37. Akoglu, H. User's Guide to Correlation Coefficients. *Turk. J. Emerg. Med.* **2018**, *18*, 91–93. [CrossRef]
38. PG-3. General Technical Specifications for Road and Bridge Construction (PG-3). 2019. Available online: https://www.transportes.gob.es/carreteras/normativa-tecnica/17-pliegos-de-prescripciones-tecnicas-generales (accessed on 15 February 2024).
39. *EN 12697-34*; Bituminous Mixtures December 2007 Test Methods for Hot Mix Asphalt Mixtures Part 34: Marshall Test. ISO: Geneva, Switzerland, 2007.
40. *EN 1426*; Bitumen and Bituminous Binders Needle Penetration Determination. ISO: Geneva, Switzerland, 2007.
41. Meiarashi, S.; Ishida, M.; Fujiwara, T.; Hasebe, M.; Nakatsuji, T. Noise Reduction Characteristics of Porous Elastic Road Surfaces. *Appl. Acoust.* **1996**, *47*, 239–250. [CrossRef]
42. Mavridou, S.; Kehagia, F. Environmental Noise Performance of Rubberized Asphalt Mixtures: Lamia's Case Study. *Procedia Environ. Sci.* **2017**, *38*, 804–811. [CrossRef]
43. Kim, H.H.; Lee, S.J. Effect of Crumb Rubber on Viscosity of Rubberized Asphalt Binders Containing Wax Additives. *Constr. Build. Mater.* **2015**, *95*, 65–73. [CrossRef]
44. Gallego Medina, J.; Tomas Raz, R.; Del Val Melus, J. Caracterizacion Mediante Ensayos Convencionales y Dinamicos de Mezclas Asfalticas Fabricadas Con Betun Modificado Con Caucho de Neumaticos. Available online: https://trid.trb.org/view/950625 (accessed on 28 October 2020).
45. Vaitkus, A.; Vorobjovas, V.; Jagniatinskis, A.; Andriejauskas, T.; Fiks, B. Peculiarity of Low Noise Pavement Design under Lithuanian Conditions. *Balt. J. Road Bridge Eng.* **2014**, *9*, 155–163. [CrossRef]
46. Irali, F.; Kivi, A.; Tighe, S.L.; Sangiorgi, C. Tire–Pavement Noise and Wearing Course Surface Characteristics of Experimental Canadian Road Pavement Sections. *Can. J. Civ. Eng.* **2015**, *42*, 818–825. [CrossRef]

47. Shatanawi, K. *The Effects of Crumb Rubber Particles on Highway Noise Reduction—A Laboratory Study*; Clemson University: Clemson, SC, USA, 2008.
48. Apaza, F.R.; Vázquez, V.F.; Paje, S.E.; Saiz, L.; Gulisano, F.; Gallego, J. The Potential Effect of Crumb Rubber on the Maximum Sound Absorption Performance of Asphalt Mixtures. *Constr. Build. Mater.* **2023**, *389*, 131789. [CrossRef]
49. Apaza, F.R.; FVázquez, V.; Gallego, J.; EPaje, S.; Perez, R. *Characterisation of the Sound Absorption of Bituminous Mixtures Modified with Rubber from End-of-Life Tires*; Jorge, S., Geroge, W., Juan, G., Eds.; RAR2022: Malaga, Spain, 29 June 2022.
50. Zhang, Y.; Li, H.; Abdelhady, A.; Yang, J. Effect of Different Factors on Sound Absorption Property of Porous Concrete. *Transp. Res. D Transp. Environ.* **2020**, *87*, 102532. [CrossRef]
51. Sachakamol, P.; Dai, L. Parametric Influence on Prediction of Sound Absorption Coefficients for Asphalt Pavements. *J. Environ. Inform.* **2011**, *18*, 1–11. [CrossRef]
52. John, T.; Kayhanian, M.; Harvey, J.T. *Optimizing Rubberising Open Graded Friction Coruse (RHMA-O)Mix Desings for Water Quality Benefits: Phase I: Literature Review*; University of California Pavement Research Center: Davis, CA, USA, 2020; p. 179.

infrastructures

MDPI

Article

Evaluation of Pigment-Modified Clear Binders and Asphalts: An Approach towards Sustainable, Heat Harvesting, and Non-Black Pavements

Gul Badin [1,*], Naveed Ahmad [2], Ying Huang [1] and Yasir Mahmood [1]

[1] Civil, Construction and Environmental Engineering Department, North Dakota State University, Fargo, ND 58102, USA; ying.huang@ndsu.edu (Y.H.); yasir.mahmood@ndsu.edu (Y.M.)

[2] Department of Civil Engineering, University of Engineering and Technology, Taxila 47050, Pakistan; n.ahmad@uettaxila.edu.pk

* Correspondence: gul.badin@ndsu.edu; Tel.: +1-701-491-0279

Abstract: Pavement construction practices have evolved due to increasing environmental impact and urban heat island (UHI) effects, as pavements, covering over 30% of urban areas, contribute to elevated air temperatures. This study introduces heat-reflective pavements, by replacing conventional black bitumen with a clear binder and pigment-modified clear binders. Titanium dioxide white, zinc ferrite yellow, and iron oxide red pigments are used to give asphalt corresponding shades. The asphalt and bitumen specimens were subjected to thermal analysis in heat sinks, under varying solar fluxes. The pigment dosage was maintained at 4%, according to the weight of the total mix, for all pigment types. The samples were heated and cooled for 3 h and 2 h, respectively. Mechanical testing was conducted to ascertain the impact of temperature variations on both the neat clear binder (C.B) and pigmented C.B and asphalt mixture samples. Wheel tracking and dynamic modulus tests were conducted to evaluate their performance under high temperatures. The results indicate that non-black asphalt mixtures exhibit significant temperature reductions, up to 9 °C, which are further enhanced by pigmented binders, up to 11 °C. It was found that asphalt with a clear or transparent binder demonstrated lower temperatures and faster heat dissipation in extreme conditions. Moreover, C.B asphalt mixtures displayed a rut resistance of 15%, with the pigmented C.B asphalt mixture showing a remarkable rut resistance of 73%, outperforming conventional asphalt. Non-black mixtures, especially C.B + zinc ferrite, showed improved resistance to permanent deformation in dynamic modulus tests.

Keywords: clear binder (C.B); UHI effect; pigmented binder; solar flux; non-black asphalt mixture; rut resistance

1. Introduction

Pavement construction practices, using advanced and innovative materials and techniques, have been on the rise in recent years. This is to lessen the impact of infrastructure on adjacent areas and to address environmental issues [1]. It is becoming increasingly common for natural ground surfaces to be covered by pavements; pavements typically account for over 30% of typical urban areas [2]. Compared to natural ground surfaces, the higher thermal inertia and lower evaporation rate of pavements contribute to the formation of urban heat islands. This phenomenon results in air temperatures in urban areas typically being higher than those of surrounding rural regions, with temperature differences ranging from 10 to 21 °C [3,4].

Asphaltenes are a class of crude oil compounds that are black in color, contributing to the dark appearance of traditional bitumen [5,6]. Conventional bitumen, which has a high tinting strength, is responsible for the black color of asphalt [7]. Studies have demonstrated that the dark color of asphalt is the leading cause of pavement high surface temperatures [8–11]. Furthermore, hot pavements significantly intensify the urban heat island effect [3]. The U.S. Environmental Protection Agency (U.S. EPA 2008) has defined low-temperature intensive or cool pavements as a range of established and emerging pavement-related solutions and technologies. These innovative pavement solutions have the potential to significantly lower pavement surface temperatures and, consequently, reduce the amount of heat released into the atmosphere when compared to conventional pavements [12]. Implementing these alternative pavement materials and technologies has the potential to mitigate the urban heat island effect and provide substantial environmental benefits.

The extent of UHI depends on various factors, such as time, city dimensions, weather conditions, urban layout, surface material composition, vegetation, and human activities [13–15]. In pavement engineering, it is important to highlight that the thermal properties of materials, including pavements, are markedly influenced by solar radiation, leading to absorbed, reflected, and stored energies [16]. Historically, numerous early studies have focused on employing thermally optimized solutions for paving, such as utilizing clear or colored materials and porous surfaces [17–19].

Several strategies for reducing the surface temperature of pavements have been developed to alleviate the urban heat island (UHI) effect. These strategies include heat-reflective pavements with surface coatings, evaporative pavements with porous surface layers, thermally modified pavements, and pavements with phase-change materials for heat storage [20–25]. As a result of the utilization of transparent bitumen by the asphalt industry, roads may be constructed in any color [26]. Producing colored asphalt mixtures for road pavements is essential to meet environmental, aesthetic, and functional requirements [27]. There are three different processes through which transparent bitumen can be produced, as follows:

- The modification of bitumen, by removing the asphaltene responsible for its dark color [6];
- Synthetic binders produced using naturally transparent and special polymer materials [28,29];
- Proper resins combined with bio-oils or organic, renewable raw materials from vegetable sources [27,28]. While these materials are not bituminous, they display rheological properties similar to bitumen, making them appropriate for the construction of roads [6].

The literature has also demonstrated that clear binders and asphalt made with clear binders perform similarly to conventional bitumen and the corresponding asphalt mixtures (HMA) [30,31]. In addition, a recent study shows that clear binders and corresponding asphalt can dissipate heat faster and absorb heat slower than conventional black bitumen [32]. Moreover, studies have also shown that traditional binders and asphalts, when mixed with metal oxide pigments, increased the thermal conductivity of the pavement, which resulted in a significant enhancement of the high-temperature performance of the wearing course [33]. Furthermore, researchers have investigated the feasibility of utilizing innovative mixtures of clear and colored materials on pavement surfaces to mitigate the UHI effect by increasing the albedo and reducing the surface temperature [34–36]. It has been shown that clear mixtures reduce the temperature significantly compared to conventional black surfaces. It has also been shown that the surface color influences the thermal response and that oxide-modified mixtures exhibit promising mechanical properties, suggesting that they may be utilized in residential areas with low traffic [37]. Hence, altering the color of conventional black binders to any non-black color, or solely using a colorless binder, could result in durable and cooler pavement structures.

Colored asphalt pavements primarily include light-colored pavements and pigmented pavements [27]. In the former approach, the aggregates are covered and bound using transparent bitumen instead of traditional black bitumen. In areas with a significant landscape or areas of historical and cultural significance, these types of transparent bitumen enhance the natural color of mineral aggregates. In the latter type, transparent bitumen is blended with pigments in asphalt mixtures or artificially colored aggregates are incorporated into the mix to create a specific aesthetic feature [27].

During hot and sunny seasons, colored pavements display higher reflectance characteristics than traditional asphalt pavements [37]. Hence, those surfaces are less likely to absorb solar radiation and remain cooler when exposed to sunlight. As a result, the air temperature in the surrounding areas decreases, because less heat is transferred from the pavement to the air [11,38]. These pavements are also highly reflective, reducing the likelihood of overheating during the summer, which increases pavement durability and reduces damage [39]. It has also been demonstrated that transparent bitumen can lessen illumination needs, thereby decreasing electricity costs [27]. As part of an effective traffic management program, colored asphalt pavements might also be used to identify bus lanes, sidewalks, crosswalks, and pedestrian areas [30].

2. Objectives and Experimental Approach

This study aims to evaluate the thermal characteristics and structural performance of various non-black asphalt binders and mixtures, compared to conventional black bitumen and asphalt. Specifically, we aim to evaluate the heat conductions, absorption, and the rate of heating and cooling of pigment-modified and unmodified clear binders (C.B.) and asphalt mixtures, in comparison to conventional black bitumen and asphalt. The structural performance will be assessed using wheel tracking and dynamic modulus tests. Temperature measurements will be taken at five different depths for all the modified and unmodified binders and asphalt to analyze their thermal characteristics. Solar flux densities will be used to calculate the corresponding power and voltage applied to a silicon plate heater. The temperature measurements are recorded using a highly sensitive data acquisition system and highly calibrated thermocouples.

3. Materials and Methods

3.1. Binders and Aggregates

For this study, a clear asphalt binder (provided in Figure 1a [1]) was imported from China, which comprises thermoplastic bicomponent resins, whose color ranged from straw yellow to amber orange [1]. While locally manufactured conventional bitumen (ARL 60/70 pen grade) was used, which is the most used bitumen grade across the country. However, in this study, it was only used for a comparison. Our recently published research provides the conventional properties of both binders [32]. The aggregates used in this research were obtained from the Margalla quarry source. The aggregates from this quarry are limestone (basic) in nature [40], as calcium carbonate is a substantial portion of this quarry. The thermal properties, like the thermal conductivity and specific heat of the various materials used in this research, are summarized in Table 1.

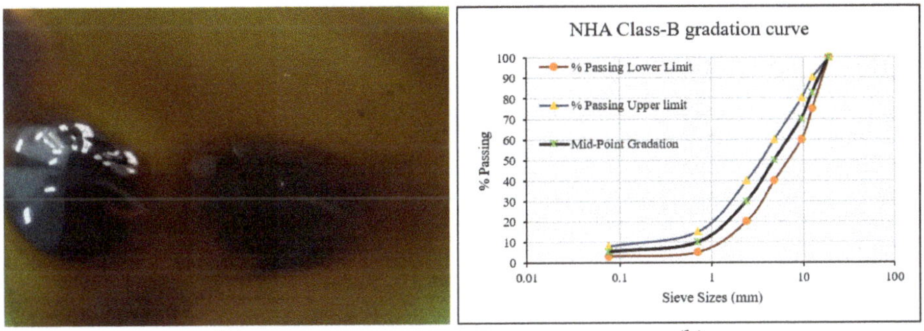

(a) (b)

Figure 1. (**a**) Thin layer of CB poured onto steel surface [1]; (**b**) mid-point gradation curve.

Table 1. Thermal properties of pigments and other materials.

Material	Specific Heat ($J\ kg^{-1}K^{-1}$)	Thermal Conductivity ($W\ m^{-1}K^{-1}$)
Iron oxide red (Fe_2O_3)	650 [41]	3 (800 °C) and 8 (200 °C) 15 at room temperature [42]
Titanium dioxide white (TiO_2)	683–697 [43]	11.7 at 25 °C [43]
Zinc ferrite yellow ($ZnFe_2O_4$)	800 [44]	3.5–4.3 at 25–85 °C [45] 1–5 at room temperature [46]
Conventional asphalt	900 [9]	1.35 at room temperature [47] 0.8–2.0 [9]
Conventional bitumen	1850–3900 at 27–127 °C [48]	0.17–0.2 [49]

3.2. Pigments

For the thermal analysis, titanium dioxide white and zinc ferrite yellow were used. However, iron oxide red was also used for mechanical testing, apart from titanium dioxide and zinc ferrite. All these pigments were imported from China. The distinct properties of all three types of pigments were extracted from their technical data sheets (TDSs) and are provided below in Table 2.

Table 2. Distinct properties of pigments extracted from their technical data sheets.

Property	Titanium Dioxide White	Zinc Ferrite Yellow	Iron Oxide Red
Heat stability (°C)	900–930	260–300	350–400
Particle size	300 nm	0.1–0.6 µm	97% ≤ 45 µm
Particle shape	Tetragonal	Acicular	Spherical
pH value	6.5–8.0	5–8	3–7
Density (g/cm^3)	4.13	5.0–5.6	0.72–1.1
Chemical formula	TiO_2	$ZnFe_2O_4$	$Fe_2O_3{\cdot}H_2O$
Tinting strength %	≥175	95–110	95–105
Moisture at 105 °C, %	0.5 Max	0.5 Max	1.0 Max
Oil absorption (mL/100 g)	19~22	25–45	15–25
Matter soluble in water %	0.5 Max	0.5 Max	0.5 Max

3.3. Material Mixing

The national highway authority's (NHA) Class-B mid-point gradation (provided in Figure 1b) was followed to prepare the asphalt mixtures, which is the finer gradation for the asphalt wearing course. The maximum particle size in this gradation is 12.5 mm (about 0.49 in). After using the Marshall mix design method, 4.4% of optimum binder content was found for both the conventional and clear binder. Previous studies recommend a pigment concentration of 3–5% for asphalt coloring purposes [50,51]; however, a mean value of 4% was adopted in this study. Hence, a ratio of 4.4:4 was used to mix the binders with the pigments to make the pigmented binder samples. This was achieved using a mechanical stirrer, equipped with a simple fan blade configuration.

Similarly, 4% pigment, according to the weight of the total aggregate mix, was used for the asphalt mixture preparation. It was also ensured that the job mix formula (JMF) was not disturbed, so the 4% pigment weight was subtracted from all the sieve size aggregates corresponding to their percentage in the total mix. The aggregates were heated at 130–140 °C, before mixing them with the pigments. The pigments were then mixed with the aggregates for 60 to 90 s. To prevent pigment lumps or accumulations, a minimum mixing time of 10 to 15 s was ensured [52].

4. Experimental Setup

This study is broadly classified into two phases of laboratory testing. The first is thermal testing, and the second one is mechanical testing.

Thermal Testing Setup

During the design and manufacturing process, two aluminum heat sinks were designed and manufactured; the larger heat sink was intended for asphalt mixture analysis, while the smaller heat sink was designed for binder analysis. The walls of each heat sink were 5 mm (about 0.2 in) thick. The heat sink walls were lined with fiber block insulation to ensure a one-dimensional heat flow. A 5 mm thick piece of acrylic glass was also mounted on the top of each heat sink and tightened with screws on each side. The base of each heat sink was provided with two slots, measuring 1.5 mm (about 0.06 in) in depth, to enable the temperature measurements. K-type thermocouples were positioned in these slots. To provide heat, silicon plate heaters from Omega Engineering USA [52] were installed at the base of the heat sinks, as the energy source. Aluminum tape was used to secure the heaters and to ensure that the entire assembly was stable and airtight. K-type thermocouples were placed at five different depths within the heat sinks to measure the temperature at various locations within the specimens. The Agilent 34972A data acquisition system was used to monitor and record the temperature at these thermocouples. A DC power supply (Keysight U8032A) was used to supply power to the silicon plate heaters. A cotton covering was applied to the extended portion of the fiber block insulation on the heat sinks to prevent unwanted heat loss. This measure was taken to ensure the accuracy and reliability of the temperature readings.

Under the Energy Sector Management Assistance Program (ESMAP), data was obtained from the World Bank via ENERGYDATA.info [53]. Three flux densities of 800, 1000, and 1200 W/m^2 were selected for analysis. After converting the flux density to power using Ohm's law, the base area of the heat sink was considered to compute the corresponding power. The conversion resulted in the respective voltage and current values for each flux. A DC power supply was used to supply the corresponding power values (V × I) to the heater and, then, to the asphalt binder/mixture sample.

The charge time was defined as three hours, during which the power was supplied. The temperature was automatically recorded every 5 s during this period. The power supply was disconnected after the 3 h charging period, and the temperature measurements were continuously recorded during the following two hours, called the discharging period. Analysis was conducted throughout the 5 h period, i.e., both the heating and cooling cycles. To ensure consistency, the heating and cooling of the samples was performed in identical conditions (room temperature). Table 3 shows the dimensions of the heat sinks, the dimensions of the silicon heaters, and the locations of the thermocouples inside the heat sinks.

Table 3. Heat sink and heater dimensions, along with thermocouple positioning.

Heat Sink \ Dimensions	Internal Dimensions (mm^3)	External Dimensions (mm^3)	Heater Dimensions (mm^3)	Thermocouple Positioning (mm)				
				T_1	T_2	T_3	T_4	T_5
Larger Heat Sink	100 × 100 × 50	110 × 110 × 55	100 × 100 × 1.5	Heat sink base	0	10	30	50
Smaller Heat Sink	60 × 60 × 25	70 × 70 × 30	60 × 60 × 1.5		0	5	15	25

The flow chart in Figure 2 illustrates the steps involved in preparing the binders and mixtures, installing the thermocouples, and applying the heat (power) to them. It is important to note that all of the pictures do not reflect the preparation of one sample type; however, the images are combined randomly to reflect the sequential steps. Furthermore, Figures 3 and 4 depict the schematics of the laboratory setup and 3D views of both heat sinks, respectively.

Figure 2. *Cont.*

(g) (h)

Figure 2. Steps involved in the preparation of binders and asphalt mixtures for thermal analysis. (**a**) TiO_2 being mixed with dry aggregates. (**b**) Binder poured into the TiO_2 blended aggregates. (**c**) Placement of a silicon heater at the base of the heat sink. (**d**) Heat sink inside the fiber block insulation, with installed thermocouples and heater. (**e**) White pigmented (TiO_2) clear binder poured into a container. (**f**) TiO_2-modified CB under experimentation. (**g**) Conventional asphalt inside the large heat sink, with installed thermocouples and heater. (**h**) White pigmented asphalt mixture before the start of the test.

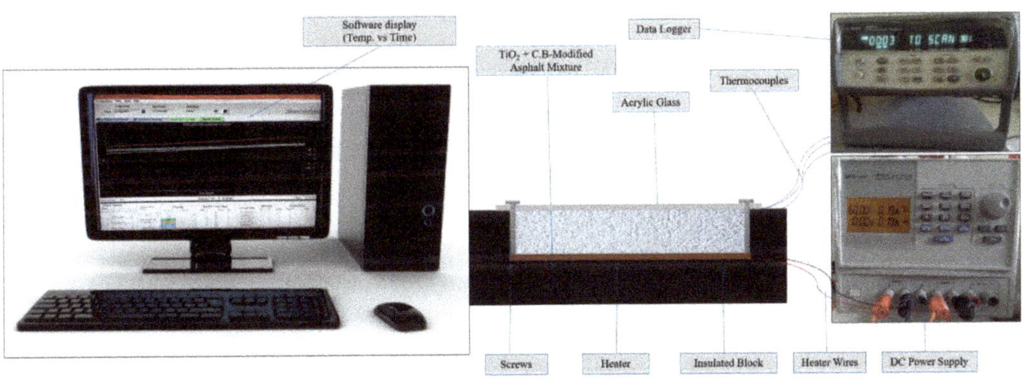

Figure 3. Schematics of laboratory setup.

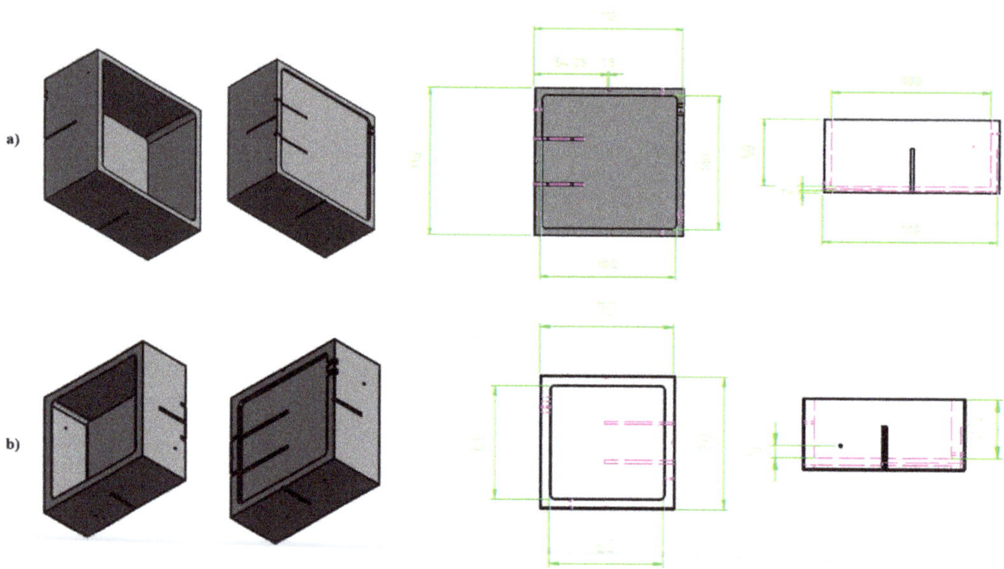

Figure 4. Three dimensional views, with labeled dimensions (mm), of (**a**) larger heat sink, and (**b**) smaller heat sink.

5. Results and Discussion

Eight combinations were prepared and evaluated for each of the three heat fluxes. The first four samples were asphalt mixture samples, while the remaining four were corresponding bitumen samples. It should be noted that all the tests were conducted at room temperature and that the cooling of all the samples was accomplished through natural convection.

5.1. Thermal Analysis of Asphalt Mixtures

Conventional asphalt, a clear binder (C.B), and the pigmented C.B-modified asphalt (white and yellow) specimens were meticulously prepared and placed into a large heat sink for thermal analysis. Each asphalt mixture was subjected to flux densities of 1200, 1000, and 800 W/m^2 for three hours (heating time). The specimens were subsequently allowed to cool for two hours (cooling time). Before applying heat, the samples were cooled to room temperature. During the 5 h temperature cycle, the heat storage and dissipation behavior of the pigmented modified and clear binder asphalt mixtures were compared to conventional black asphalt. The goal was to observe whether there was any enhancement in the heat dissipation and diminutions in the heat storage behavior between the conventional and modified asphalt mixtures.

Figure 5a–d shows the temperature curves for the five-hour test period of all four asphalt mixtures at 1200 W/m^2. T1-T5 represents the temperature at the five thermocouples from the heat sink's base to the top (50 mm), respectively. Also, the average curve "Avg" was drawn, showing the average temperature of the entire 50 mm sample. As the heat source is applied to the mixtures, the temperature increases w.r.t time. Figure 5a shows three significantly different temperature readings at the five thermocouples. The T4 and T5 readings almost overlap until half of the heating time has elapsed. Similarly, there is no significant difference in the temperature readings for T1 and T2 in the first 20 to 30 min. The overlapping region indicates similar heat absorption at that specific time. However, the temperature curve for thermocouple T3 significantly differs from the four neighboring thermocouples. It is also worth noting that, although the curves are parabolic, they are more inclined towards linear behavior. The maximum surface temperature of asphalt (T2) is close to 65 °C, which is almost the same as the actual pavement surface temperature during peak summer [24], and the temperatures closer to 70 °C are responsible for the UHI effect [54]. The higher temperature curves, seen in Figure 5a, also indicate lower thermal conductivity and higher specific heat of conventional asphalt, which traps the heat in the asphalt for extended periods, elevating the overall temperature of the structure. The average "Avg" curve represents the average temperature of the 50 mm thick asphalt mixture sample and, in the first case, it almost overlaps with the temperature curve for T3. Hence, we can conclude that the average temperature of the 50 mm thick conventional asphalt lies at a depth of 10 mm (about 0.39 in).

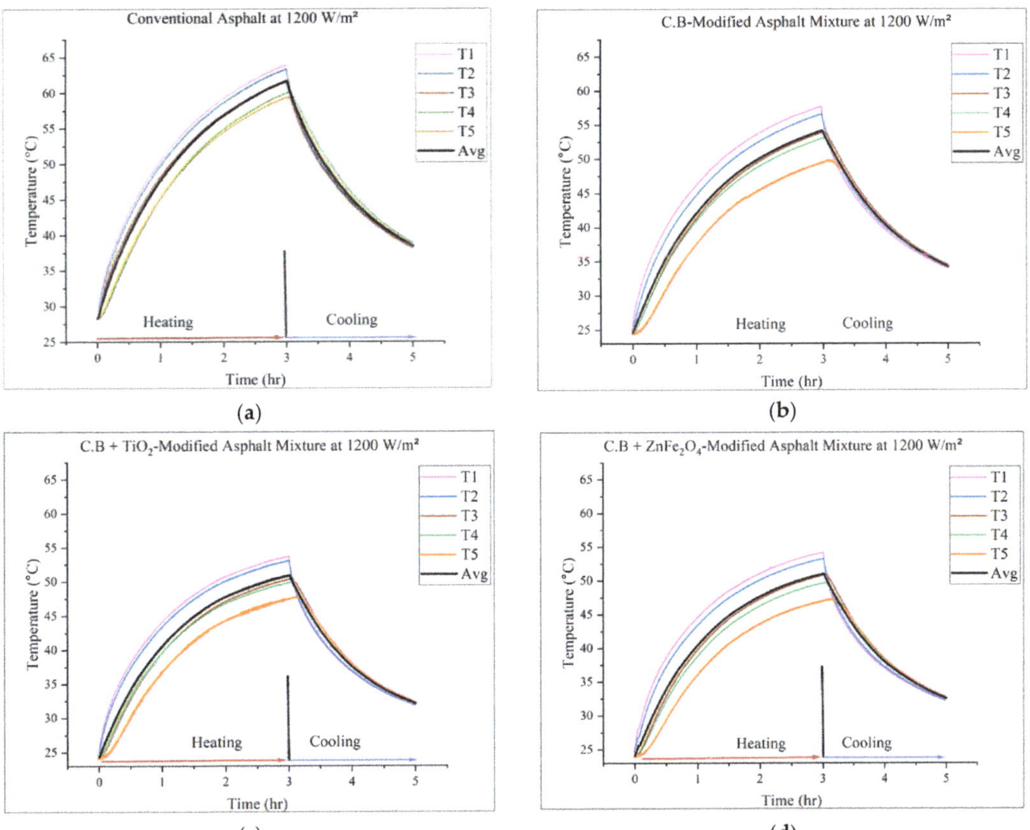

Figure 5. Heating and cooling temperature curves at 1200 flux for: (**a**) conventional asphalt mixture; (**b**) asphalt mixture prepared with neat clear binder; (**c**) asphalt mixture prepared with titanium dioxide-blended clear binder; and (**d**) asphalt mixture prepared with zinc ferrite-blended clear binder.

The heating trend is pretty much different in the rest of the three asphalt mixture combinations, as seen in Figure 5b–d. Unlike conventional asphalt, the curves overlap only between T3 and T4 during almost the whole first hour of heating. However, in part (d) of Figure 5, there is no significant overlapping of the temperature curves. It means that all the modified asphalt mixture combinations distribute the heat evenly down the structure. Hence, the chances of heat accumulation are less. Moreover, the average curve for all the modified combinations is just above the T3 curve; thus, the average temperature of modified asphalt mixtures lies at a depth of less than 10 mm. This makes heat dissipation easier and faster for these mixtures, than for conventional asphalt. It is also worth noting that the gap between the T2 and T5 curves, representing the surface and 50 mm (about 1.97 in) temperatures, respectively, is significant for all the modified combinations compared to conventional asphalt. The gap between the T2 and T5 curves represents around a 4 °C gap in the temperature for the traditional black asphalt, while it is 7–8 °C for the other modified combinations. The maximum temperatures for all three combinations are around 52–53 °C in the modified mixtures. Therefore, it is easier for them to dissipate heat quickly.

Also, these combinations are less likely to participate in UHI augmentation, as their maximum absorbed temperature is far less than 70 °C. Many researchers have studied UHI mitigation by introducing pigments into asphalt mixtures. There was a maximum difference of 11 °C and an average of 7.5 °C between yellow pigment-modified asphalt pavement samples and conventional asphalt mixtures [55]. Similarly, pigments have also been found to cool the internal structure and surface of the pavement, making solar heating reflective coating layers (SHRCLs). Studies have reported a 10 ± 2.5 °C difference between unmodified asphalt and SHRCL surfaces [56]. Also, the internal temperature of asphalt was reduced by 11.5 °C and 13 °C with pigment G and pigment Y blends, respectively [35].

Figure 6a,b represents the mean temperature of all four asphalt mixture combinations at 1000 and 800 W/m^2. The starting temperatures of the titanium dioxide and zinc ferrite C.B-modified samples are a little higher than room temperature; however, it does not affect the objectives of this study, as we aim to observe the maximum temperature absorbed by a sample at the end of the heating phase. Again, there is a massive difference between the conventional and modified mixtures at the end of the heating phase. The pigmented C.B mixtures have almost the same temperature throughout the heating and cooling phases at both fluxes. However, the neat C.B. asphalt mixture temperature is between that of the conventional and pigmented mixtures.

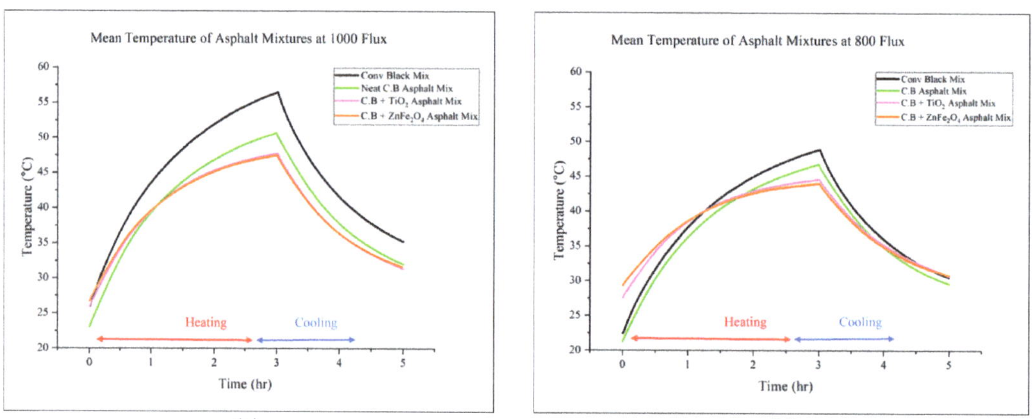

(a) (b)

Figure 6. Mean temperature of 50 mm thick asphalt mixtures at (**a**) 1000 W/m^2 and (**b**) 800 W/m^2.

5.1.1. Thermal Analysis of Binders

Like Figure 6, a similar discussion applies to Figure 7a–d, which illustrates the heating and cooling behavior of conventional black bitumen, the neat clear binder, the titanium dioxide-modified clear binder, and the zinc ferrite-modified clear binder, respectively. Compared to Figure 6a, Figure 7a exhibits an almost identical pattern of overlapping curves; however, in the second half of the heating phase, the average curve does not overlap with T3, indicating a lower average temperature than the asphalt mixtures. The "Avg" curve indicates the mean temperature of a 25 mm (about 0.98 in) thick binder sample, which is close to the temperature at a depth of 6–7 mm. The most significant difference between all four samples is the temperature gap between T1 and T2. Unlike conventional asphalt, which allows the surface base of the heater to reach 72 °C (approx.), the C.B. and pigmented C.B. binder absorb and dissipate heat effectively, limiting T1 to 62 °C and below. As an additional point of interest, both virgin binders (black and transparent) exhibit similar temperature distributions between T2 and T5 (except for the actual temperatures). However, both pigmented C.B binders resist heat transfer and storage at the bottommost part (25 mm) of the sample. This could be attributed to the efficient heat dissipation capability of the pigments. Overall, the lower specific heat and higher thermal conductivity of pigments enable the corresponding samples to quickly dissipate heat back into the walls of the container (environment). This results in the samples staying at lower temperatures for an elongated period, thus decreasing the time required to cool down.

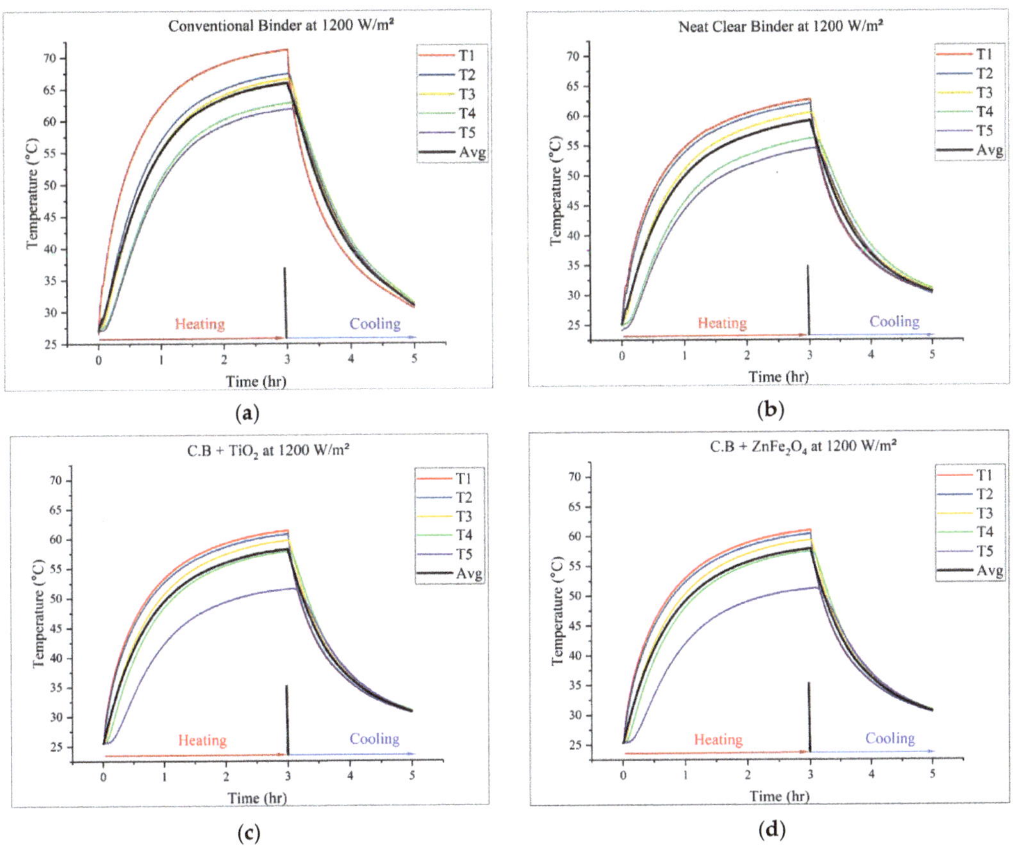

Figure 7. Heating and cooling temperature curves at 1200 flux for: (**a**) conventional black bitumen, (**b**) neat clear binder, (**c**) clear binder-modified with titanium dioxide pigments, and (**d**) clear binder-modified with zinc ferrite pigments.

The shape of the curves in Figure 7 appear to be more bent during the heating phase, as opposed to the heating pattern of the asphalt mixture. Consequently, the change in heat over time is not linear, since binders, being viscoelastic, are highly susceptible to temperature changes [57]. The graphs demonstrate that the samples with pigment modifications tend to retain heat more in areas close to the heat source, leading to a faster dissipation of heat (when the heat supply is turned off). However, neat binders evenly distribute heat, resulting in higher temperatures at the end of the heating phase. The cooling behavior in almost all the graphs is identical; however, for conventional binders, T1 shows an abrupt cooling pattern similar to that of heating.

Figure 8a,b illustrates the mean temperature of all four binder combinations at 1000 and 800 W/m^2. The starting temperature of the zinc ferrite C.B-modified sample is a little higher than the other samples; however, it does not affect the objectives of this study, as we aim to observe the amount of heat absorbed and dissipated at the end of the heating and cooling phase, which is independent of a difference of 2–3 °C in the initial temperatures. As is evident from the above comparison, there is a significant difference between the modified and conventional binders in terms of the absorbed temperature at the end of the heating phase. Pigmented C.B binders maintain, approximately, the same temperature at both flux levels throughout the heating and cooling phases. However, the neat C.B temperature is slightly higher than the pigmented binder temperature. On the other hand, the black binder appears to have absorbed the maximum temperature. The results of a similar study, in which the heat dissipation and energy storage capabilities of nano-modified binders were studied using the same test conditions, also support our findings. Nano-modified binders were reported to stay 8–10 °C cooler than unmodified binders [58].

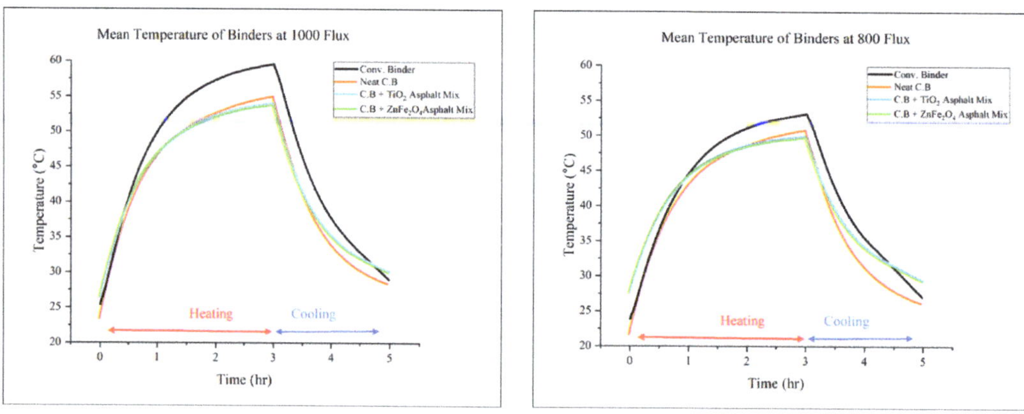

(a) (b)

Figure 8. Mean temperature of 25 mm thick bituminous samples at (**a**) 1000 W/m^2 and (**b**) 800 W/m^2.

5.1.2. Heating and Cooling Trends

Another factor that requires discussion in this study is the variation in the heating rate, in addition to the change in the elevated temperature. Since we are primarily concerned with asphalt mixtures as the actual materials used for road paving, the scope of this analysis is limited to asphalt mixtures only. We set a target temperature range of 45–50 °C (heating time) and 50–45 °C (cooling time) for every sample. The calculations are shown for a flux density of 1200 W/m^2 for the heating and cooling phases, which is the extreme case in this study. The heat maps show the time each sample took to attain the designated temperature. As a point of clarification, both the heating and cooling rates were measured against the average curve, which reflects the overall temperature of a 50 mm sample.

Figure 9a shows the variation in heat intensity over time and the time each sample took to reach the designated temperature, i.e., 50 °C. The heat map of the heating time (Figure 9a) shows that the conventional asphalt mixture reached the target temperature in 0.4 h, while the neat clear binder took slightly less than twice the time to achieve 50 °C. Moreover, both C.B-pigmented asphalt mixtures took over an hour to attain the target temperature. However, the titanium dioxide-modified mixture proved to be the most resistant asphalt mixture to temperature absorption. Pigmented mixtures are highly thermally conductive, so they conduct heat to the outside and take longer to reach the same temperature. This property results in a faster heat dissipation rate in pigmented mixtures, resulting in cooler pavement structures. In addition, it also contributes to reducing the UHI effect and increasing the resistance to permanent deformation.

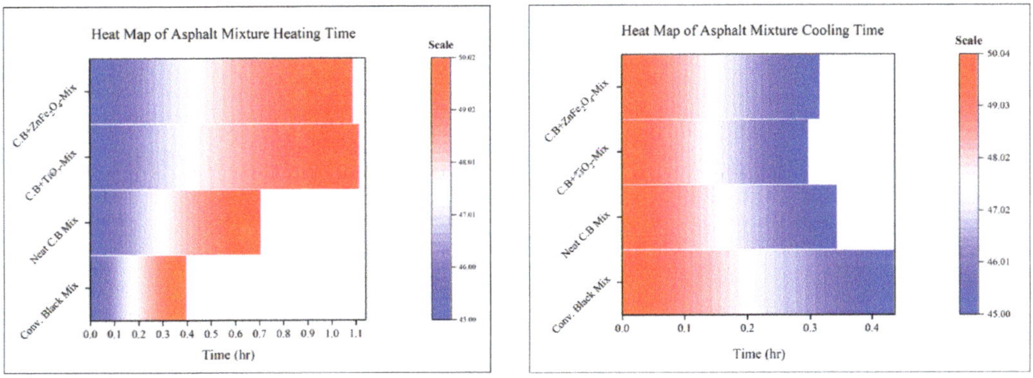

(**a**) (**b**)

Figure 9. Heat map against Avg. curve of the asphalt mixtures, showing (**a**) heating time from 45–50 °C and (**b**) cooling time from 50–45 °C.

Similarly, Figure 9b shows the comparative cooling time of the asphalt mixture combinations. The time taken by all the mixtures (from 50 °C to 45 °C) was observed and used to draw heat maps. Also, the heat maps indicate the change in temperature for each mixture over time. The heat maps display a significant difference in the time taken by all the asphalt mixtures to attain 45 °C. Conventional asphalt took almost 0.45 h to drop its temperature to 5 °C. Similarly, asphalt prepared with a neat clear binder took nearly 0.35 h to drop its temperature. Adding pigments to the C.B asphalt mixtures made the cooling process even quicker. Hence, titanium dioxide C.B-prepared asphalt took the shortest time to reduce its temperature, while zinc ferrite took around 0.31 h to drop to 5 °C. Therefore, the heating and cooling analysis reflects the material's thermal conductivity and specific heat values (from Table 1). Consequently, non-black asphalt mixtures are preferred over conventional asphalt due to their cooling behavior. Additionally, the cooling analysis has demonstrated that pigmented asphalt mixtures conduct more heat and effectively dissipate heat. This property could be directly attributed to UHI mitigation and a lower pavement temperature. The heat energy dissipation, in a similar study, also reported 8% to 9% of heat being dissipated by nano-modified bitumen, while the corresponding asphalt mixtures dissipated 14–16% of the absorbed heat [58].

The rate of heating and cooling of the binder combinations was also studied apart from observing the heating and cooling trends of the asphalt mixtures. Figure 10a demonstrates the time all the four binder combinations took to get from 45 °C to 55 °C. The time at 45 °C is considered the starting time, while the time corresponding to 55 °C is regarded as the final time. As binders are more temperature sensitive, a temperature bracket of 10 °C was considered to observe the heating and cooling time. The conventional binder reached the target temperature in less than half an hour, while the neat clear binder took twice the time, i.e., one hour. Adding pigments to the clear binder further increased the heating time, making it more resistant to absorbing higher temperatures. There is no relatively significant difference in the cooling time between the binders, as seen in Figure 10b. However, the neat C.B and pigmented binders took less time than conventional black binders to cool down from 55 °C to 45 °C.

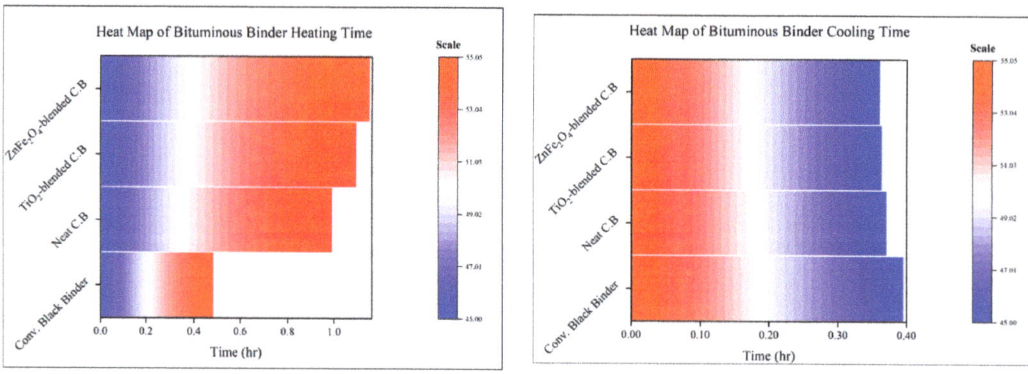

(a) (b)

Figure 10. Heat map against Avg. curve of bituminous binders, showing (**a**) heating time from 45–55 °C and (**b**) cooling time from 55–45 °C.

5.1.3. Heat Maps

Although previously discussed graphs present the temperature behaviors of different asphalt binders and mixtures, it is essential to visualize the relative heating and cooling behaviors over time. Figure 11 illustrates the heat maps of all four combinations at 1200 W/m^2, and these graphs depict the temperature intensities of the average curves, which is the overall temperature of the 50 mm (mixtures) and 25 mm (binders) thick samples. The most significant difference, as seen in both the images in Figure 11a,b, for the asphalt mixtures and bituminous binders, respectively, is the temperature intensity. However, in the asphalt mixtures, the conventional black asphalt maintains the highest temperature level absorbed for up to 20–30 min, even after turning off the power supply (cooling time). Moreover, at 45 °C and above, the temperature is expanded from the second to the fourth hour of the test duration.

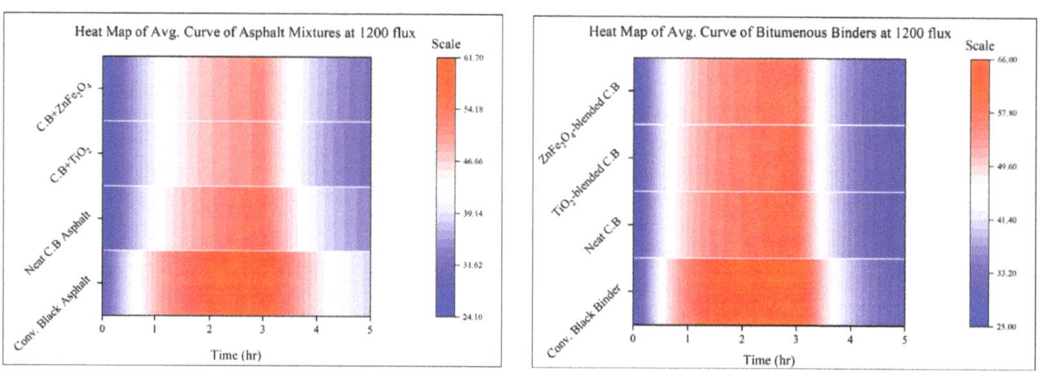

(a) (b)

Figure 11. Heat map showing the temperature variations over time for (**a**) asphalt mixtures and (**b**) bituminous binders.

In contrast, apart from the much lower temperature intensities, the neat clear binder and pigment-modified clear binder mixtures depict a higher temperature expansion for a maximum of two hours. The pigment-modified mixtures have nearly the same temperature pattern over the 5 h test cycle, except for titanium dioxide, which is more efficient during the cooling period. Furthermore, from the scale used, the difference in the maximum absorbed temperature between the black and non-black mixtures is around 15 °C.

Maintaining the highest temperature absorbed during the start of the cooling period can also be seen in the case of conventional black bitumen, as seen in Figure 11b. However, the clear binder and pigmented clear binders showed immediate cooling after switching off the power supply. The final cooling temperature is much higher for the clear and pigmented binders than for black bitumen. The difference in the maximum temperature between the black and non-black binders is also around 12–13 °C. The heat maps indicate that conventional black asphalt stores the highest temperature and stays at elevated temperatures for extended periods, which significantly aids the UHI effect [59].

5.2. Performance Analysis of Asphalt Mixtures

Performance tests were also performed in this study to observe the reflectance of thermal enhancement on the functional/structural performance of the corresponding asphalts.

5.2.1. Rut Resistance of Asphalt

According to the BS EN 12697-25 [60] standard, the Cooper wheel tracking test (CWTT) was performed to determine the resistance of various asphalt mixtures to permanent deformation or rutting. A slab measuring 300 × 300 × 50 mm^3 was prepared and compacted using a Cooper roller compactor to perform the test. A wheel with a load between 700 and 740 N was used to pass each sample 10,000 times, and the rut depth was measured. A wheel with a diameter of 8 inches and a thickness of 2 inches was used. In addition, the machine ran at 26.5 rpm at 55 °C. Figure 12 shows the results of the CWTT. The rutting potential of five asphalt mixture combinations was measured, including conventional black asphalt, asphalt prepared with the neat clear binder, and pigment (red, white, orange)-modified C.B asphalts.

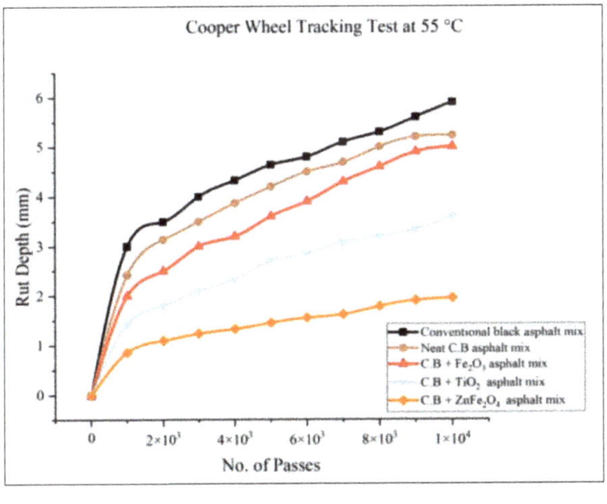

Figure 12. Cooper wheel tracking test (CWTT) of asphalt mixtures at 55 °C.

From the graph (Figure 12), it is clear that the neat C.B asphalt and all pigmented C.B asphalt mixtures demonstrated a significant increase in rut resistance over conventional black asphalt. It is evident that the rut value has decreased by 5 mm, which means that the rut resistance has increased. The maximum rut depth of around 5.8 mm was observed for conventional asphalt. In contrast, rut depths of 5.1 mm, 5 mm, 3.5 mm, and 1.8 mm were recorded for the neat C.B, iron oxide red, titanium dioxide white, and zinc ferrite orange asphalt mixtures. Hence, it can be concluded that the maximum rut depth was observed for conventional asphalt; however, the neat C.B, red, white, and orange-pigmented asphalt mixtures showed lesser rut depths. The zinc ferrite orange asphalt mixture was the most resistant to ruts, followed by titanium dioxide. Furthermore, one of the objectives of this research was to enhance the high-temperature performance by adding pigments and replacing the black binder, which was also achieved during the thermal analysis.

As demonstrated by the permanent deformation analysis, the increase in stiffness of the asphalt mixtures at higher temperatures indicates an improvement in the colored binder's elastic response. This improves the performance of the asphalt mixtures at high temperatures. Additionally, the observed reduction in the pavement temperature, even by as little as 5 °C, can notably enhance the dynamic stability and rut resistance of asphalt, particularly as temperatures rise [61], apart from the reduction in heat islands. Furthermore, the high viscosity of colored asphalt is identified as a contributing factor to increased rut values [62]. This highlights the multifaceted nature of colored asphalt's performance characteristics.

5.2.2. Dynamic Modulus Test

Various asphalt mixtures, including conventional asphalt and asphalt mixtures prepared with different combinations of clear binders (C.B) and pigments, namely, iron oxide red, titanium dioxide white, and zinc ferrite orange, were tested for dynamic modulus to determine their resistance to permanent deformation. A Superpave gyratory compactor was used to prepare cylindrical specimens, with a diameter of 150 mm and a height of 170 mm. Moreover, 600 kPa of pressure at 160 °C was applied to achieve the specified specimen height. From the compacted samples, specimens with a diameter of 101.6 mm and a height of 150 mm were extracted. The dynamic modulus was measured at frequencies ranging from 25 to 0.1 Hz, under various loading conditions, namely 1050, 525, 195, and 52.5 kPa. The AASHTO TP 62 [63] guidelines were followed for sample preparation and testing. To account for the regional climatic conditions, the tests were conducted at both 40 °C and 55 °C. Figure 13 illustrates the dynamic modulus test results. Also, Figure 14 shows the five colored cylindrical samples before and after cutting.

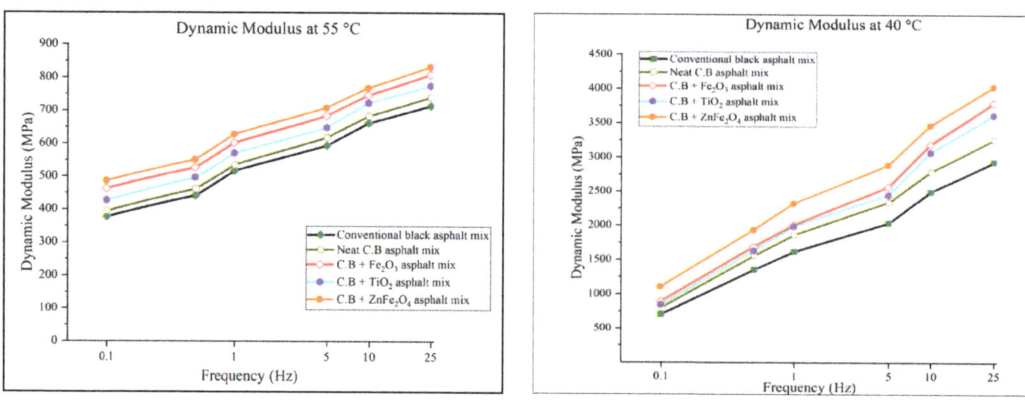

(a) (b)

Figure 13. Dynamic modulus of asphalt mixtures at (**a**) 55 °C and (**b**) 40 °C.

The dynamic modulus directly indicates a material's resistance to permanent deformation, commonly known as rut resistance [64]. It can be seen from the graph (Figure 13) that the conventional asphalt has a lower dynamic modulus than asphalt mixtures prepared with a clear binder only and those containing red iron oxide pigments, white titanium dioxide pigments, and orange zinc ferrite pigments, particularly at higher temperatures. The dynamic modulus of pigmented asphalt mixtures appears to be higher for a range of frequencies, indicating a better performance when compared to pigmented asphalt mixtures.

(a) (b)

Figure 14. Dynamic modulus samples (**a**) after core cutting and (**b**) before core cutting.

In particular, the zinc ferrite orange-pigmented clear binder (C.B) asphalt exhibits the highest dynamic modulus and, therefore, the highest rutting resistance of all combinations. Asphalt prepared with C.B modified with Fe_2O_3 takes second place, while TiO_2+C.B-pigmented asphalt takes third place. On the other hand, the neat clear binder (C.B) asphalt exhibits the lowest dynamic modulus and rut resistance compared to the colored mixtures. However, it still shows increased resistance than conventional black asphalt mixtures. It is evident that the higher the value of dynamic modulus, the higher the resistance against permanent deformation and vice versa, which means that it is a direct measure of rut resistance [64].

All the mixtures generally exhibited a higher dynamic modulus at lower temperatures (i.e., 40 °C). However, it was lower at higher temperatures (55 °C). This indicates that asphalt is a highly temperature-dependent material, indicating greater rut susceptibility at higher temperatures [65]. Overall, these dynamic modulus test results suggest that the addition of pigments, specifically zinc ferrite orange, significantly improves asphalt resistance to permanent deformation at higher temperatures, demonstrating the potential benefits of these pigment additives when it comes to improving the durability and performance of asphalt pavements. It is evident from these findings that pigment additives are essential for enhancing rut resistance in asphalt pavements under a variety of climatic conditions. These findings have significant implications for the design and construction of asphalt pavements. Furthermore, it has also been reported that adding pigments to asphalt mixtures improves the mixture's high-temperature performance by reducing its surface and internal temperatures, while not compromising its fatigue life [33].

5.3. ANOVA (Analysis of Variance) Analysis

Based on the thermal testing results, the ANOVA analysis in Table 4 shows significant differences between conventional and other binders. The F-statistic indicates a statistically significant difference between the binder groups compared to a p-value of 0 (rounded). A significant F-statistic demonstrates different heating rates between the binder groups. The ANOVA results do not explicitly provide statistics about the cooling behavior; if the heating behavior differs significantly between the groups, we can infer that binders will also display different cooling behavior. We can examine the ANOVA results between the binder groups to understand how the clear binder (C.B) compares to the conventional binder (conv. black asphalt). The thermal test results indicate that the clear binder is slower to heat up than the conventional black asphalt (i.e., average for clear binder = 41.1069, average for conv. black asphalt = 49.0344).

Table 4. Statistical and ANOVA analysis of thermal testing.

Statistics						
Groups	Count	Sum	Average		Variance	
Conv. Black Asphalt	3601	176,572.9325	49.0344		69.7382	
Neat C.B Asphalt	3601	155,958.9918	43.3099		54.0948	
C.B + TiO_2	3601	148,025.9923	41.1069		47.0287	
C.B + $ZnFe_2O_4$	3601	148,432.8170	41.2199		46.0895	
ANOVA Results for thermal testing						
Source of Variation	SS	df	MS	F	*p*-Value	F crit
Between Groups	149,366.1305	3	49,788.7102	917.9705	0	2.6055
Within Groups	781,024.4668	14,400	54.2378			
Total	930,390.5973	14,403				

Similarly, conventional asphalt (conv. black asphalt) may also cool down slower than clear asphalt, indicating a similar trend for cooling. These binder groups can be applied differently in different scenarios because of their significant differences in heating and cooling behaviors. A clear binder may be better than a conventional binder if slow heating and rapid cooling rates are desired. Considering the specific temperature requirements and performance expectations of various engineering projects, these findings can assist in optimizing the selection of binder materials.

The ANOVA results for measuring the rut depth in Table 5, using a wheel tracker, for the five binders (clear binder, red, white, yellow, and conventional) clearly show a different rut depth for at least two binders. Statistically significant differences in the rut depth of the binders are indicated by an F-statistic of 9.631 with a *p*-value of 6.271×10^{-6} (remarkably close to 0). Based on the small *p*-value, at least two binder groups perform differently, suggesting that the null hypothesis should be rejected. As we intend to compare the performance of the clear binder with that of the conventional binder, we will focus on the two relevant groups of binders. The wheel tracking test result for the conventional binder was 4.2781, while for the clear binder it was 3.7627. Compared to the conventional binder, the clear binder performs better, on average, due to its lower average penetration by the wheel tracker. Based on the ANOVA results, the clear binder performs significantly better than conventional binders in terms of its structural performance.

Table 5. Statistical and ANOVA analysis of rutting test.

Statistics						
Groups	Count	Sum	Average		Variance	
Conventional	12	47.0586	4.2781		3.1557	
Clear binder	12	41.3900	3.7627		2.4111	
C.B + Fe_2O_3	12	42.6500	3.5542		2.1059	
C.B + TiO_2	12	27.3700	2.2808		1.0498	
C.B + $ZnFe_2O_4$	12	15.8300	1.3192		0.3082	
ANOVA Results for rutting analysis						
Source of Variation	SS	df	MS	F	*p*-value	F crit
Between Groups	68.1594	4	17.0398	9.6311	0.0000	2.5463
Within Groups	93.7708	53	1.7693			
Total	161.9301	57				

The normal probability plot, as shown in Figure 15a, clearly shows that our sample percentile is an evenly distributed pattern of data points compared to the conventional binder. The data points follow some skewness for the initial points. Still, the remaining points are perfectly normally distributed, which indicates a normally distributed dataset that validates the reliability of our analysis. Figure 15b–d shows the line fit plots of neat C.B asphalt, C.B + $ZnFe_2O_4$, and C.B + TiO_2 with conventional black asphalt. By observing the line fit plots closely, it is clear that the conventional binder tends to attain higher temperatures than the clear and pigmented binders and asphalts, indicating its ability to dissipate heat faster.

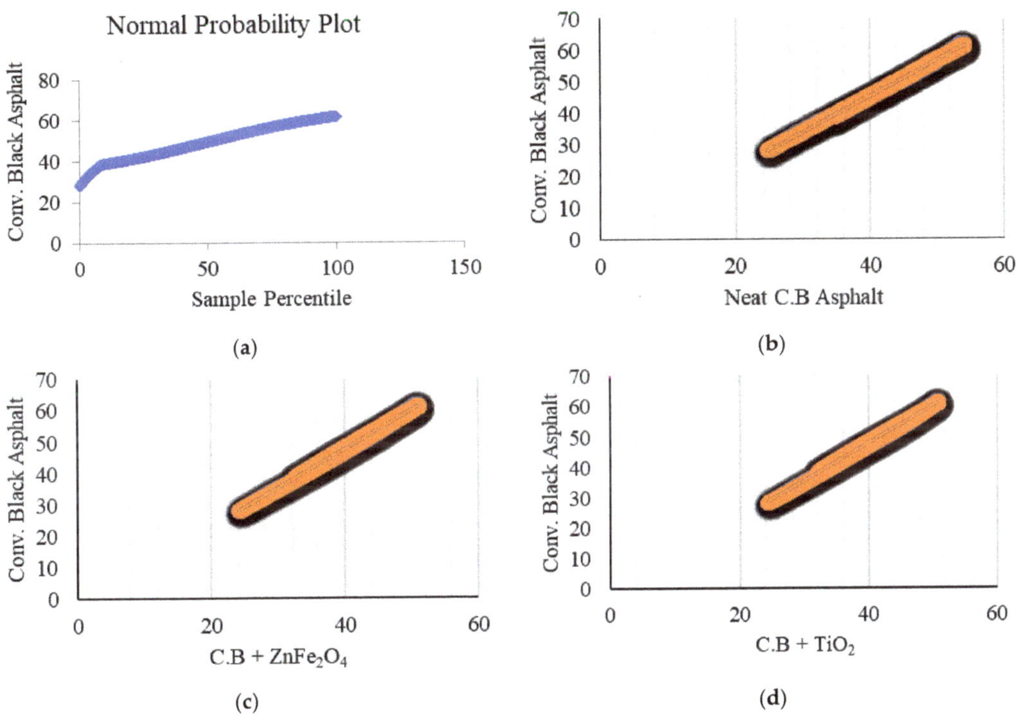

Figure 15. (**a**) Normal probability plot. (**b**) Line fit plot for neat C.B asphalt. (**c**) Line fit plot for C.B + $ZnFe_2O_4$. (**d**) Line fit plot for C.B + TiO_2 with conventional black asphalt.

Table 6 shows the correlation matrix of all four binders, by showing the effectiveness of the modified binders compared to the conventional black binder. The high correlation coefficients observed, among the variables, indicate strong linear relationships between the asphalt mixtures. By conducting correlation analysis, we aim to pinpoint which characteristics of the modified binders, such as the pigment content or clear binder composition, contribute most significantly to the observed differences in thermal performance compared to conventional black asphalt. This understanding will provide valuable insights for optimizing asphalt mixture design and performance in future applications.

Table 6. Correlation matrix of all four mixtures.

	Conv. Black Asphalt	Neat C.B Asphalt	C.B + TiO_2	C.B + $ZnFe_2O_4$
Conv. Black Asphalt	1			
Neat C.B Asphalt	0.9994	1		
C.B + TiO_2	0.9976	0.9965	1	
C.B + $ZnFe_2O_4$	0.9984	0.9978	0.9997	1

6. Conclusions

This study aimed to introduce a clear binder and pigment-modified clear binders to replace black bitumen in the asphalt mixture, with a non-black and solar-reflective pavement surface that could help reduce the UHI effect. The following are the key findings in this study:

- Asphalt prepared with neat C.B displayed an average 9 °C lower temperature than the conventional black asphalt mixture. The temperature was further reduced by 2 °C by adding pigments, resulting in an overall temperature difference of 11 °C.
- Significant temperature differences between black and non-black asphalt mixtures were observed at 1000 and 800 W/m^2, with gaps of 7 °C and 10 °C at 1000 flux and 2 °C and 5 °C at 800 flux, indicating a higher efficiency of non-black pavements at extreme temperatures.
- The average temperature absorbed by the conventional black binder at 1200 W/m^2 was just above 65 °C, while the neat clear binder displayed a 7 °C lower temperature. Similarly, pigmented binders further lowered the temperature to 1–2 °C. At the lower fluxes, i.e., 1000 and 800, the difference in the average absorbed temperature between the black bitumen and clear binder (C.B, pigmented C.B) was 5–7 °C and 2–3 °C, respectively.
- The neat C.B asphalt mixture took 75% longer to reach the target temperature than conventional black asphalt, with the pigmented mixtures taking an additional 170% (zinc ferrite mix) to 180% (titanium dioxide mix) longer. It indicates a higher resistance to heat absorption and prolonged cooling.
- The cooling rate revealed that conventional asphalt took the longest (approx. 0.45 h) to cool down to 45 °C, while the neat C.B mix took 27% less time (approx. 0.33 h) to return to the target temperature. The pigmented mixture further lessened the cooling time up to 33% (0.3 h) and 31% (0.31 h) for TiO_2 and $ZnFe_2O_4$ mixtures, respectively.
- The heat maps indicated significant heat accumulation in conventional asphalt mixtures and binders over a wide period. However, the neat and pigment-modified C.B mixtures appeared more relaxed over the entire heat cycle.
- The neat C.B asphalt mixture exhibited around 15% more resistance to rut depth than conventional asphalt, with red, white, and orange-pigmented mixtures showing 20%, 43%, and 73% rut resistance, respectively.
- All the non-black mixtures performed slightly better than conventional black mixtures in the dynamic modulus test. A maximum resistance to permanent deformation of 15% was noted in the C.B + $ZnFe_2O_4$ mixture compared to conventional black asphalt mixtures.

Overall, thermally conductive materials with a lower specific heat are preferred for heat harvesting and cool pavement applications, since they enhance the efficiency of heat transfer and storage. This helps harness thermal energy for various purposes, creating more excellent surfaces and mitigating urban heat issues.

Thermal analysis, in this study, was carried out assuming fixed solar flux densities (800, 1000, and 1200 W/m^2) to simulate solar radiation; however, solar radiation varies considerably over time and geographical locations. Moreover, despite the efforts to ensure a one-dimensional heat flow and minimize transverse heat dissipation during the charging phase, it is recognized that achieving perfect insulation and heat confinement within heat sinks is challenging. This could introduce slight inaccuracies in the observed temperatures, particularly at higher heat flux levels. Additionally, variations in sensor placement (i.e., inaccurate height, other than specified) could impact the reliability of our temperature readings, or result in inaccurate temperature measurements at certain depths. Furthermore, the continuous heat dissipation during the charging phase and subsequent cooling period could slightly impact the observed maximum temperatures.

The authors recommend the strategic application of these mixtures in urban areas with high percentages of paved surfaces, specifically in densely populated streets, from a UHI mitigation perspective. They can be utilized in bicycle lanes, pedestrian walkways, tennis courts, running tracks, and bus stops, for thermal comfort. Colored pavements can also reduce lighting requirements, so their application in tunnels is also suggested. From a structural performance perspective, the authors suggest performing thermal analysis for at least 1–2 ft^2 slabs under direct sunlight over 24 h to compare the performance of asphalt prepared with pigmented clear binders. After that, some correlations should be made for future reference to lab tests. Additionally, the adhesion properties of C.B and pigmented asphalt are worth investigating to understand their interaction and potential impact on pavement performance. Moreover, investigating the cracking resistance of these mixtures to assess their durability and long-term performance under various environmental conditions would be an interesting study for the future. After all these findings, constructing test tracks of up to a few hundred feet long of such materials for normal traffic to assess their durability and rut resistance at higher temperatures would be worth testing before scaling up to larger roadway constructions of such materials.

Author Contributions: Conceptualization, G.B. and N.A.; methodology, G.B. and N.A.; software, G.B. and Y.M.; validation, G.B., Y.H. and Y.M.; formal analysis, G.B. and Y.M.; investigation, G.B. and N.A.; resources, N.A. and Y.H.; data curation, G.B.; writing—original draft preparation, G.B.; writing—review and editing, G.B. and Y.M.; visualization, G.B.; supervision, N.A. and Y.H. All authors have read and agreed to the published version of the manuscript.

Funding: This research received no external funding.

Data Availability Statement: The data supporting the findings of this study are available within the article.

Conflicts of Interest: The authors declare that they have no known competing financial interests or personal relationships that could have appeared to influence the work reported in this paper.

References

1. Asif, S.A.; Ahmad, N. Comparative Study of Various Properties of Clear Binder with Traditional Black Binder. *Arab. J. Sci. Eng.* **2022**, *47*, 12979–12991. [CrossRef]
2. Akbari, H.; Kolokotsa, D. Three decades of urban heat islands and mitigation technologies research. *Energy Build.* **2016**, *133*, 834–842. [CrossRef]
3. Sreedhar, S.; Biligiri, K.P. Comprehensive laboratory evaluation of thermophysical properties of pavement materials: Effects on urban heat island. *J. Mater. Civ. Eng.* **2016**, *28*, 04016026. [CrossRef]
4. Chen, J.; Chu, R.; Wang, H.; Zhang, L.; Chen, X.; Du, Y. Alleviating urban heat island effect using high-conductivity permeable concrete pavement. *J. Clean. Prod.* **2019**, *237*, 117722. [CrossRef]
5. Bai, Y.; Bai, Q. *Subsea Engineering Handbook*; Gulf Professional Publishing: Houston, TX, USA, 2018. [CrossRef]
6. Lesueur, D. The colloidal structure of bitumen: Consequences on the rheology and on the mechanisms of bitumen modification. *Adv. Colloid Interface Sci.* **2009**, *145*, 42–82. [CrossRef] [PubMed]
7. Badin, G.; Ahmad, N.; Ali, H.M.; Ahmad, T.; Jameel, M.S. Effect of addition of pigments on thermal characteristics and the resulting performance enhancement of asphalt. *Constr. Build. Mater.* **2021**, *302*, 124212. [CrossRef]
8. Higashiyama, H.; Sano, M.; Nakanishi, F.; Takahashi, O.; Tsukuma, S. Field measurements of road surface temperature of several asphalt pavements with temperature rise reducing function. *Case Stud. Constr. Mater.* **2016**, *4*, 73–80. [CrossRef]
9. US DOT (FHWA). Pavement Thermal Performance and Contribution to Urban and Global Climate. Available online: https://www.fhwa.dot.gov/pavement/sustainability/articles/pavement_thermal.cfm (accessed on 7 August 2023).
10. Li, H.; Harvey, J.T.; Holland, T.J.; Kayhanian, M. The use of reflective and permeable pavements as a potential practice for heat island mitigation and stormwater management. *Environ. Res. Lett.* **2013**, *8*, 015023. [CrossRef]
11. Synnefa, A.; Karlessi, T.; Gaitani, N.; Santamouris, M.; Assimakopoulos, D.N.; Papakatsikas, C. Experimental testing of cool colored thin layer asphalt and estimation of its potential to improve the urban microclimate. *Build. Environ.* **2011**, *46*, 38–44. [CrossRef]
12. U.S. EPA. *Reducing Urban Heat Islands: Compendium of Strategies (Cool Pavements)*; U.S. EPA: Washington, DC, USA, 2008; pp. 1–23. Available online: https://www.epa.gov/sites/default/files/2017-05/documents/reducing_urban_heat_islands_ch_5.pdf (accessed on 20 February 2024).
13. Chang, H.-T. A temporal and spatial analysis of urban heat island in basin city utilizing remote sensing techniques. *Int. Arch. Photogramm. Remote Sens. Spat. Inf. Sci.* **2016**, *41*, 165–170. [CrossRef]
14. Miner, M.J.; Taylor, R.A.; Jones, C.; Phelan, P.E. Efficiency, economics, and the urban heat island. *Environ. Urban.* **2017**, *29*, 183–194. [CrossRef]

15. Robaa, E.-S. Effect of urbanization and industrialization processes on outdoor thermal human comfort in Egypt. *Atmos. Clim. Sci.* **2011**, *1*, 100. [CrossRef]
16. Gutzler, D.S. Ecological climatology: Concepts and applications. *Nat. Resour. J.* **2003**, *43*, 1291–1296.
17. Coseo, P.; Larsen, L. Cooling the heat island in compact urban environments: The effectiveness of Chicago's green alley program. *Procedia Eng.* **2015**, *118*, 691–710. [CrossRef]
18. Correia, D.; Ferreira, A. Energy Harvesting on Airport Pavements: State-of-the-Art. *Sustainability* **2021**, *13*, 5893. [CrossRef]
19. Wang, Y.; Berardi, U.; Akbari, H. The urban heat island effect in the city of Toronto. *Procedia Eng.* **2015**, *118*, 137–144. [CrossRef]
20. Qin, Y.H. A review on the development of cool pavements to mitigate urban heat island effect. *Renew. Sustain. Energy Rev.* **2015**, *52*, 445–459. [CrossRef]
21. Santamouris, M.; Gaitani, N.; Spanou, A.; Saliari, M.; Giannopoulou, K.; Vasilakopoulou, K.; Kardomateas, T. Using cool paving materials to improve microclimate of urban areas—Design realization and results of the flisvos project. *Build. Environ.* **2012**, *53*, 128–136. [CrossRef]
22. Chen, J.; Yin, X.J.; Wang, H.; Ding, Y.M. Evaluation of durability and functional performance of porous polyurethane mixture in porous pavement. *J. Clean. Prod.* **2018**, *188*, 12–19. [CrossRef]
23. Chen, J.; Li, J.H.; Wang, H.; Huang, W.; Sun, W.; Xu, T. Preparation and effectiveness of composite phase change material for performance improvement of Open Graded Friction Course. *J. Clean. Prod.* **2019**, *214*, 259–269. [CrossRef]
24. Shi, X.J.; Rew, Y.; Ivers, E.; Shon, C.S.; Stenger, E.M.; Park, P. Effects of thermally modified asphalt concrete on pavement temperature. *Int. J. Pavement Eng.* **2019**, *20*, 669–681. [CrossRef]
25. Du, Y.F.; Liu, P.S.; Wang, J.C.; Wang, H.; Hu, S.W.; Tian, J.; Li, Y.T. Laboratory investigation of phase change effect of polyethylene glycolon on asphalt binder and mixture performance. *Constr. Build. Mater.* **2019**, *212*, 1–9. [CrossRef]
26. Sengoz, B.; Bagayogo, L.; Oner, J.; Topal, A. Investigation of rheological properties of transparent bitumen. *Constr. Build. Mater.* **2017**, *154*, 1105–1111. [CrossRef]
27. Bocci, M.; Grilli, A.; Cardone, F.; Virgili, A. Clear asphalt mixture for wearing course in tunnels: Experimental application in the province of bolzano. *Procedia-Soc. Behav. Sci.* **2012**, *53*, 115–124. [CrossRef]
28. Partal, P.; Martinez-Boza, F.; Conde, B.; Gallegos, C. Rheological characterisation of synthetic binders and unmodified bitumens. *Fuel* **1999**, *78*, 1–10. [CrossRef]
29. Airey, G.D.; Mohammed, M.H.; Fichter, C. Rheological characteristics of synthetic road binders. *Fuel* **2008**, *87*, 1763–1775. [CrossRef]
30. Lee, H.; Kim, Y. Laboratory evaluation of color polymer concrete pavement with synthetic resin binder for exclusive bus lanes. *Transp. Res. Rec.* **2007**, *1991*, 124–132. [CrossRef]
31. Santagata, F.; Canestrari, F.; Ferrotti, G.; Graziani, A. Experimental characterization of transparent synthetic binder mixes reinforced with cellulose fibres. In Proceedings of the 4th International SIIV Congress, Palermo, Italy, 12–14 September 2007. Available online: https://iris.univpm.it/handle/11566/52988 (accessed on 23 August 2023).
32. Badin, G.; Huang, Y.; Ahmad, N. Comparative Analysis of Thermally Investigated Pigment-Modified Asphalt Binders. In Proceedings of the ASCE Airfield and Highway Pavements 2023 Conference, Austin, Texas, USA, June 14−17 2023; pp. 174–184.
33. Badin, G.; Ahmad, N.; Ali, H.M. Experimental investigation into the thermal augmentation of pigmented asphalt. *Phys. A Stat. Mech. Its Appl.* **2020**, *551*, 123974. [CrossRef]
34. Cheela, V.R.S.; John, M.; Biswas, W.; Sarker, P. Combating Urban Heat Island Effect-A Review of Reflective Pavements and Tree Shading Strategies. *Buildings* **2021**, *11*, 93. [CrossRef]
35. Jiang, L.; Wang, L.C.; Wang, S.Y. A novel solar reflective coating with functional gradient multilayer structure for cooling asphalt pavements. *Constr. Build. Mater.* **2019**, *210*, 13–21. [CrossRef]
36. Karlessi, T.; Santamouris, M.; Apostolakis, K.; Synnefa, A.; Livada, I. Development and testing of thermochromic coatings for buildings and urban structures. *Sol. Energy* **2009**, *83*, 538–551. [CrossRef]
37. Pasetto, M.; Pasquini, E.; Giacomello, G.; Baliello, A. Innovative pavement surfaces as urban heat islands mitigation strategy: Chromatic, thermal and mechanical characterisation of clear/coloured mixtures. *Road Mater. Pavement Des.* **2019**, *20*, S533–S555. [CrossRef]
38. Gaitani, N.; Spanou, A.; Saliari, M.; Synnefa, A.; Vassilakopoulou, K.; Papadopoulou, K.; Pavlou, K.; Santamouris, M.; Papaioannou, M.; Lagoudaki, A. Improving the microclimate in urban areas: A case study in the centre of Athens. *Build. Serv. Eng. Res. Technol.* **2011**, *32*, 53–71. [CrossRef]
39. Pomerantz, M.; Akbari, H.; Harvey, J. *The Benefits of Cooler Pavements on Durability and Visibility*; Report No. LBNL-43443; Lawrence Berkeley National Laboratory: Berkeley, CA, USA, 2000.
40. Jameel, M.S.; Ahmad, N.; Badin, G.; Khan, A.H.; Ali, B.; Raza, A. Comparison of hydrated lime, paraffin wax and low-density polyethylene modified bituminous binders: A perspective of adhesion and moisture damage. *Int. J. Pavement Eng.* **2023**, *24*, 2168659. [CrossRef]
41. Elements, A. Iron(III) Oxide Properties (Theoretical). Available online: https://www.americanelements.com/iron-iii-oxide-1309-37-1 (accessed on 8 August 2023).
42. Takeda, M.; Onishi, T.; Nakakubo, S.; Fujimoto, S. Physical Properties of Iron-Oxide Scales on Si-Containing Steels at High Temperature. *Mater. Trans.* **2009**, *50*, 2242–2246. [CrossRef]

43. Corporation, G. Titanium Dioxide—Titania (TiO2). Available online: https://www.azom.com/properties.aspx?ArticleID=1179 (accessed on 7 August 2023).
44. TDK. Mn-Zn Ferrite, Material Characteristics. 2022. Available online: https://product.tdk.com/system/files/dam/doc/product/ferrite/ferrite/ferrite-core/catalog/ferrite_mn-zn_material_characteristics_en.pdf (accessed on 8 August 2023).
45. Ferrite Materials. MAGNETICS. Available online: https://www.mag-inc.com/Media/Magnetics/File-Library/Products/Ferrite/Magnetics-Ferrite-Materials-Web-8-17a.pdf (accessed on 7 August 2023).
46. Overview of the Ferrite, in FERRITES—TDK. 2021. Available online: https://product.tdk.com/en/system/files?file=dam/doc/product/ferrite/ferrite/ferrite-core/catalog/ferrite_summary_en.pdf (accessed on 7 August 2023).
47. Abbas, F.A.; Alhamdo, M.H. Enhancing the thermal conductivity of hot mix asphalt. *Results Eng.* **2023**, *17*, 100827. [CrossRef]
48. Lindberg, W.R.; Thomas, R.R.; Christensen, R.J. measurements of specific-heat, thermal-conductivity and thermal-diffusivity of Utah tar sands. *Fuel* **1985**, *64*, 80–85. [CrossRef]
49. Pan, P.; Wu, S.P.; Hu, X.D.; Liu, G.; Li, B. Effect of Material Composition and Environmental Condition on Thermal Characteristics of Conductive Asphalt Concrete. *Materials* **2017**, *10*, 218. [CrossRef]
50. Del Carpio, J.A.V.; Marinoski, D.L.; Triches, G.; Lamberts, R.; de Melo, J.V.S. Urban pavements used in Brazil: Characterization of solar reflectance and temperature verification in the field. *Sol. Energy* **2016**, *134*, 72–81. [CrossRef]
51. LANXESS. Colouring of Bituminous Mixes, in "Inorganic Pigments—Technical Information Competence Center Construction. LANXESS Deutschland GmbH Business Unit, Inorganic Pigments D-51368 Leverkusen. 2002. Available online: https://lanxess.com/en/Products-and-Brands/Brands/Bayferrox/Construction-Applications (accessed on 7 July 2023).
52. E. Inc. Silicone Rubber Fiberglass Flexible Heater. Available online: https://www.omega.com/en-us/industrial-heaters/surface-heaters/flexible-heaters/srfra-srfga/p/SRFGA-404-10-P (accessed on 9 July 2023).
53. Affolter, R. Pakistan—Solar Radiation Measurement Data. 2017. Available online: https://energydata.info/dataset/pakistan-solar-radiation-measurement-data (accessed on 9 July 2023).
54. Abbas, F.A.; Alhamdo, M.H. Thermal performance of asphalt solar collector by improving tube and slab characteristics. *Int. J. Thermofluids* **2023**, *17*, 100293. [CrossRef]
55. Kyriakodis, G.; Santamouris, M. Using reflective pavements to mitigate urban heat island in warm climates-Results from a large scale urban mitigation project. *Urban Clim.* **2018**, *24*, 326–339. [CrossRef]
56. Sha, A.; Liu, Z.; Tang, K.; Li, P. Solar heating reflective coating layer (SHRCL) to cool the asphalt pavement surface. *Constr. Build. Mater.* **2017**, *139*, 355–364. [CrossRef]
57. Porto, M.; Caputo, P.; Loise, V.; Eskandarsefat, S.; Teltayev, B.; Rossi, C.O. Bitumen and Bitumen Modification: A Review on Latest Advances. *Appl. Sci.* **2019**, *9*, 742. [CrossRef]
58. Qureshi, F.A.; Ahmad, N.; Ali, H.M. Heat dissipation in bituminous asphalt catalyzed by different metallic oxide nanopowders. *Constr. Build. Mater.* **2021**, *276*, 122220. [CrossRef]
59. Jabbar, H.K.; Hamoodi, M.N.; Al-Hameedawi, A.N. Urban heat islands: A review of contributing factors, effects and data. *IOP Conf. Ser. Earth Environ. Sci.* **2023**, *1129*, 012038. [CrossRef]
60. *CSN EN 12697-25*; Bituminous Mixtures-Test Methods-Part 25: Cyclic Compression Test. European Standards: Released 01 October, 2017. Available online: https://www.en-standard.eu/csn-en-12697-25-bituminous-mixtures-test-methods-part-25-cyclic-compression-test/ (accessed on 25 February 2024).
61. Zheng, M.L.; Han, L.L.; Wang, F.; Mi, H.C.; Li, Y.F.; He, L.T. Comparison and analysis on heat reflective coating for asphalt pavement based on cooling effect and anti-skid performance. *Constr. Build. Mater.* **2015**, *93*, 1197–1205. [CrossRef]
62. Xin, Z.G. Research application of colored asphalt mixture pavement. *Adv. Mater. Res.* **2014**, *900*, 459–462. [CrossRef]
63. *AASHTO TP 62*; Standard Method of Test for Determining Dynamic Modulus of Hot Mix Asphalt (HMA). GlobalSpec: New York, NY, USA. Available online: https://standards.globalspec.com/std/1283471/aashto-tp-62 (accessed on 25 February 2024).
64. Hafeez, M.; Ahmad, N.; Kamal, M.A.; Rafi, J.; Ul Haq, M.F.; Jamal; Zaidi, S.B.A.; Nasir, M.A. Experimental Investigation into the Structural and Functional Performance of Graphene Nano-Platelet (GNP)-Doped Asphalt. *Appl. Sci.* **2019**, *9*, 686. [CrossRef]
65. Li, P.; Zheng, M.L.; Wang, F.; Che, F.; Li, H.Y.; Ma, Q.L.; Wang, Y.H. Laboratory Performance Evaluation of High Modulus Asphalt Concrete Modified with Different Additives. *Adv. Mater. Sci. Eng.* **2017**, *2017*, 7236153. [CrossRef]

infrastructures MDPI

Article

Effect of Olive Waste Ash as a Partial Replacement of Cement on the Volume Stability of Cement Paste

Safa Ghazzawi [1], Hassan Ghanem [1,*], Jamal Khatib [1,2], Samer El Zahab [3] and Adel Elkordi [1,4]

1 Faculty of Engineering, Beirut Arab University, 12-5020 Beirut, Lebanon; s.ghazzawi@bau.edu.lb (S.G.); j.khatib@bau.edu.lb (J.K.); a.elkordi@bau.edu.lb (A.E.)
2 Faculty of Engineering, University of Wolverhampton, Wolverhampton WV1 1LY, UK
3 College of Engineering and Technology, American University of the Middle East, Egaila 54200, Kuwait; samer.el-zahab@aum.edu.kw
4 Faculty of Engineering, Alexandria University, Alexandria 5423021, Egypt
* Correspondence: h.ghanem@bau.edu.lb

Abstract: Over the last decades, concrete has been excessively prone to cracks resulting from shrinkage. These dimensional changes can be affected by the incorporation of supplementary cementitious materials. This work used olive waste ash (OWA), which could substantially tackle this problem and achieve sustainability goals. For this issue, five cement paste mixes were prepared by replacing cement with OWA at different percentages varying from 0 to 20% by weight with a constant increment of 5%. The water-to-cement ratio was 0.45 for all mixes. Compressive strength and flexural strength were investigated at 7, 28, and 90 days. In addition, three shrinkage tests (drying, autogenous, and chemical) and expansion tests were also conducted for each mix and measured during 90 days of curing. The experimental findings indicated that there was a loss in compressive and flexural strength in the existence of OWA. Among all mixes containing OWA, the samples incorporating 10% OWA exhibited maximum strength values. Furthermore, the chemical and autogenous shrinkage decreased with the incorporation of OWA. However, the drying shrinkage decreased at lower levels of substitutions and increased at higher replacement levels. In addition, there was a growth in expansion rates for up to 10% of OWA content, followed by a decrease at higher levels (beyond 10%). Additionally, correlations between these volumetric stability tests were performed. It was shown that a positive linear correlation existed between chemical shrinkage and autogenous and drying shrinkage; however, there was a negative relationship between chemical shrinkage and expansion.

Keywords: cement paste; olive waste ash; chemical shrinkage; autogenous shrinkage; drying shrinkage; expansion; compressive strength; flexural strength

1. Introduction

Volume stability is an intrinsic engineering property that directly affects the serviceability and long-term durability of various cement-based materials. This has motivated researchers to widely focus their studies on the shrinkage performance of concrete structures, including chemical-, drying-, and autogenous-shrinkage as well as expansion. Thus, concrete can expand or contract excessively in light of changes in the moisture content, temperature, and chemical reactions [1–9]. By definition, drying shrinkage is the contraction behavior of the concrete during the hardening period, which is likely to be associated with the internal loss of water by evaporation [1]. It is affected by many parameters such as humidity, the size of the specimen, water-to-cement ratio, and paste volume [1]. Autogenous shrinkage is defined as a reduction in the dimensions (volume or length) of cementitious materials without any transfer of moisture to the surrounding environment [2]. As cement hydration progresses, the matrix enters the stage of self-desiccation independently of external conditions; therefore, this type of shrinkage can also be known as self-desiccation shrinkage [3]. Chemical shrinkage is well-defined as the internal early change in the volume of cement-based materials due to the hydration process [4,5]. The cement composition is the principal influence factor on the development of chemical shrinkage [6]. Three methods are mainly applied to measure the chemical shrinkage rate: dilatometry, pycnometry, and gravimetry [5]. The dimensional change that occurs when the specimen is exposed to excess moisture or water is known as "expansion" [7]. Generally, the moisture content and the change in temperature of the relative humidity have significant effects on the mechanism of the expansion [8,9].

Extensive studies have assessed the effect of supplementary cementitious materials (SCMs) on shrinkage magnitudes, and its trend has changed depending on the quantity of the SCM included. It was displayed that concrete made with clay and limestone fines presents high autogenous shrinkage strains compared to the control concrete [10]. The influence of limestone fines (LF) replacing cement in both mortar and paste on the four types of shrinkage has been examined as well [11–14]. The outcomes revealed an improvement in chemical shrinkage for 15% of LF for both cement paste and mortar, and, for up to 10% inclusion, an increase in autogenous shrinkage occurred [11–14]. However, the drying shrinkage increased as LF content increased, and noticeable changes in expansion were observed with replacements surpassing 15% of LF [11–14]. Moreover, it was stated that fly ash replacing ordinary cement in concrete had a good performance in terms of lowering drying shrinkage [15]. Another experimental work reported that in the presence of nano-silica, the chemical shrinkage gradually rose in the first curing days [16]. The effects of furnace bottom ash (FBA) used as a natural sand substitute on the drying shrinkage of concrete were also investigated: the results indicated that as FBA content increased, the rate of drying shrinkage increased [17]. Additionally, the normal cement was partially replaced by combining fly ash (FA), blast furnace slag (BFS), and metakaolin (MK) to produce green concrete with lower drying shrinkage, reducing structural cracking [18].

Lebanon is famous for its olive tree agricultural sector and the olive oil production process, which produces huge amounts of olive oil residues. One of these wastes is "pomace", which is burned to obtain OWA. Actually, this local material contains complex organic components that are difficult to decompose when disposed of in landfills, thereby, causing harm to the ecosystem. The literature review presents a lack of using OWA as a cement substitute in cement-based materials for volumetric change investigation for sustainability purposes. Recently, OWA has been gaining increased interest in light of its efficient properties and environmental friendliness, which is characterized by its pozzolanic reaction, filler effect, and low-cost effectiveness [19,20]. In general, pozzolanic materials are natural or industrial resources essentially consisting of silicon (SiO_2) and aluminum (Al_2O_3) oxides [21]. This material, when mixed with water, reacts with the hydrate products to form a calcium–silicate–hydroxide (C-S-H) gel responsible for strength-development and durability purposes. The stronger the pozzolanic action is, the more attractive it is for diverse applications including building construction, soil stabilization, and production of geopolymers. A previous study reported that OWA was beneficial in enhancing the workability of cement paste [19]. Another research carried out on rammed earth blocks showed that OWA possessed cementitious and pozzolanic behavior and improved its mechanical performance [20]. Additionally, olive waste biomass ash could behave as a filler in self-compacting concrete, resulting in a compressive strength similar to that of the reference mix [21].

In addition to the described mechanical functionality of OWA, shrinkage parameters using different levels were previously examined. A prior investigation was made on non-structural recycled concrete with low replacement percentages of natural sand with olive biomass bottom ash (0, 3, and 6%) yielding a higher drying shrinkage value [21]. On the other hand, the paste expansion tended to drop with the addition of OWA at various levels ranging from 3 to 15% [22]. Another assessment aimed to manufacture cement mortar using different treated biomass bottom ash (BBA) [23]. The laboratory results indicated that OWA contributed to the occurrence of high dimensional changes compared to the control mix [23]. In self-consolidating concrete production, this effect could also be explained by the filler effect of the ash, which inhibited water evaporation [24]. On the other hand, in brick manufacturing, the specimens made with 20% olive pomace bottom ash (OPBA) exhibited the highest linear shrinkage [25]. Olive biomass bottom ash (BBA) was used as an economic agent for expansive soil stabilization in the construction of road embankments due to its capacity to reduce the expansion of soils [26]. Furthermore, BBA was used in mortar specimens to examine its effect on mechanical performance using varying factors [27]. These factors included the type and content of cement implemented (515, 485, and 450 g/L) and the percentage of replacement of sand or cement with BBA (0, 10, and 20%). The findings indicated that the increase in BBA content decreased porosity, density and compressive and flexural strengths. Also, the decline in porosity and density was strongly correlated to the high absorption water of this ash [27]. Another study used pomace and olive kernel as alternatives to cement with percentages of 10 and 30% by mass to fabricate mortars with better stiffness [28]. Based on the evaluation tests, adding 10% of these two alternatives approximately maintained the compressive strength of the control mixture (around 50 MPa). Additionally, the influence of particle size was investigated. As a result, after two hours of milling, the residual for ordinary cement on a 45 µm sieve opening was 20%, and replacing cement with 30% of both agricultural wastes contributed to a larger Blaine-specific surface relative to traditional cement [28]. Moreover, several types of sustainable materials, such as rice husk ash (RHA) and OWA were simultaneously applied to produce high-strength concrete [29]. The substitution levels of RHA used varied from 0 to 25% with an increment of 5%, while OWA levels ranged from 0 to 7.5% with an addition of 2.5%. The tests conducted included the slump test, compressive, splitting tensile, flexural strength, modulus of elasticity, and bond strength. The outcomes emphasized that replacing cement with 20% RHA and 5% OWA significantly enhanced all the mechanical strengths due to the pozzolanic activity of these alternative materials [29]. A previous study was conducted to fabricate self-compacting concrete with two green materials such as sea sand powder (SS) and olive waste bottom ash (OW) instead of traditional filler (limestone powder) [30,31]. To investigate the effect of OW filler on the performance of concrete, different properties including compressive and tensile strengths as well as volume stability were evaluated. As a result, OW contributed to a loss in compressive strength as well as a decrease in workability; however, there was an augmentation in shrinkage values mostly associated with the high porosity [30,31]. Table 1 provides a summary of the key information from all prior studies using OWA or a combination with other SCMs.

To the best of the Authors' knowledge, there is little information in the literature about the influence of OWA as a partial cement replacement on the volume stability of cement paste. For this purpose, five partial replacement compositions of 0, 5, 10, 15, and 20% of OWA were implemented in this research. The mechanical performance, including compressive and flexural strength, was investigated. In addition, this paper provided insights into how OWA impacted autogenous shrinkage, chemical shrinkage, drying shrinkage, and expansion. The correlations between these volumetric stability parameters were also elucidated. It is expected, at the end of this study, that the use of OWA could lead to the development of sustainable infrastructures.

Table 1. Summary of previous studies using OWA with/without other SCMs.

References	Materials (%)	Tests Conducted	Results
[21]	OWA (0, 3, and 6%) replacing natural sand	-Drying shrinkage for non-structural recycled concrete	-Higher drying shrinkage
[22]	OWA (0 to 15% with a constant addition of 3%) replacing cement	-Setting time -Expansion for cement paste	-Increase in setting time, thus, retarded hydration -Gradual decrease in expansion values with increasing OWA content

Table 1. *Cont.*

References	Materials (%)	Tests Conducted	Results
[23]	BBA replacing cement	-Drying shrinkage for mortar	-High dimensional changes related to the increased porosity
[25]	OPBA (0, 5, 10, 15, and 20% by weight)	For clay bricks: -Bulk density -Linear shrinkage -Water absorption -Porosity -Compressive strength -Thermal conductivity	-Greater compressive strength in the sample with 5% of OPBA (11.5 MPa) -Highest linear shrinkage and porosity as well as lowest density for 20% OPBA sample -Best thermal insulation
[26]	BBA (0, 15, 50, and 100%)	In soil stabilization for road embankment	-Enhancement in mechanical properties -Reduction in soil expansion
[27]	-BBA (0, 10, and 20%) replacing cement and natural sand -Two types of cement (CEM-I and CEM-II) -amount of cement (515, 485, and 450 g/L	For mortars: -Compressive strength -Flexural strength -Density -Porosity	-Decline in the mechanical properties -Decrease in density and porosity due to high water absorption of BBA
[28]	Olive pomace and olive kernel replacing cement with 10 and 30% by mass	For cement mortars: -Compressive strength -influence of particle size -Heat evolution -Setting time	-Similar compressive strength to that of the control mixture -After two hours of milling, a larger Blaine-specific surface relative to traditional cement with the addition of 30% of pomace and olive kernel -Greater heat evolution and hydration
[29]	-OWA (0, 2.5, 5, and 7.5%) replacing cement -RHA (0, 5, 10, 15, 20, and 25%) replacing cement	For high-strength concrete: -Slump test -Compressive strength -Splitting tensile strength -Flexural strength -Modulus of elasticity -Bond strength	-Decrease in slump records and workability -Increase in compressive strength by 58% for replacements of 20% RHA with 5% OWA -Improvement in all mechanical strengths due to the pozzolanic activity of these ashes -Larger surface areas of binder, thus, leading to high water absorption -Lower amount of pores related to the densification of the matrix
[30,31]	-SS filler (10 and 15% of total aggregate content) -OW filler (5, 10, and 15% of total aggregate content)	-Compressive strength -Volume stability	-Detrimental effect on workability and compressive strength -Increase in volumetric shrinkage and porosity -The usage of OW must be less than 5% -No segregation and bleeding in OW mixes

2. Experimental Test

2.1. Materials

Ordinary Portland Cement CN PA-L 42.5-type obtained from Sabaa plant, Tripoli, Lebanon and OWA with a density of 950 kg/m^3 were used in this work. The OWA was delivered from an olive press located in Zgharta, North Lebanon, and obtained after burning large amounts of olive residue generated from the olive oil extraction in a boiler for 8 h. The OWA was milled using a Los Angeles abrasion machine for two hours to obtain finer particles. Finally, OWA was sieved through a No. 200 sieve. A brief description of the process of olive waste ash OWA production is clarified in Figure 1. The chemical composition and the particle size distribution of OWA are shown in Table 2 and Figure 2, respectively. The chemical composition showed that OWA mainly consists of CaO (36.13%), SiO_2 (24.73%), and K_2O (9.56%).

Figure 1. Process of OWA preparation.

Table 2. Chemical composition of OWA and cement.

Oxide	SiO_2	Al_2O_3	Fe_2O_3	CaO	MgO	SO_3	K_2O	Na_2O	LOI	Other
OWA	24.73	3.41	3.83	36.13	2.81	0.03	9.56	1.42	14.7	3.38
Cement	18.53	3.93	3.06	61.78	1.74	2.92	0.47	0.18	6.3	1.09

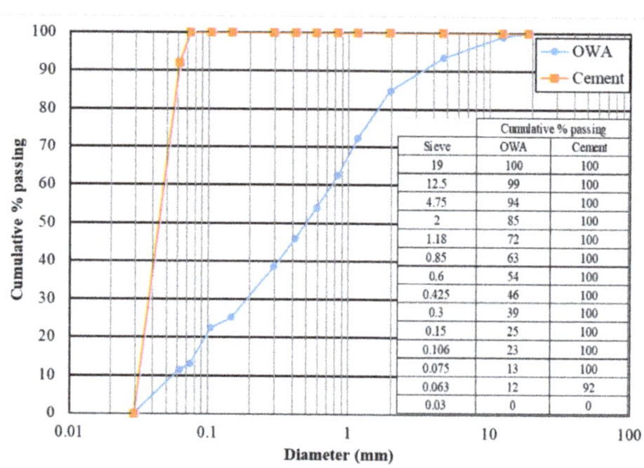

Figure 2. Particle size distribution of OWA and cement.

2.2. Mixture Proportions

Five paste mixtures were prepared with five percentages of 0, 5, 10, 15, and 20% of OWA replacing the cement. For each type of shrinkage test (chemical, drying, autogenous shrinkage, and expansion), ten specimens were cast so the total number of specimens was forty. The water-to-cement ratio used was 0.45. Table 3 summarizes all mixture proportions and component amounts.

Table 3. Mix proportions and component quantities of paste.

	Proportions			Amount (kg/m^3)		
Paste Code	**Cement**	**OWA**	**W/C Ratio**	**Cement**	**OWA**	**Water**
P0%	1	0	0.45	1303.5	0.0	586.2
P5%	0.95	0.05	0.45	1260.0	66.2	566.9
P10%	0.9	0.1	0.45	1215.1	134.7	546.9
P15%	0.85	0.15	0.45	1168.5	205.6	526.2
P20%	0.8	0.2	0.45	1120.1	280.6	504.1

2.3. Testing Procedure and Specimen Preparation

2.3.1. Compressive and Flexural Strengths

To determine the compressive strength, 50 × 50 × 50 mm cubes were tested following ASTM C109 guidelines [32]. According to the flexural strength evaluation, beams of 40 × 40 × 160 mm were used as per ASTM C348 standards [33]. The data were measured at 7, 28, and 90 days.

2.3.2. Chemical Shrinkage

Chemical shrinkage was tested according to ASTM 1608 by measuring the internal volume change in cement paste due to the hydration of the cementing materials [34]. The equipment employed in this test was a graduated pipette of 2 mL, a bottle of 250 mL, a spatula, and a rubber stopper. The procedure started by placing 30 g of each mix at the bottom of the bottle corresponding to a depth of 1.8 cm. Then, the water was slowly added to the top of the matrix until the bottle was filled. The pipette was inserted into the bottle through the stopper, filled with water, and sealed at the top with a drop of oil to prevent water evaporation, as seen in Figure 3. The reading of the drop in the level of water in the pipette refers to the chemical shrinkage (change in volume) [35,36] and is computed every one hour for the first 24 h and then every two days until reaching 90 days. Two replicate samples were tested for each mix. The chemical shrinkage value was monitored by calculating the average of two readings, with the first reading considered as zero. The calculation of chemical shrinkage is expressed by the following equation:

$$\Delta V/V = 3\ \Delta L/L \quad (1)$$

where ΔV = change in volume in the pipette (mL), V = initial volume of the sample (mL), ΔL = change in length in the pipette (μm), L = initial length of the sample (m).

2.3.3. Drying Shrinkage

Drying shrinkage was tested following ASTM C157 [37]. For this test, molds of 25 × 25 × 300 mm^3 were used. After 24 h in the curing chamber (25 °C), the samples were demolded, and two Demec points separated by a distance of 200 mm were placed on each side of the specimen. The displacement of the Demec points was measured using a dial gauge and the length was reported every 2 days for 90 days while the specimens were in the chamber. The average of four readings from two replicated specimens refers to the drying shrinkage (change in length). The specimens are shown in Figure 4.

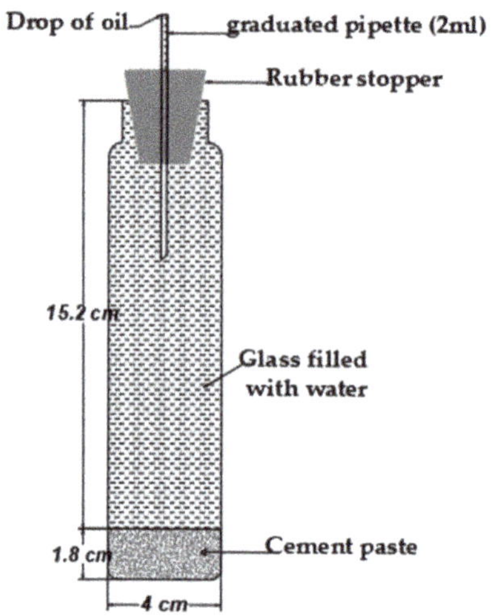

(a)

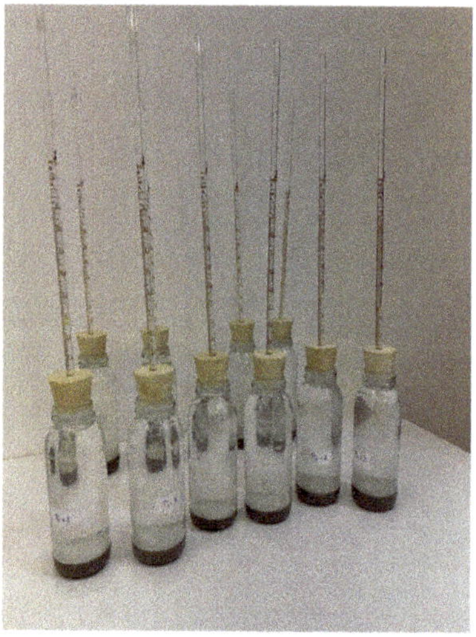

(b)

Figure 3. (**a**) Chemical shrinkage setup; (**b**) chemical shrinkage specimens.

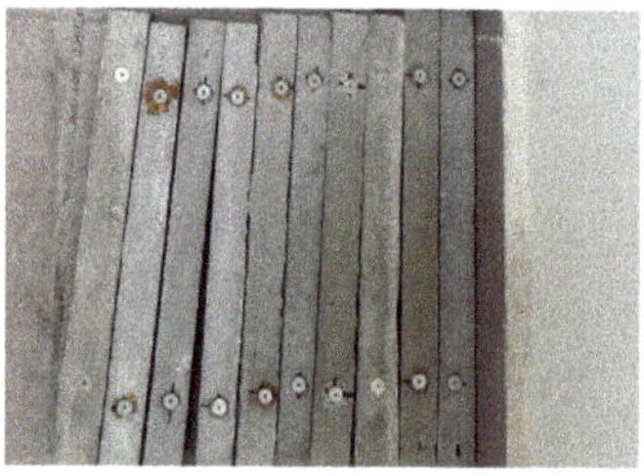

(a)

(b)

Figure 4. Specimens for (**a**) drying shrinkage; and (**b**) autogenous shrinkage.

2.3.4. Autogenous Shrinkage

Paste specimens of 25 × 25 × 300 mm^3 were tested according to ASTM C192 [38]. To record autogenous shrinkage (starting after 24 h), each specimen was immediately covered with a plastic bag as illustrated in Figure 4. After 24 h, the mean values of four readings represent the starting point (reference data). The same process for measuring the drying shrinkage records was employed for autogenous shrinkage.

2.3.5. Expansion

The specimens were cast in 25 × 25 × 300 mm^3 molds. After 24 h, the formwork was removed, and each specimen was totally immersed in water at a constant temperature (20 ± 1 °C) as illustrated in Figure 5. The same procedure for monitoring the drying shrinkage data was applied for monitoring expansion as well.

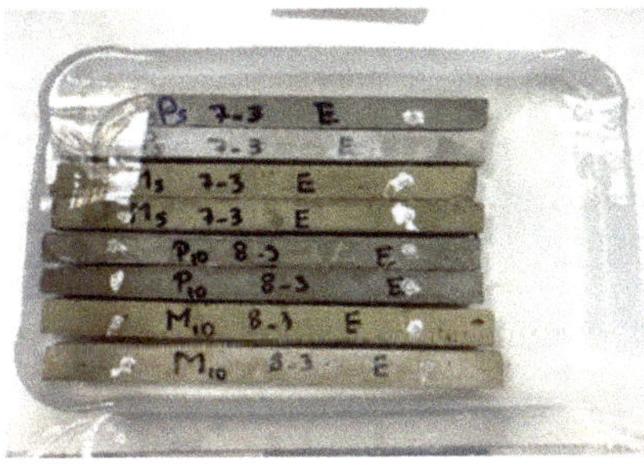

(a)

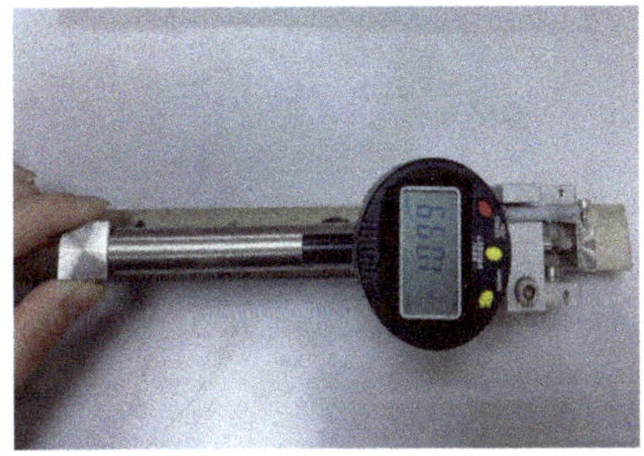

(b)

Figure 5. (**a**) Expansion specimens; (**b**) strain gauge.

3. Analysis of the Results

3.1. Compressive and Flexural Strength

The effects of OWA induced at various percentages on compressive and flexural strengths are indicated in Table 4. As can be seen from the results, as the curing time increases, there is a significant increase in compressive strength values. For example, the control mixture displays compressive strength values of 27, 47, and 50 MPa at 7, 28, and 90 days, respectively. However, the incorporation of OWA decreases these measurements independently of the amount of OWA induced. More precisely, among all mixtures containing OWA, there is a negligible loss in compressive strength in the P10% sample with a rate of 8%, displaying a magnitude of 46 MPa after 90 days of curing. Therefore, 10% of OWA can be considered as the optimal substitution level. This behavior could be attributed to the fact that the chemical composition—with 36.1% CaO and 24.7% SiO_2—indicated that OWA was a cementitious material with good pozzolanic properties; this leads to the formation of additional C-S-H gel, which is the major contributor to strengthening the matrix [20,28]. Compared to the reference mixture, the decrease in compressive strength continues for the addition of 15 and 20% of OWA, achieving drops of 38 and 59%, respectively. There are two potential reasons for this: First, there was a decrease in hydrate products (C_2S and C_3S) resulting from the loss of cement [20,28]. Second, the higher porosity took place with the increase in OWA content [23,25].

Table 4. Compressive and flexural strengths of all cement pastes.

Paste	Compressive Strength (MPa)			Flexural Strength (MPa)		
	7 Days	28 Days	90 Days	7 Days	28 Days	90 Days
P0%	27.2	47	50	2.6	3.7	4
P5%	23	31.5	32.2	2.4	3.4	3.6
P10%	25.4	42.5	45.4	2.5	3.6	3.8
P15%	23	28.7	31	2.3	2.6	2.8
P20%	14.4	18.9	20.3	2	2.3	2.5

The flexural strength results show an identical trend to the compressive strength pattern. For instance, the 90-day flexural strength is 4 MPa for the paste without OWA. The sample containing 10% of OWA displays the lowest rates of 3 and 5% after 28 and 90 days, respectively; however, this decline peaks with the addition of 20% of OWA, achieving 37%. The causes of this negative effect of OWA are as follows: first, the reduction in cement content in the matrix, and second, OWA seemed to weaken the bonding between OWA and the cement particles associated with the augmentation of the number of voids in the paste [27,29].

3.2. Chemical Shrinkage

Figure 6 depicts the results for the chemical shrinkage (μm/m) measured over 90 days. The time immediately following placement (t = 0) is considered the starting point for data collection. It is noteworthy that in the early period of curing, all curves exhibit high chemical shrinkage. Conversely, during the late ages, there is no alteration in chemical shrinkage values, which are observed in a steady state. These findings are consistent with a prior study [16]. For instance, on day 1, the chemical shrinkage for P0% was 0 μm/m, and this value rises to magnitudes of 660 and 811 μm/m at 28 and 90 days, corresponding to an increase of 23%. As previously mentioned, chemical shrinkage is referred to as the reduction in the total volume of hydrate products during the hydration process of cement. With the increase in curing time, this decrease gradually progresses during the hardening stages, thus leading to an increase in chemical shrinkage [12–14]. On the other hand, it is noticed that 10% of OWA partially replacing cement extremely reduces the chemical shrinkage to a measurement of 680 μm/m after 90 days of curing, showing a great reduction rate of 16%. However, the samples containing different OWA content do not show the same significant decline as observed in the sample P10%; for example, P5%, P15%, and P20% show chemical shrinkage values of 740, 750, and 780 μm/m, respectively. This slight drop is equivalent to 8, 8, and 3% compared to the control paste.

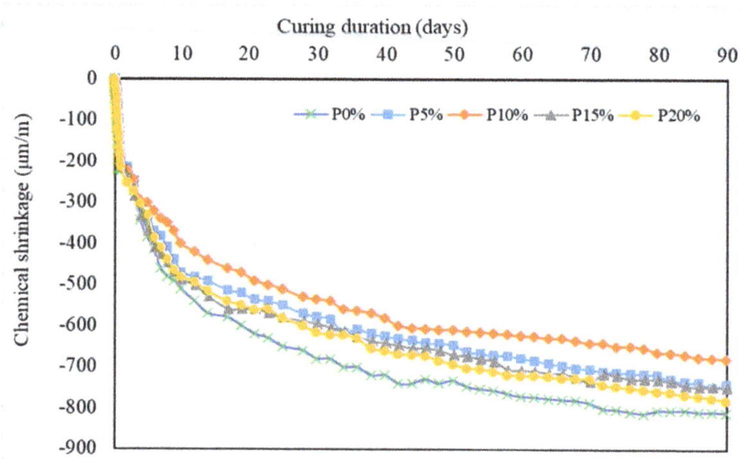

Figure 6. Chemical shrinkage of pastes with different levels of OWA measured over 90 days.

The efficiency of implementing OWA in cement paste was referred to as its potential to reduce chemical shrinkage. This effect could be mostly explained by the fact that OWA contains large amounts of SiO_2, and Al_2O_3, indicating that OWA was likely linked to the pozzolanic activity, as stated in the literature review [21,28,39]. Therefore, the decline in chemical shrinkage could be attributed to two factors: first, the loss of cement in the paste, thereby the reduction in clinker phases responsible for chemical shrinkage occurrence; and second, the slower pozzolanic reactivity of OWA with the calcium hydroxide (CH) obtained from cement hydration. The two factors formed a combined effect in delaying the early-age hydration mechanism, therefore lowering chemical shrinkage measurements [23]. Specifically, among all mixes incorporating OWA, the mix containing 10% OWA shows the optimal reduction level as this percentage indicates a good balance between the loss of cement quantity replaced with OWA and the sufficiency of the pozzolanic reaction to continuously form hydrate products, maximizing the decrease in chemical shrinkage.

3.3. Drying Shrinkage

Figure 7 displays the drying shrinkage values of all specimens measured over 90 days. As a result, the curing time shows a major effect on the magnitude of drying shrinkage [40,41]. For example, on day 1 after placing, P5% is recorded as 0 µm/m; then, this value rises, reaching 1125 and 1850 µm/m at 28 and 90 days, respectively. Meanwhile, the drying shrinkage rises by 64% from 28 to 90 days. In addition, the drying shrinkage measurements show a logical variation with varying proportions of OWA. At lower substitution levels of 5 and 10% of OWA content, the drying shrinkage exhibits a considerable decrease, followed by a notable increase at higher levels of 15 and 20% of OWA content. For example, after 90 days, the drying shrinkage for P0% is 2015 µm/m; then, this value sharply drops to reach 1730 µm/m for 10% OWA content (P10%), achieving a decline of 14%. However, inducing higher percentages of OWA increases the drying shrinkage for samples P15% and P20% to 2275 and 2350 µm/m, representing increases of 13% and 16% over the same period, respectively.

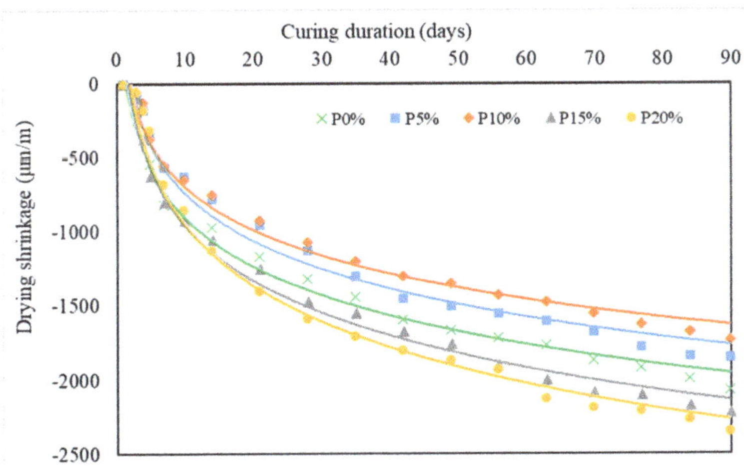

Figure 7. Drying shrinkage of pastes with different levels of OWA measured over 90 days.

The test results demonstrate that the drying shrinkage behaved differently in the existence of OWA [12,21]. These changes are attributed to a variety of factors like the material's stiffness, pore structure, and internal pressures. Thus, with the addition of 5 and 10% (low levels), the pozzolanic reaction of OWA might play a significant role in decreasing this type of shrinkage. Hence, as additional C-S-H gel developed, the matrix became denser, leading to a notable reduction in porosity, meaning that less water was needed for evaporation when the paste dried, mitigating the drying shrinkage [41]. However, with the addition of 15 and 20% (high levels), OWA might incompletely compensate for the increased loss of cement amount, resulting in a rise in capillary pores, which facilitates the loss of water [1,42]. Accordingly, the high and continuous evaporation of water within the paste contributed to a reduction in the material's stiffness and, consequently, a weakness in resisting the internal pressures caused by moisture escape, leading to an amplification of the deformation of the beam under drying conditions [17]. Furthermore, another relevant explanation for the increase in drying shrinkage could be the disjoining pressure that exists between the pores promoting the contraction of the solid body [9,22,43,44]. In light of the above outcomes, it is highly recommended to avoid using high replacement levels in the manufacturing of more durable cement-based materials, as well-demonstrated in a prior study [23]. It is inferred that the optimal percentage of cement replacement with OWA can be achieved at 10% due to its substantial effect on mitigating drying shrinkage, powerfully adequate for improving long-term durability.

3.4. Autogenous Shrinkage

The data for autogenous shrinkage of all cement pastes during 90 days are illustrated in Figure 8. As seen from the curve, there is a considerable impact of the curing time on the development of self-desiccation, particularly since all values increase as the curing time increases [45,46]. For instance, P5% shows 0 μm/m on day 1 of curing; this value rises to 600 and 950 μm/m at 28 and 90 days, respectively. In addition, incorporating OWA improves the autogenous shrinkage of the paste at various OWA content. For instance, after 90 days, P0% records the highest autogenous shrinkage value of 1125 μm/m. The inclusion of 10% of OWA content ultimately reduces autogenous shrinkage to 788 μm/m, showing a decrease of 30%. However, this drop is less pronounced with the incorporation of 15 and 20% of OWA, showing magnitudes reaching 950 and 975 μm/m, respectively. This is equivalent to a drop of 15 and 13% compared to the reference paste.

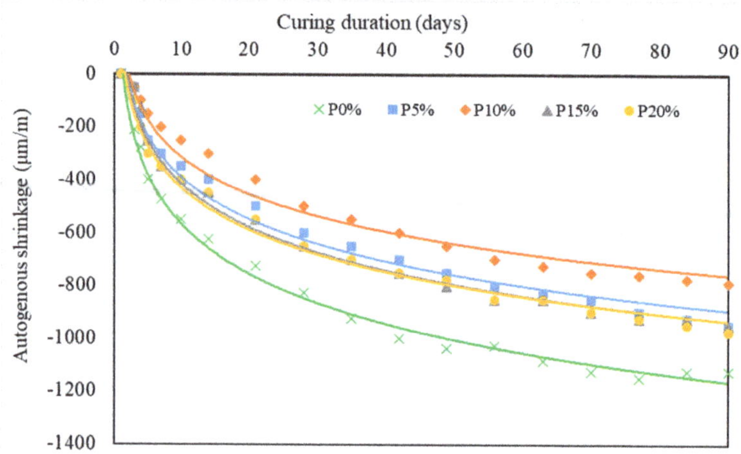

Figure 8. Autogenous shrinkage of pastes with different levels of OWA measured over 90 days.

Based on the findings, as autogenous shrinkage was directly dependent on the early-age hydration mechanism, the drop in autogenous shrinkage values could be explained by the fact that the additional OWA retarded the hydration process in the early stages, lowering the magnitude of autogenous shrinkage [47,48]. In this context, the pozzolanic properties of OWA potentially slowed the reaction between OWA and CH, reducing the early-age hydration, and mitigating, by this action, the self-desiccation of the paste [20,48,49]. On the other hand, the filler effect of OWA led to the refinement of the pores, which slowed down the internal water movement into the paste, inhibiting by this behavior the development of self-desiccation as the fine particles could accommodate the voids [19,21]. It is crucial to note that OWA is an effective material in the improvement of the relative humidity, which in turn helps the mix to reduce the self-desiccation changes [3,50–52]. As expected, at the level of 10% OWA, there is a perfect balance between the reduced amount of cement and the suitable percentage of OWA for pozzolanic reactivity, thus controlling the porosity in a way to not be exceeded and maximizing the decrease in autogenous shrinkage. Accordingly, it is recommended to use 10% to prevent the volumetric instability resulting from autogenous shrinkage.

3.5. Expansion

Figure 9 presents the measurements for expansion of all the pastes with the varying OWA content in μm/m measured over 90 days. As displayed, all mixes show a continuous expansion along with the curing time. It can also be observed that the addition of up to 10% increases the expansion values, while inducing 15% and beyond declines the expansion.

For instance, the reference mixture (P0%) has an expansion level of 2075 μm/m at 90 days. This value slightly rises for P5% and P10%, reaching 2250 and 2175 μm/m, corresponding to a growth of 8 and 5%, respectively. Then, these magnitudes drop for P15% and P20%, achieving 1900 and 1550 μm/m, respectively; this is equivalent to 9 and 25% increases relative to the control paste.

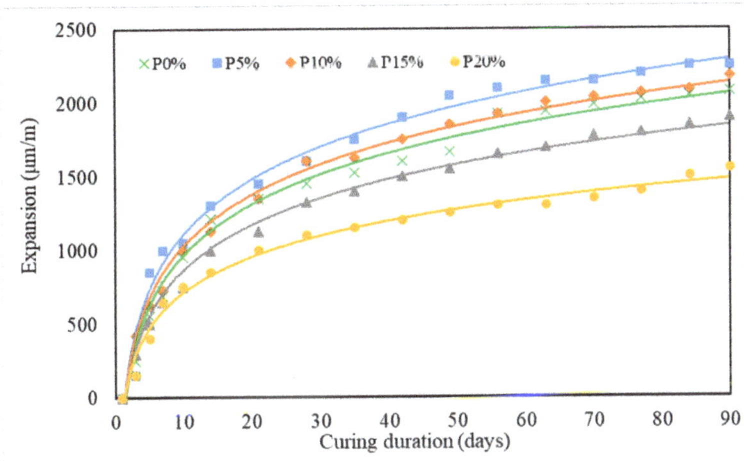

Figure 9. Expansion of pastes with different levels of OWA measured over 90 days.

The different behavior of expansion in the presence of OWA was principally attributed to the pozzolanic reactions of OWA. For OWA content up to 10%, the expansion occurred because the amount of OWA induced was insufficient to react with CH resulting from hydration This remaining quantity of CH could subsequently react with some available constituents in OWA, contributing to the generation of expansive agents [53,54]. On the contrary, exceeding 10% of OWA content led to a further reduction in expansion measurements. This was mostly based on several reasons: First, the higher reduction in CH, the less production of expansive agents, making the paste more stable against expansion [55,56]. Second, this stability could also be the result of the filler effect of OWA, allowing it to accommodate the expansion volume [19,49]. Finally, the chemical composition of OWA also provided a reasonable indication of expansion performance. Particularly, as OWA replacement levels increase, the quantity of CaO decreases, which in turn affects the swelling of the specimens [57]. In addition, the limited quantity of SO_3 present in OWA (below the 3.5% threshold) reacted with CaO to form small amounts of ettringite, which is an expansive agent, which implied expansion limitations [58–60]. For this reason, it was beneficial to use higher replacement levels of cement with OWA to withstand expansion, showing a pronounced decrease in swelling. In particular, among all mixes, the paste containing 20% OWA possesses the optimal percentage of replacement in terms of lowering expansion.

3.6. Relationships between Length Change Parameters

Figure 10 displays the length change parameters (drying, autogenous, chemical shrinkage, and expansion) among all mixes during the 90 days, which evolve in the early ages and are prolonged for the long term [52,61]. It appears, from the obtained results, that OWA substantially improves various types of shrinkage, which are key indicators of the durability and serviceability of cement-based materials. The magnitude of each type of volumetric stability parameter could be interpreted by the hydration process and the development of pores within the blended cement paste [47,53]. This study reveals that drying shrinkage records the highest volumetric variances compared to other types of shrinkage, reaching

2350 μm/m in the P20% sample. This was primarily caused by moisture diffusion in drying conditions and probably could last for future years after the curing of the paste [62–64]. This type is the best criterion for longer durability and, thereby, the lifespan of the structure [65].

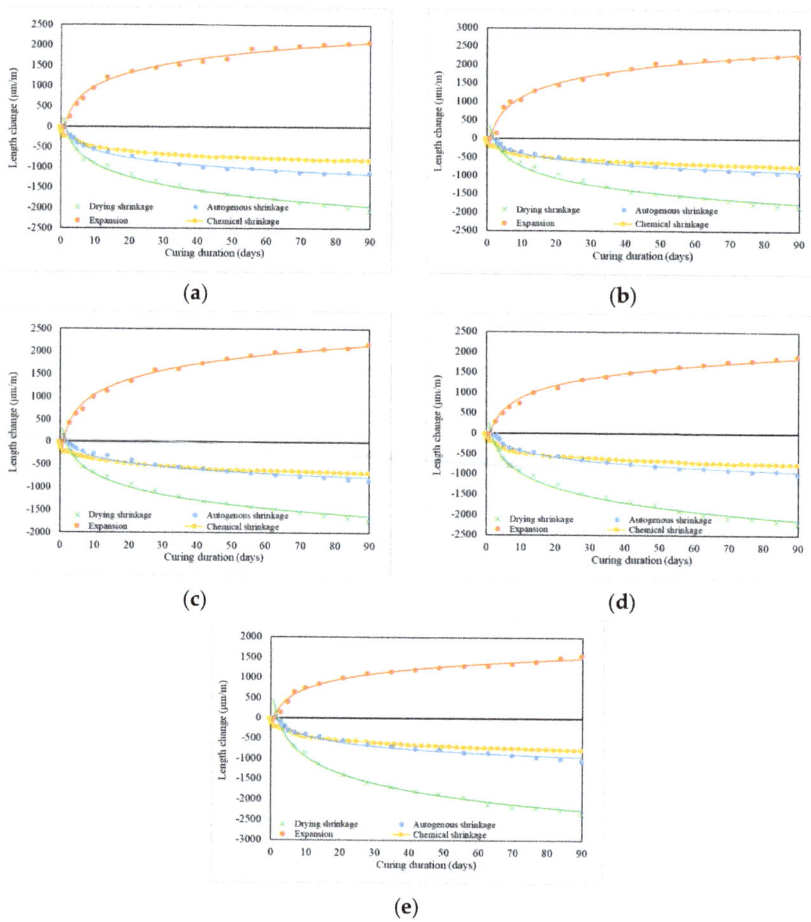

Figure 10. Length change in pastes containing (**a**) 0% OWA; (**b**) 5% OWA; (**c**) 10% OWA; (**d**) 15% OWA; (**e**) 20% OWA.

Furthermore, two points can be observed in these figures: First, the chemical and autogenous shrinkage are equal in the early ages (the first few days) [45]. Second, the autogenous shrinkage values are higher than those for chemical shrinkage as time increases. For instance, at 90 days, for 5% addition, the chemical and autogenous readings are 740 and 950 μm/m, indicating a difference rate of 28%, as seen in Figure 10b. The inclusion of 10% OWA exhibits the lowest chemical and autogenous shrinkage values of 680 and 788 μm/m, respectively, as illustrated in Figure 10c. This is equivalent to a difference rate of 16%. However, the incorporation of 15% OWA raises the difference rate to 27%, as shown in Figure 10d. Similarly, the rate of difference between the chemical and autogenous shrinkage is around 25% for P20% samples, as illustrated in Figure 10e. The cause of these findings could be explained by the fact that both mechanisms are directly linked to the hydration process and occurred in the early periods [45,66,67].

To further visualize the relationships of the shrinkage performance properties, a regression equation is used to determine the coefficient of determination R^2 as displayed in

Figures 11–13 for days 1, 7, 28, and 90. As can be seen, a positive linear correlation exists between chemical shrinkage and drying shrinkage, with a high R^2 of 0.97, 0.99, 0.99, 0.97, and 0.99 for P0%, P5%, P10%, and P20%, respectively. This is because the reduction in volume resulting from the chemical reactions could induce additional internal stresses and pores, leading to higher shrinkage as the paste dries [68–70].

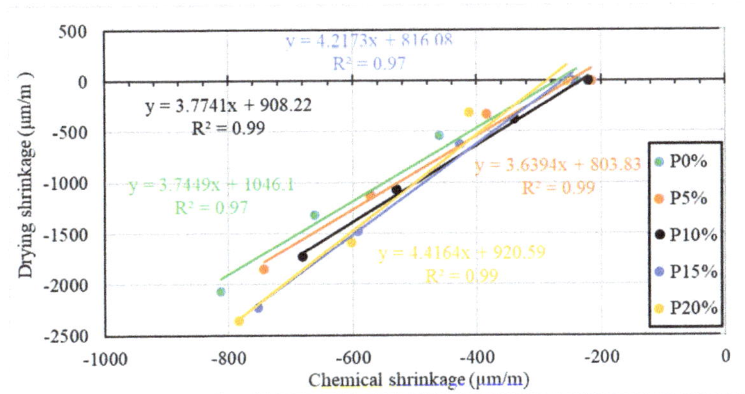

Figure 11. Correlation between drying and chemical shrinkage for all levels of OWA.

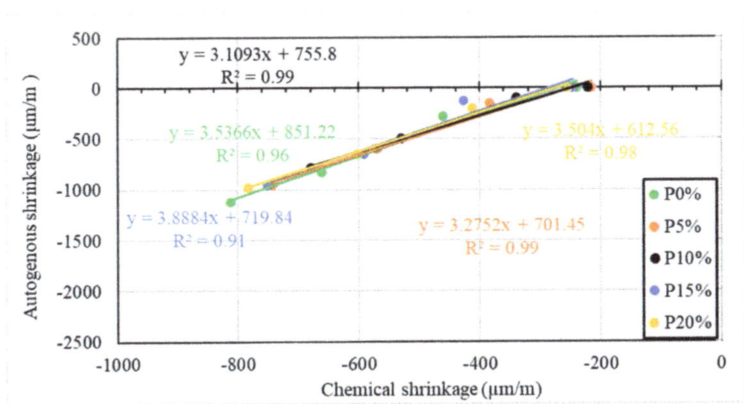

Figure 12. Correlation between autogenous and chemical shrinkage for all levels of OWA.

Furthermore, there is a strong dependence between chemical and autogenous shrinkage, showing a significant coefficient of correlation R^2 equal to 0.96, 0.99, 0.99, 0.91, and 0.98 for P0%, P5%, P10%, P15%, and P20%, respectively. One possible explanation of this link is that chemical shrinkage can be called "internal autogenous shrinkage" while autogenous shrinkage can be referred to as "external chemical shrinkage" in some cases [63,67]. Hence, in the early stage, after reaching the stiffening point and the formation of the structure, the matrix enters the self-desiccation process due to the retention of moisture in the capillary pores; in this case, both autogenous and chemical shrinkage cannot be considered interchangeable [3,67,68].

Conversely, the expansion indicates an inverse relationship with chemical shrinkage, as evidenced by a high R^2. This correlation may be interpreted to mean that the chemical shrinkage can continuously occur even if the paste is in an expansion state, especially when, in both cases, the paste is still immersed in cured water [71]. Overall, it is fundamentally justified that chemical shrinkage is the driving force for other types of shrinkage.

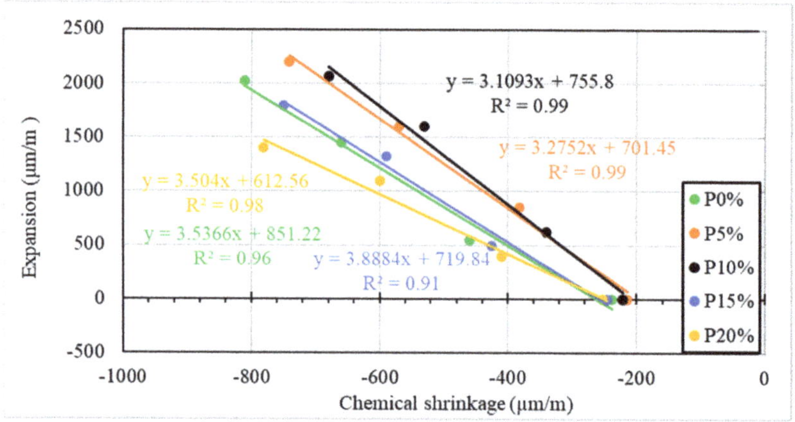

Figure 13. Correlation between expansion and chemical shrinkage for all levels of OWA.

4. Conclusions

It is interesting to develop a sustainable management strategy to save natural resources and promote their utilization in construction. To achieve this target, olive waste ash (OWA) is used as a partial replacement for cement to test the volume stability parameters including chemical, autogenous, and drying shrinkage as well as expansion. Based on this study, several points can be outlined:

- The presence of OWA adversely affects the mechanical properties of the pastes. For the addition of 10% OWA, the rate of reduction in both compressive and flexural strength was around 8 and 5%, respectively, compared to the control mixture. Beyond this level, at the 15 and 20% OWA levels, the compressive and flexural strengths significantly decreased, reaching 59 and 36%, respectively, for P20% samples;
- OWA had a positive impact on chemical shrinkage. At day 90, the chemical shrinkage P0% exhibited the highest level of chemical shrinkage. However, adding 10% of OWA sharply reduced the chemical shrinkage to a rate of 16%. The drop in chemical shrinkage was found to be less pronounced for the incorporation of 15% and 20% of OWA, indicating a reduction of 8 and 3%, respectively, compared with the free OWA paste;
- The variation in drying shrinkage depended on the percentages of OWA content. Hence, in the comparison with the control mix, the drying shrinkage ultimately decreased, reaching a rate of 14% for P10% after 90 days of curing. However, there was an increase in drying shrinkage for P15% and P20% at the rates of 13 and 16% over the same period;
- Regarding autogenous shrinkage, there was a decrease in autogenous measurements with varying OWA content. This drop was optimal for the incorporation of 10%, showing a percentage of decline of 30%. On the other hand, this reduction was less pronounced in the P15% and P20% mixes, which exhibited decreases of 15 and 13% compared to the reference paste;
- The expansion behaved differently with varying OWA levels. Hence, P5% and P10% exhibited a slight increase in expansion measurements, reaching rates of 8 and 5%. Contrarily, a notable decrease was observed for P15% and P20%, achieving drops of 9 and 25% relative to the control paste;
- A positive linear relationship was observed between chemical shrinkage and drying and autogenous shrinkage. In addition, a negative linear correlation was found between chemical shrinkage and expansion. These correlations confirmed that chemical shrinkage was the driving force for the other types of shrinkage.

These conclusions suggest that olive waste ash has a substantial impact on the shrinkage and expansion of cement paste, highlighting its effectiveness as a partial replacement for cement in the development of more durable and sustainable construction materials.

Author Contributions: Conceptualization, J.K. and H.G.; methodology, S.G. and H.G.; formal analysis, S.G. and H.G.; writing—original draft preparation, S.G. and H.G.; writing—review and editing, H.G. and S.E.Z.; supervision, J.K. and A.E.; project administration, H.G. All authors have read and agreed to the published version of the manuscript.

Funding: This research received no external funding.

Data Availability Statement: Data are included in this paper.

Acknowledgments: The authors express their gratitude for the assistance provided by the staff and technicians at BAU Laboratories.

Conflicts of Interest: The authors declare no conflicts of interest.

References

1. Bissonnette, B.; Pierre, P.; Pigeon, M. Influence of key parameters on drying shrinkage of cementitious materials. *Cem. Concr. Res.* **1999**, *29*, 1655–1662. [CrossRef]
2. Holt, E. Contribution of mixture design to chemical and autogenous shrinkage of concrete at early ages. *Cem. Concr. Res.* **2005**, *35*, 464–472. [CrossRef]
3. Wu, L.; Farzadnia, N.; Shi, C.; Zhang, Z.; Wang, H. Autogenous shrinkage of high performance concrete: A review. *Constr. Build. Mater.* **2017**, *149*, 62–75. [CrossRef]
4. Tazawa, E.I.; Miyazawa, S.; Kasai, T. Chemical shrinkage and autogenous shrinkage of hydrating cement paste. *Cem. Concr. Res.* **1995**, *25*, 288–292. [CrossRef]
5. Zhang, T.; Gao, P.; Luo, R.; Guo, Y.; Wei, J.; Yu, Q. Measurement of chemical shrinkage of cement paste: Comparison study of ASTM C 1608 and an improved method. *Constr. Build. Mater.* **2013**, *48*, 662–669. [CrossRef]
6. Yodsudjai, W.; Wang, K. Chemical shrinkage behavior of pastes made with different types of cements. *Constr. Build. Mater.* **2013**, *40*, 854–862. [CrossRef]
7. Van Breugel, K.; Van Tuan, N. Autogenous Shrinkage of HPC and Ways to Mitigate it. *Key Eng. Mater.* **2015**, *629*, 3–20. [CrossRef]
8. Sellevold, E.J.; Bjøntegaard, Ø. Coefficient of thermal expansion of cement paste and concrete: Mechanisms of moisture interaction. *Mater. Struct.* **2006**, *39*, 809–815. [CrossRef]
9. Thomas, J.J.; Jennings, H.M. Changes in the size of pores during shrinkage (or expansion) of cement paste and concrete. *Cem. Concr. Res.* **2003**, *33*, 1897–1900. [CrossRef]
10. Nguyen, Q.D.; Afroz, S.; Zhang, Y.; Kim, T.; Li, W.; Castel, A. Autogenous and total shrinkage of limestone calcined clay cement (LC3) concretes. *Constr. Build. Mater.* **2022**, *314*, 125720. [CrossRef]
11. Khatib, J.; Ramadan, R.; Ghanem, H.; ElKordi, A. Effect of using limestone fines on the chemical shrinkage of pastes and mortars. *Environ. Sci. Pollut. Res.* **2023**, *30*, 25287–25298. [CrossRef] [PubMed]
12. Khatib, J.; Ramadan, R.; Ghanem, H.; Elkordi, A. Volume stability of cement paste containing limestone fines. *Buildings* **2021**, *11*, 366. [CrossRef]
13. Khatib, J.M.; Ramadan, R.; Ghanem, H.; Elkordi, A.; Sonebi, M. Effect of limestone fines as a partial replacement of cement on the chemical, autogenous, drying shrinkage and expansion of mortars. *Mater. Today Proc.* **2022**, *58*, 1199–1204. [CrossRef]
14. Khatib, J.M.; Ramadan, R.; Ghanem, H.; Elkordi, A.; Baalbaki, O.; Kırgız, M. Chemical shrinkage of paste and mortar containing limestone fines. *Mater. Today Proc.* **2022**, *61*, 530–536. [CrossRef]
15. Ruengsillapanun, K.; Udtaranakron, T.; Pulngern, T.; Tangchirapat, W.; Jaturapitakkul, C. Mechanical properties, shrinkage, and heat evolution of alkali activated fly ash concrete. *Constr. Build. Mater.* **2021**, *299*, 123954. [CrossRef]
16. Wang, J.; Cheng, Y.; Yuan, L.; Xu, D.; Du, P.; Hou, P.; Zhou, Z.; Cheng, X.; Liu, S.; Wang, Y. Effect of nano-silica on chemical and volume shrinkage of cement-based composites. *Constr. Build. Mater.* **2020**, *247*, 118529. [CrossRef]
17. Bai, Y.; Darcy, F.; Basheer, P.A.M. Strength and drying shrinkage properties of concrete containing furnace bottom ash as fine aggregate. *Constr. Build. Mater.* **2005**, *19*, 691–697. [CrossRef]
18. Liu, K.; Shui, Z.; Sun, T.; Ling, G.; Li, X.; Cheng, S. Effects of combined expansive agents and supplementary cementitious materials on the mechanical properties, shrinkage and chloride penetration of self-compacting concrete. *Constr. Build. Mater.* **2019**, *211*, 120–129. [CrossRef]
19. Cuenca, J.; Rodríguez, J.; Martín-Morales, M.; Sánchez-Roldán, Z.; Zamorano, M. Effects of olive residue biomass fly ash as filler in self-compacting concrete. *Constr. Build. Mater.* **2013**, *40*, 702–709. [CrossRef]
20. Ghanem, H.; Bouz, C.E.; Ramadan, R.; Trad, A.; Khatib, J.; Elkordi, A. Effect of Incorporating Cement and Olive Waste Ash on the Mechanical Properties of Rammed Earth Block. *Infrastructures* **2024**, *9*, 122. [CrossRef]
21. Beltrán, M.G.; Agrela, F.; Barbudo, A.; Ayuso, J.; Ramirez, A. Mechanical and durability properties of concretes manufactured with biomass bottom ash and recycled coarse aggregates. *Constr. Build. Mater.* **2014**, *72*, 231–238. [CrossRef]

22. Alkheder, S.; Obaidat, Y.T.; Taamneh, M. Effect of olive waste (Husk) on behavior of cement paste. *Case Stud. Constr. Mater.* **2016**, *5*, 19–25. [CrossRef]
23. Rosales, J.; Cabrera, M.; Beltrán, M.G.; López, M.; Agrela, F. Effects of treatments on biomass bottom ash applied to the manufacture of cement mortars. *J. Clean. Prod.* **2017**, *154*, 424–435. [CrossRef]
24. Şahmaran, M.; Yaman, İ.Ö.; Tokyay, M. Transport and mechanical properties of self consolidating concrete with high volume fly ash. *Cem. Concr. Compos.* **2009**, *31*, 99–106. [CrossRef]
25. El Boukili, G.; Ouakarrouch, M.; Lechheb, M.; Kifani-Sahban, F.; Khaldoune, A. Recycling of olive pomace bottom ash (by-product of the clay brick industry) for manufacturing sustainable fired clay bricks. *Silicon* **2021**, *14*, 4849–4863. [CrossRef]
26. Cabrera, M.; Rosales, J.; Ayuso, J.; Estaire, J.; Agrela, F. Feasibility of using olive biomass bottom ash in the sub-bases of roads and rural paths. *Constr. Build. Mater.* **2018**, *181*, 266–275. [CrossRef]
27. Beltrán, M.G.; Barbudo, A.; Agrela, F.; Jiménez, J.R.; de Brito, J. Mechanical performance of bedding mortars made with olive biomass bottom ash. *Constr. Build. Mater.* **2016**, *112*, 699–707. [CrossRef]
28. Kerrai, L.; Belaadi, S.; Solimando, R.; Zirour, F.R. Valorisation of organic waste: Use of olive kernels and pomace for cement manufacture. *J. Clean.* **2020**, *277*, 123703. [CrossRef]
29. Hakeem, I.Y.; Agwa, I.S.; Tayeh, B.A.; Abd-Elrahman, M.H. Effect of using a combination of rice husk and olive waste ashes on high-strength concrete properties. *Case Stud. Constr. Mater.* **2022**, *17*, e01486. [CrossRef]
30. Cheraghalizadeh, R.; Akcaoglu, T. Utilization of Olive Waste Ash and Sea Sand Powder in Self-compacting Concrete. *Iran. J. Sci. Technol. Trans. Civ. Eng.* **2019**, *43*, 663–672. [CrossRef]
31. Akçaoğlu, T.; Cheraghalizadeh, R. *Effects of Limestone Powder, Olive Waste Ash and Sea Sand Powder on Properties of Self Compacting Concrete*; Eastern Mediterranean University, Faculty of Engineering, Dept. of Civil Engineering: Gazimağusa, Cyprus, 2019.
32. *ASTM, C109/C109M-16a*; Standard Test Method for Compressive Strength of Hydraulic Cement Mortars (Using 2-in. or [50-mm] Cube Specimens). ASTM International: West Conshohocken, PA, USA, 2016.
33. *ASTM, C348-02*; Standard Test Method for Flexural Strength of Hydraulic-Cement Mortars. ASTM International: West Conshohocken, PA, USA, 2002.
34. *ASTM, C. 1608*; Standard Test Method for Chemical Shrinkage of Hydraulic Cement Paste. ASTM International: West Conshohocken, PA, USA, 2007; pp. 667–670.
35. Sarkar, S.; Halder, A.; Bishnoi, S. Shrinkage in concretes containing fly ash. In Proceedings of the UKIERI Concrete Congress 2013, Jalandhar, India, 5–8 March 2013.
36. Al-Massri, G.; Ghanem, H.; Khatib, J.; Kırgız, M.S.; Elkordi, A. Chemical shrinkage, autogenous shrinkage, drying shrinkage, and expansion stability of interfacial transition zone material using alkali-treated banana fiber for concrete. *J. Struct. Integr. Maint.* **2024**, *9*, 2390650. [CrossRef]
37. *ASTM, C 157*; Standard Test Method for Length Change of Hardened Hydraulic-Cement Mortar and Concrete. ASTM International: West Conshohocken, PA, USA, 2008.
38. *ASTM, C 192*; Standard Practice for Making and Curing Concrete Test Specimens in the Laboratory. ASTM International: West Conshohocken, PA, USA, 2014.
39. Lu, T.; Li, Z.; Huang, H. Effect of supplementary materials on the autogenous shrinkage of cement paste. *Materials* **2020**, *13*, 3367. [CrossRef] [PubMed]
40. Bartojay, K.; Lucero, C. *Comparative Analysis on Reducing Concrete Shrinkage and Cracking. Research and Development Office*; US Department of the Interior, Bureau of Reclamation: Washington, DC, USA, 2018.
41. Ye, H.; Radlińska, A. A review and comparative study of existing shrinkage prediction models for portland and non-portland cementitious materials. *Adv. Mater. Sci. Eng.* **2016**, *2016*, 2418219. [CrossRef]
42. Al-Akhras, N.M.; Al-Akhras, K.M.; Attom, M.F. Performance of olive waste ash concrete exposed to elevated temperatures. *Fire Saf. J.* **2009**, *44*, 370–375. [CrossRef]
43. Beltzung, F.; Wittmann, F.H. Role of disjoining pressure in cement based materials. *Cem. Concr. Res.* **2005**, *35*, 2364–2370. [CrossRef]
44. Lura, P.; Jensen, O.M.; Van Breugel, K. Autogenous shrinkage in high-performance cement paste: An evaluation of basic mechanisms. *Cem. Concr. Res.* **2003**, *33*, 223–232. [CrossRef]
45. Ghanem, H.; Ramadan, R.; Khatib, J.M.; Elkordi, A. A Review on Chemical and Autogenous Shrinkage of Cementitious Systems. *Materials* **2024**, *17*, 283. [CrossRef]
46. Wang, L.; Zhou, S.; Shi, Y.; Huang, Y.; Zhao, F.; Huo, T.; Tang, S. The influence of fly ash dosages on the permeability, pore structure and fractal features of face slab concrete. *Fractal Fract.* **2022**, *6*, 476. [CrossRef]
47. Siddique, R. Performance characteristics of high-volume Class F fly ash concrete. *Cem. Concr. Res.* **2004**, *34*, 487–493. [CrossRef]
48. Saha, A.K.; Sarker, P.K. Sustainable use of ferronickel slag fine aggregate and fly ash in structural concrete: Mechanical properties and leaching study. *J. Clean. Prod.* **2017**, *162*, 438–448. [CrossRef]
49. Marušić, E.; Štirmer, N. Autogenous shrinkage and expansion related to compressive strength and concrete composition. *J. Adv. Concr. Technol.* **2016**, *14*, 489–501. [CrossRef]
50. Mohamed, A.M.; Tayeh, B.A.; Aisheh, Y.I.A.; Salih, M.N.A. Utilising olive-stone biomass ash and examining its effect on green concrete: A review paper. *J. Mater. Res. Technol.* **2023**, *24*, 7091–7107. [CrossRef]

51. Barcelo, L.; Moranville, M.; Clavaud, B. Autogenous shrinkage of concrete: A balance between autogenous swelling and self-desiccation. *Cem. Concr. Res.* **2005**, *35*, 177–183. [CrossRef]
52. Jun, Z.; Dongwei, H.; Haoyu, C. Experimental and theoretical studies on autogenous shrinkage of concrete at early ages. *J. Mater. Civ. Eng.* **2011**, *23*, 312–320. [CrossRef]
53. Gorospe, K.; Booya, E.; Ghaednia, H.; Das, S. Effect of various glass aggregates on the shrinkage and expansion of cement mortar. *Constr. Build. Mater.* **2019**, *210*, 301–311. [CrossRef]
54. Lian, C.; Zhuge, Y.; Beecham, S. The relationship between porosity and strength for porous concrete. *Constr. Build. Mater.* **2011**, *25*, 4294–4298. [CrossRef]
55. Figueira, R.B.; Sousa, R.; Coelho, L.; Azenha, M.; De Almeida, J.M.; Jorge, P.A.S.; Silva, C.J.R. Alkali-silica reaction in concrete: Mechanisms, mitigation and test methods. *Constr. Build. Mater.* **2019**, *222*, 903–931. [CrossRef]
56. Achour, T.; JARRAYA, E.; Jamel, N.E.J.I. Apports combinés de l'expérimentation et de la modélisation à la compréhension des propriétés mécaniques des bétons. *Rev. Des Compos. Et Des Mater. Av.* **2017**, *27*, 123. [CrossRef]
57. Thomas, M.; Dunster, A.; Nixon, P.; Blackwell, B. Effect of fly ash on the expansion of concrete due to alkali-silica reaction–Exposure site studies. *Cem. Concr. Compos.* **2011**, *33*, 359–367. [CrossRef]
58. Horkoss, S.; Escadeillas, G.; Rizk, T.; Lteif, R. The effect of the source of cement SO3 on the expansion of mortars. *Case Stud. Constr. Mater.* **2016**, *4*, 62–72. [CrossRef]
59. Wang, D.; Shi, C.; Farzadnia, N.; Shi, Z.; Jia, H.; Ou, Z. A review on use of limestone powder in cement-based materials: Mechanism, hydration and microstructures. *Constr. Build. Mater.* **2018**, *181*, 659–672. [CrossRef]
60. Nobre, J.; Ahmed, H.; Bravo, M.; Evangelista, L.; De Brito, J. Magnesia (MgO) production and characterization, and its influence on the performance of cementitious materials: A review. *Materials* **2020**, *13*, 4752. [CrossRef]
61. Dey, A.; Vastrad, A.V.; Bado, M.F.; Sokolov, A.; Kaklauskas, G. Long-term concrete shrinkage influence on the performance of reinforced concrete structures. *Materials* **2021**, *14*, 254. [CrossRef] [PubMed]
62. Yang, Y.; Sato, R.; Kawai, K. Autogenous shrinkage of high-strength concrete containing silica fume under drying at early ages. *Cem. Concr. Res.* **2005**, *35*, 449–456. [CrossRef]
63. Jiang, Z.; Sun, Z.; Wang, P. Internal relative humidity distribution in high-performance cement paste due to moisture diffusion and self-desiccation. *Cem. Concr. Res.* **2006**, *36*, 320–325. [CrossRef]
64. Tran, N.P.; Gunasekara, C.; Law, D.W.; Houshyar, S.; Setunge, S.; Cwirzen, A. A critical review on drying shrinkage mitigation strategies in cement-based materials. *J. Build. Eng.* **2021**, *38*, 102210. [CrossRef]
65. Sirtoli, D.; Wyrzykowski, M.; Riva, P.; Lura, P. Autogenous and drying shrinkage of mortars based on Portland and calcium sulfoaluminate cements. *Mater. Struct.* **2020**, *53*, 1–14. [CrossRef]
66. Gagné, R.; Aouad, I.; Shen, J.; Poulin, C. Development of a new experimental technique for the study of the autogenous shrinkage of cement paste. *Mat. Struct.* **1999**, *32*, 635–642. [CrossRef]
67. Harald, J. Chemical shrinkage of cement pastes with plasticizing admixtures. *Nord. Concr. Res.* **2000**, *24*, 39–44.
68. Al-Massri, G.; Ghanem, H.; Khatib, J.; El-Zahab, S.; Elkordi, A. The Effect of Adding Banana Fibers on the Physical and Mechanical Properties of Mortar for Paving Block Applications. *Ceramics* **2024**, *7*, 1533–1553. [CrossRef]
69. Holt, E.E. Where did these cracks come from? *Concr. Int.* **2000**, *22*, 57–60.
70. Valcuende, M.; Marco, E.; Parra, C.; Serna, P. Influence of limestone filler and viscosity-modifying admixture on the shrinkage of self-compacting concrete. *Cem. Concr. Res.* **2012**, *42*, 583–592. [CrossRef]
71. Zhutovsky, S.; Kovler, K. Chemical shrinkage of high-strength/high-performance cementitious materials. *Int. Rev. Civ. Eng* **2010**, *1*, 110–118. [CrossRef]

infrastructures

MDPI

Article

Conceptual Design of Public Charging Stations for Freight Road Transport

Jakub Hospodka *, Jindřich Sadil, Helena Bínová, Kekula František , Hykš Oldřich, Hykšová Magdalena and Neubergová Kristýna

Faculty of Transportation Sciences, Czech Technical University in Prague, 166 36 Prague, Czech Republic; sadilj@fd.cvut.cz (J.S.); kekulfra@fd.cvut.cz (K.F.)

* Correspondence: hospodka@fd.cvut.cz

Abstract: We present a comprehensive methodology for a two-step approach to address the task at hand. The first step involves the optimal placement of charging stations, while the second step focuses on determining the necessary capacity of the charging stations based on traffic factors. This methodology is applicable to countries, states, or specific areas where the placement and optimization of charging stations for truck road transport are being considered. We identify the key inputs required for solving such a task. In the results section, we demonstrate the outcomes using a model example for the Czech Republic.

Keywords: electric freight transport; charging station; infrastructure; freight transport charging

1. Introduction

This article is focused on developing a methodology for the placement of charging stations for battery electric freight trucks (CHS). The specific task at hand involves designing charging stations with parameters different from those used for personal transportation, where demands on parking spaces and charging power of charging stands are lower, while facilities and services are planned for a higher number of people. Principles for the planning of public charging infrastructure for personal battery electric vehicles (BEV), including a literature review, are discussed, e.g., in [1,2]. Detailed technical guidelines are provided in [3].

Given that freight transportation entails distinct traffic patterns and typically covers larger distances, the allocation problem for locating these charging stations becomes a novel and specific challenge. Currently, almost no publicly accessible infrastructure for heavy-duty electric vehicles is available in the European Union. The most advanced exception is a 600-km road stretch between the Rhine-Neckar and Rhine-Ruhr metropolitan areas in Germany, built to boost the logistical sector along one of the busiest freight routes in Europe. Till the end of the year 2023, it is supposed to contain eight ultra-fast public charging locations.

Considering the commitment of countries to reduce CO_2 emissions, a shift in traction within the freight truck industry can help achieve these goals. However, the "chicken-and-egg" problem persists in implementing electric freight transportation. This problem arises from carriers' reluctance to invest in electric freight vehicles due to the lack of an adequate charging network. Simultaneously, private operators hesitate to invest in charging stations as there is currently no demand. Thus, the role of the government appears indispensable in facilitating the establishment of charging stations. Since the impetus for change comes from a higher level, namely the state, it is possible to plan and execute it optimally.

Consequently, in this case, the placement of charging stations should not be random, based solely on available space, but rather systematically designed to ensure maximum coverage of existing transportation flows in the automotive freight industry. This entails locating the minimum number of stations with sufficient capacity to achieve the desired coverage. Recently, several studies have been published aimed at the design of future charging infrastructure for electric road freight transport. Speth et al. [4] use traffic count data as input and combine them with on-site queueing models to obtain a fast-charging network in Germany with a 100 km distance between locations. Speth et al. [5] define a network of stations on a European highway network based on synthetic transport flow data. However, the location selection does not take into account the suitability of the location for a charging area. We find it important to incorporate more inputs, including parking area availability, power grid connection, and other aspects of the analysis.

To address the task at hand, we present a comprehensive methodology for a two-step approach. The first step involves the optimal placement of charging stations in three stages corresponding to the years 2025, 2030, and 2035. The first two stages are based on the Alternative Fuel Infrastructure Regulation (AFIR) [6], which requires charging stations for freight vehicles in urban nodes and at regular intervals along the TEN-T road network. The last stage, corresponding to the original proposal of AFIR by the European Commission [7,8] and the broadened version proposed by the European Parliament for trilogue negotiations), concludes the basic coverage of the whole network.

The second step of the methodology focuses on determining the necessary capacity and other parameters of the charging stations based on data related to traffic in the catchment area, the specific electricity consumption of considered types of vehicles under the given conditions, the ratio of BEV and the ratio of public charging, parameters of BEV, charging outlets and data on the electrical network. Sources for this data depend on available information in the investigated country/region. For the model example of the Czech Republic discussed in Section 2.2, data on traffic were obtained from the national traffic census [9], which provides traffic intensities of various types of vehicles (including five different categories of freight vehicles above 3.5 t) on work/weekend days, day/night-time, peak periods etc. This census data also includes geographic information on individual segments, allowing for the computation of traffic output (vehicle-kilometers per day) on main and other routes in catchment areas of particular stations, used subsequently for the assessment of the number of charging vehicles and their output [10]. Further, the document [11] provides data on daily, weekly, and yearly variations of traffic intensities.

To assess the specific electricity consumption of different types of vehicles in the conditions of the Czech Republic, the set of open simulation programs SUMO (Simulation of Urban Mobility), together with its extension PHEM, was used for modeling a substantial part of the traffic network in the Czech Republic, simulation of traffic flows and calculation of energy and fuel consumption for present types of vehicles and for various possible future scenarios [12]. Other approaches can also be used, as described e.g., in [13]. The assessment of spatial requirements was based on data from Technical Conditions 171 [14], including dimensions of vehicles from considered categories and the norm [15] specifying minimal distances between parked vehicles. The estimation of the ratio of BEV and the need for public charging is based on scenarios discussed in studies [16,17] that also take into account statistics of trip lengths in different countries. This ratio includes the fact that vehicles use the energy earned from depot chargers to get to the highway network, and vice versa; returning to the depot can be connected with a battery depletion even below a level that is safe for travel on highways. Data on the electric network and available capacity of transformer stations were obtained from electricity providers (ČEZ distribuce, PRE distribuce, and EG.D.).

The proposed methodology determines the course of the required power, load diagrams, numbers and occupancy of charging outlets, and space requirements of the charging infrastructure stations. It also evaluates free distribution capacity and specifies the choice of the station battery. It is applicable to countries, states, or specific areas where the placement and optimization of charging stations for truck road transport are being considered. We identify the key inputs required for solving such a task. In the results section, we demonstrate the outcomes using a model example for the Czech Republic.

2. Methods

The conceptual design of truck charging stations presented in this paper consists of two main steps: Expert design of the location of charging stations in the network (see Section 2.1) and power and spatial needs calculation based on traffic demands for individual charging stations (see Figure 1 and Section 2.2).

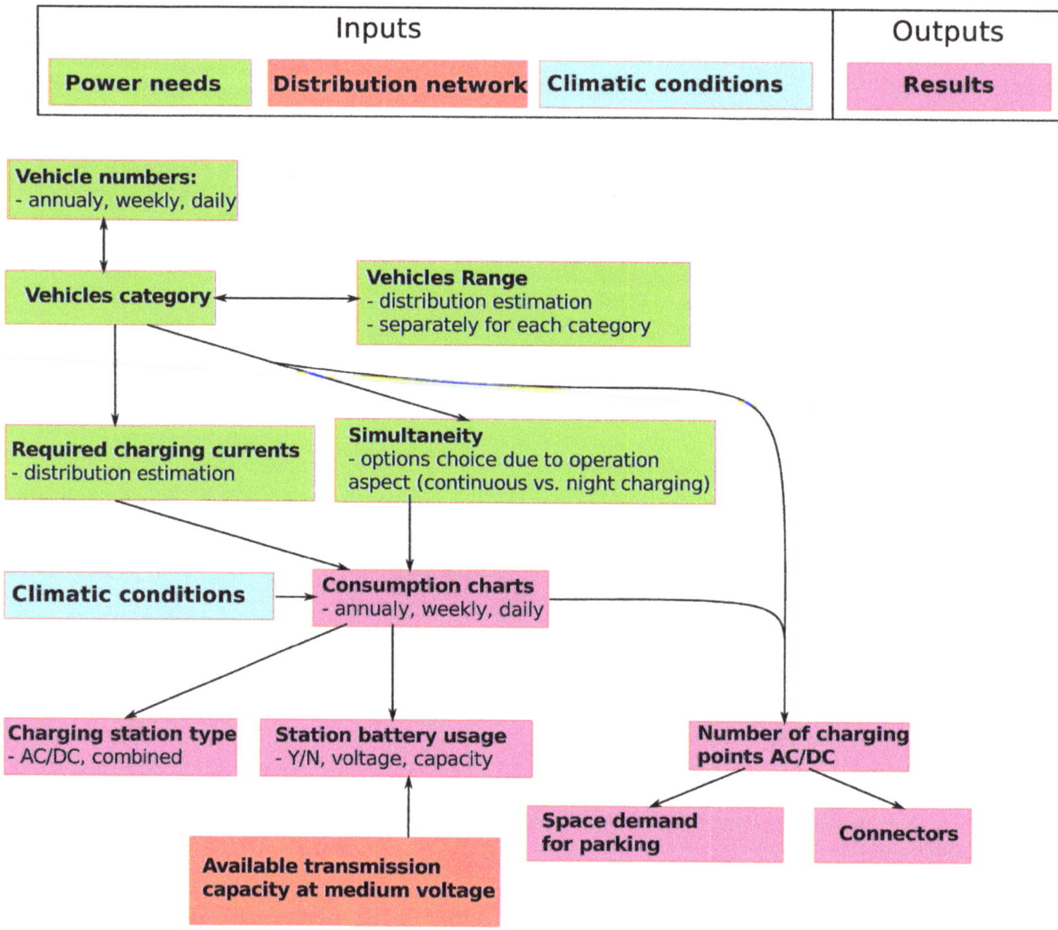

Figure 1. Architecture of the proposed methodology—Step 2: Calculation.

The fundamental aspect of a charging station is its location. In a given area, the total traffic in all directions is considered. Free transformer capacity of the nearest very high/high-voltage transformer station is then found.

The methodology uses vehicle categories monitored by the national traffic census [9]. The relevant categories for heavy road transport are SN, SNP, TN, TNP, and NSN; these categories cover the categories N2 and N3 according to European directive 2007/46/ES.

2.1. Location of Charging Stations

The placement of charging stations is based on the legislative package of the European Commission Fit for 55, in particular, the amendments to the legislation in the field of freight transport on land communications, which obliges individual member states to create a basic network of charging stations for freight transport. According to the proposal of the European Commission for the year 2021 [7], publicly accessible charging stations for freight transport are to be deployed on the roads of the Core TEN-T network at maximum distances of 60 km and in important urban nodes by the end of the year 2025. In the second stage, by the end of the year 2030, charging stations on the Comprehensive TEN-T is to be built at a maximum distance of 100 km, and the performances of the stations on the Core TEN-T are to be strengthened. In the third stage, the performances of the stations on the Comprehensive TEN-T network are also to be strengthened. The new version of AFIR [6] accepted in 2023 requires the deployment of charging stations along at least 15% of the length of the TEN-T road network till 2025 and 50% till 2027.

To satisfy the requirements as well as to cover the area, the target distances of ca. 50 km are considered. We are convinced that we cannot rely on the expectation of a significant breakthrough in the volume and mass capacity of charging cells in the near future, and a higher density of charging stations is necessary to ensure truly safe charging for freight transport throughout the country.

Besides the legislative conditions, the methodology is based on a detailed overview of all rest stops on the European TEN-T network (in the considered country/region), including stationing, the capacity of parking spaces for trucks and possibilities of its extension, the connection of directions, the existence, the size and the possibility of extension of electrical connection, distances from nearby charging stations, including those on other roads, and the existence of facilities for the driver or the possibility of their additional construction.

Charging stations are lined up at indicated distances on the respective main roads, with the lining up starting at the main communication hubs of city centers, especially Prague and Brno. This also corresponds to the requirement of deploying charging stations in important urban nodes in the first stage. Charging stations for the city center and the first charging points of the TEN-T network are located on city bypasses. These stations often overlap or are located at minimum distances. For example, Prague is the main communication hub of the republic; the stations are located at the exits from Prague at the rest stops on the D1, D11, D8, and D5 highways as Core TEN-T roads, as close as possible to the Prague Ring Road D0. Additional charging stations on the Core TEN-T follow every 50 km from these first charging stations. For the location of charging stations on the TEN-T comprehensive, the length of the Prague Ring Road, approx. 88 km, must be considered; the circuit is divided by stations into four approximately equal, 22 km long sections. This distance must be taken into account when leaving the circuit on the radials of the Comprehensive TEN-T network. The first charging stations on the Comprehensive TEN-T starting from Prague are, therefore, approximately 30 km from the Prague Ring Road. Furthermore, they are, of course, at distances of 100 and 50 km, respectively. On the other hand, the stations have to cover the network up to the frontiers, facilitating the transfer between neighboring countries (in any case, the distance of the last station from the frontiers should be smaller than 30 km); in the model example of the Czech Republic, this distance was minimized.

2.1.1. Phase 1: Urban Nodes and Core TEN-T

The location of charging stations on the Core TEN-T network (marked in blue) is based on urban nodes where the charging stations (marked in red in Figure 2) are expected to be built by 2025.

Due to the closeness of the deadline of the first phase, the uncertainty regarding the supply of electric vehicles and their price, and the uncertainty regarding the availability and price of electricity, it is proposed to split the construction of stations on the Core network into the first two stages. Thus, in the first stage, a network of charging stations for Core TEN-T would be built, but at distances of 100 km, which still meets the requirements of [6].

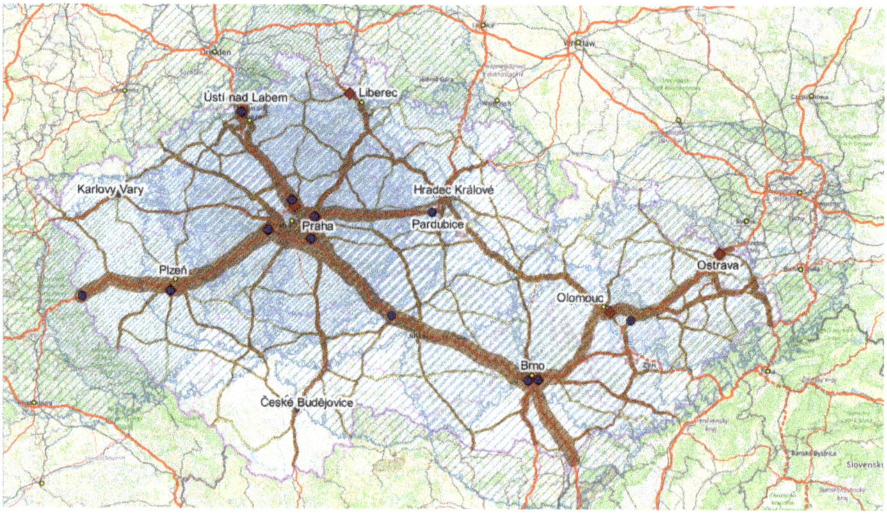

Figure 2. Phase 1, Urban nodes (red) and basic coverage of the Core TEN-T (dark blue).

2.1.2. Phase 2: Densification of the Core TEN-T Network and Basic Coverage of the Comprehensive TEN-T Network

For the second phase until 2030, the Core TEN-T network charging station distances will be densified to 50 km. See in Figure 3. In addition, the construction of a network of charging stations on the Comprehensive TEN-T is planned. Here, according to European recommendations, the distances between stations should be a maximum of 100 km.

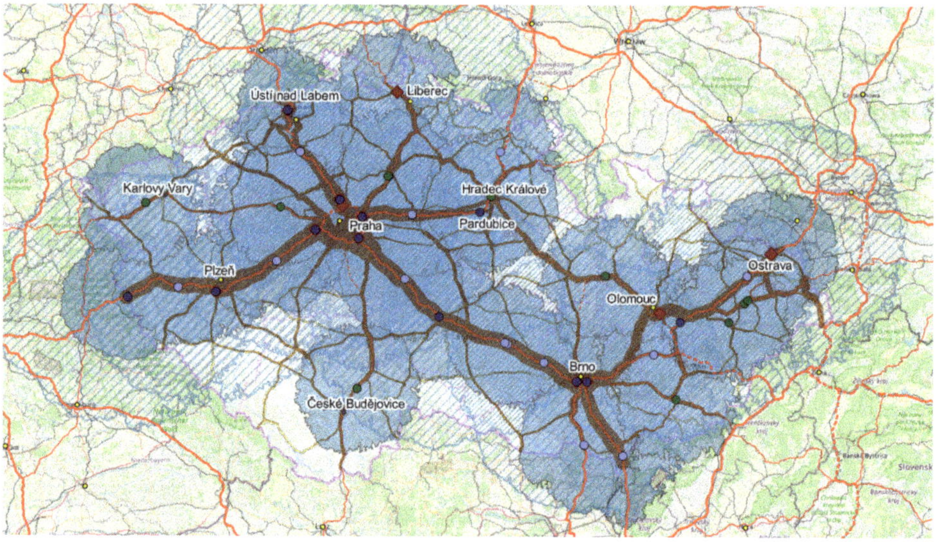

Figure 3. Phase 2, Densification of the Core TEN-T network (light blue) and basic coverage of the Comprehensive TEN-T network (dark green).

2.1.3. Phase 3 (Figure 4): Densification of the Comprehensive TEN-T Network, Additional Coverage of the Czech Republic

For the third phase until 2035, the Comprehensive TEN-T network charging station distances would be densified to 50 km to ensure truly safe charging for freight transport throughout the entire territory of the Czech Republic, without the need to rely on a significant breakthrough in the volume and energy density of traction batteries.

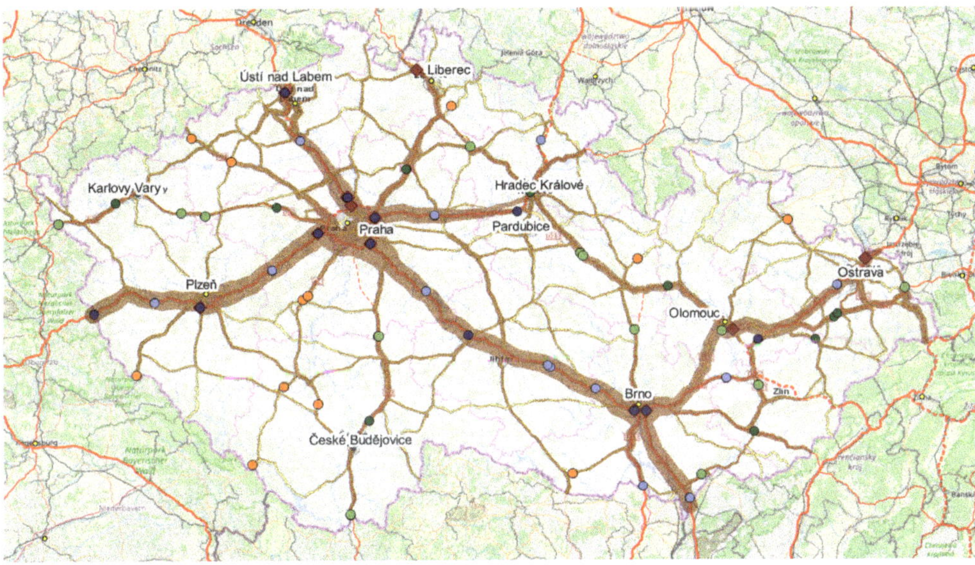

Figure 4. Phase 3, Densification of the Comprehensive TEN-T network (light green) and additional coverage of the traffic network of the Czech Republic (orange).

2.2. *Methodology of Choosing the Concept of Charging Stations for Road Freight Transport*

This is the second step in the process. This step takes place after the task "Location of charging stations in the network" described in Section 2.1 is completed. The schematics of this step are described in Figure 1. The aim of this methodology is to systematically establish suitable concepts of charging stations for road freight transport for the given location of the charging station. The methodology will determine the course of the required power, load diagrams, numbers and occupancy of charging outlets, and space requirements of the charging infrastructure stations; it will evaluate free distribution capacity and specify the choice of the station battery.

2.2.1. Input Parameters

The methodology assumes the following input parameters, and we provided the sources which we used, for example, the case of the Czech Republic:

- Traffic intensity in the area on workdays [9],
- Data from Technical Conditions 189: daily, weekly, and yearly variations in traffic intensity [11],
- Data from Technical Conditions 171: dimensions of vehicles in each category [14],
- Data from ČSN 73 6056: minimum distances between vehicles in each category in perpendicular parking [15],
- Traffic output of main and other routes in the catchment area [10],
- Electricity consumption [12],

- The ratio of battery electric vehicles (BEV) and the ratio of public charging, in the form of scenarios (sets of parameters),
- Probability of overnight charging of BEV, if the vehicle arrives at night,
- BEV parameters: range, charging current of the traction battery in xC units, Current stated in units of a multiple of the battery capacity.

The methodology assumes the following input parameters of the charging station:

- Selected power of charging outlets (e.g., 43, 170, 350 kW),
- Free capacity of the nearest transformer station (MVA),
- Distance to high-voltage power lines,
- High-voltage level,
- Dimensions of the charging stand structure.

2.2.2. Calculation

The basic time step in the calculation of daily courses of the quantities is 1 min. One day corresponds to 1440 samples (discrete time intervals).

2.2.3. Load Diagrams

Daily, weekly, and yearly variations of traffic intensity are determined for each vehicle category based on data from [11]. Based on daily traffic intensities corrected for weekly and yearly variations and based on the number of charged BEVs in the catchment area of the charging station, moments of arrival of individual vehicles at the charging station, the required charging distance (Distance of BEV ride within its range and which corresponds to the respective charge of the traction battery at the charging station), charge, charging current and the duration of discharge are determined. The course of the power of individual BEVs is summed up to determine the necessary momentary power and the reserved capacity (Reserved capacity is the power determined as the energy per a period of 15 min divided by 15 min). The results are determined for the worst-case day in the year. Figure 5 shows an example of the course of the power required together with the reserved capacity during the day in the given month.

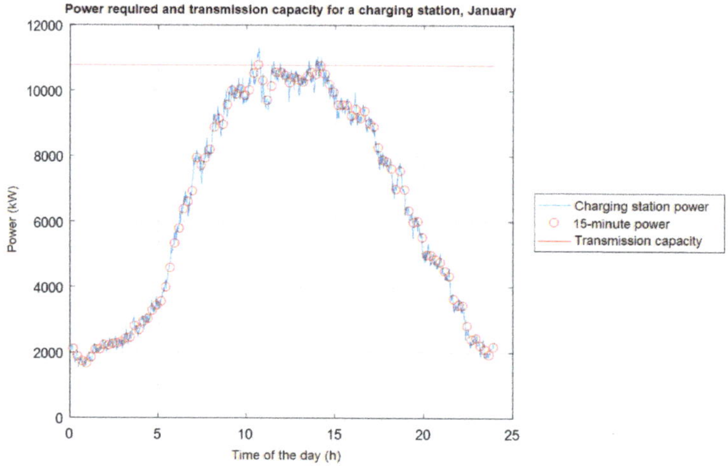

Figure 5. An illustrative example of required power and reserved capacity of a specific charging station of a specific month.

2.2.4. Occupancy

In each time step, it is found out which new vehicles have arrived to be charged, and they are allocated to a charging outlet according to the required power. In the case of a lack of sufficient charging outlets of the required power level, a charging outlet is added. The occupancy of charging outlets of each power level and the usability of the power of the charging outlet is monitored.

2.2.5. Climatic Conditions

Temperatures measured at the meteorological station in the Clementinum, Prague, are considered [18]. The average minimum values of average daily temperatures over the past 20 years are determined for each month in the year. It is assumed that the range of a BEV is lower and the consumption is higher due to heating in lower temperatures. A decrease in battery capacity is not considered; it is assumed that the station battery is placed in an air-conditioned space.

Load diagrams (Figure 5) are determined for individual months in the year while respecting climatic conditions. Occupancy calculation is done separately for each month in the year. The highest occupancy of all months of the year is determined for each 15 min interval.

2.2.6. Estimation of Reserved Capacity

Values of reserved capacity for individual months (Figure 6) are determined. In the case of variance of the parameters of the uniform distribution, the calculation is repeated. Power and occupancy are calculated repeatedly, and the 90th percentile of the results is determined. The reserved capacity for the worst-case day in the year is determined and compared to the free distribution capacity in the very high/high-voltage transformer station, which is closest in terms of power lines.

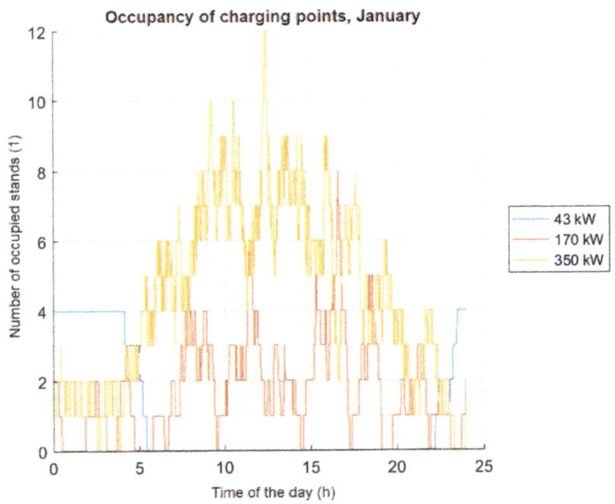

Figure 6. An illustrative example of occupancy of charging points of power levels of a specific charging station in a specific month.

2.2.7. Station Battery

After evaluating the free distribution capacity, it is determined whether it is advantageous to use a station battery in the given month. If the use of the station battery is recommended, the basic parameters of the battery are designed.

2.2.8. Space Requirements

The calculation of space requirements is based on the dimensions of vehicles in each category, the smallest distance between vehicles in each category, and the size of the charging outlet structure. Only perpendicular parking is considered. The calculation does not include the space requirements of the high- and low-voltage substations, the space for safe operation of the stands, the dimensions of the pavement and access roads, and the turning envelope of the vehicles. The required area of parking is assigned to the charging outlets of the given power level based on the knowledge of their occupancy by each category of vehicle. A maximalist approach is considered when calculating the parking area, going from the category with the largest vehicle dimensions to the smallest ones. The area required for the charging stand is calculated from the knowledge of their total number and the dimensions of their structures.

3. Results

After the two aforementioned steps, the following results are achieved.

3.1. Results of Location of Charging Stations (Step 1)

This section contains the results of the placement of charging stations for a model example of the Czech Republic.

The phasing is proposed by authors and goes as follows:

3.1.1. Year 2025: Phase 1

Core TEN-T network with maximum distances of 100 km (near-term, uncertain BEV availability) in significant urban nodes, see in Figure 7.

Figure 7. Map depicting 14 charging stations in Phase 1.

3.1.2. Year 2030: Phase 2 (Figure 3)

Core TEN-T network with maximum distances of 50 km.

Comprehensive TEN-T network with maximum distances of 100 km. See in Figure 8.

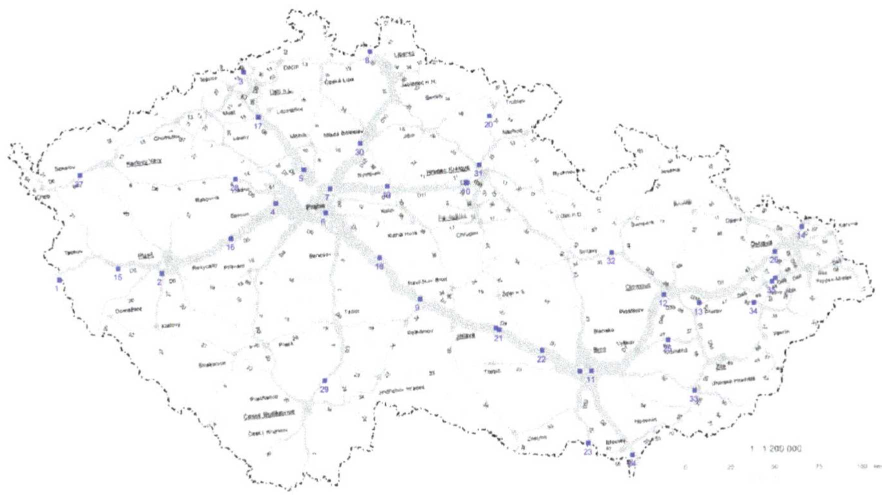

Figure 8. Map depicting 35 charging stations in Phase 2.

3.1.3. Year 2035: Phase 3

Comprehensive TEN-T network with maximum distances of 50 km.

Coverage of the transport network in the Czech Republic outside the TEN-T network.

Determination of P [MW] and n_(≥350 kW) for each phase based on the methodology results [19].

Phase 2 corresponds to the spacing requirements for charging stations (CHS) as specified by AFIR for the Core TEN-T and Comprehensive TEN-T networks. Phase 3 introduces additional CHS beyond the requirements set by AFIR. See in Figure 9.

Figure 9. Map depicting 57 charging stations in Phase 3.

3.1.4. Charging Stations Overview

The following Table 1 summarizes the list of proposed charging stations for phases 1 to 3 described above.

Table 1. Summary of the main attributes of the proposed charging stations for phases 1 to 3. Shortcuts used in the table: CHS—charging station, TR—transformer station, Lon.—Longitude, Lat.—latitude, capac.—capacity, mv—medium voltage, comp—comprehensive.

Phase	Station ID	Type	Road Nr.	CHS Name	CHS Lon. (°N)	CHS Lat. (°E)	TR Lon. (°N)	TR Lat. (°E)	TR Name	TR Capac. (MVA)	Missing mv Line (km)	Parking Places (Pcs.)
Phase 1	1	core	D5	Rozvadov	49.6490	12.5321	49.7920	12.6604	Tachov	5	0.6	218
	2	core	D5	Šlovice	49.6787	13.3289	49.6848	13.4239	Černice	24	0.6	64
	3	core	D8	Varvažov	50.7118	13.9702	50.6866	14.0320	Ustí Sever	10	0.3	71
	4	core	D5	Rudná	50.0336	14.2163	50.0543	14.2120	Chýně	15	0.4	50
	5	core	D8	Klíčany	50.2072	14.4359	50.1340	14.4439	Praha Bohnice	5	0.2	28
	6	core	D1	Nupaky	49.9841	14.6038	49.9952	14.6366	Říčany	1	0.3	120
	7	core	D11	Beranka	50.1083	14.6389	50.0867	14.6145	Běchovice	39	0.2	49
	8	comp	I/35	Chrastava	50.8175	14.9498	50.7895	15.0531	Hrádek n.Nisou	14	0.2	10
	9	core	D1	Humpolec	49.5366	15.3324	49.5462	15.3376	Humpolec	17	0.2	28
	10	core	D11	Osice	50.1371	15.6990	50.1841	15.8235	Hradec Králové	5	1.1	60
	11	core	D1	Brno	49.1625	16.6615	49.1671	16.6313	Komárov	5	1	122
	12	comp	D35	Olomouc	49.5538	17.2282	49.5633	17.2043	Hněvotín	10	1.6	18
	13	core	D1	Osek	49.5105	17.5002	49.5030	17.4981	Prosenice	5	0.6	80
	14	city	D1	Antošovice	49.9024	18.3069	49.8950	18.3367	Bohumín	10	1.1	53
Phase 2	15	core	D5	Kladruby	49.7029	12.9876	49.7640	12.9980	Stříbro	18	0.2	30
	16	core	D5	Záluží	49.8548	13.8713	49.8309	13.8587	Hořovice	1	0	101
	17	core	D8	Siřejovice	50.4808	14.0839	50.4159	14.0485	Libochovice	25	0.2	39
	18	core	D1	Střechov	49.7520	15.0208	49.7017	14.9503	Římovice	5	0.8	63
	19	core	D11	Vrbova Lhota	50.1204	15.0829	50.0343	15.1724	Kolín západ	7	1.1	160
	20	core	I/37	Výšinka	50.4829	15.8752	50.5766	15.9582	Poříčí	15	0	0
	21	core	D1	Kochánov	49.3775	15.9464	49.3468	15.9977	Velké Meziříčí	11	2.9	42
	22	core	D1	Devět Křížů	49.2698	16.2786	49.2774	16.2348	Velká Bíteš	20	1.1	53
	23	core	D52	Mikulov	48.7895	16.6358	48.8151	16.6238	Mikulov	18	0.2	0
	24	core	D2	Lanžhot	48.7267	16.9842	48.7812	16.9061	Břeclav	2	0.7	122
	25	core	D1	Křenovice	49.3202	17.2610	49.3149	17.4494	Kojetín	1,5	2	48
	26	core	D1	Klimkovice	49.7753	18.0976	49.8107	18.1519	Ostrava Poruba	20	1	166
	27	comp	D6	Staré Sedlo	50.1847	12.6948	50.1607	12.6750	Vítkov	15	0.2	30
	28	comp	D6	Nové Strašecí	50.1611	13.9052	50.1421	13.9751	Tuchlovice	18	0	5
	29	comp	I/3	Švamberk	49.1171	14.5929	49.2097	14.7207	Veselí n Lužnicí	14	0.3	0
	30	comp	D10	Brodce	50.3424	14.8735	50.3076	14.8482	Dražice	1	0.3	19
	31	comp	I/35	HradecKrálové	50.2290	15.7972	50.2511	15.7674	Všestary	5	0	10
	32	comp	I/35	Mohelnice	49.7737	16.8216	49.7496	16.6500	Mor. Třebová	10	0.9	10
	33	comp	D55	Uher. Hradiště	49.0592	17.4665	49.0731	17.4627	Uher.Hradiště	2	0.3	25
	34	comp	I/35	Lešná	49.5109	17.9307	49.4775	17.9580	ValašskéMeziříčí	5	0.3	15
	35	comp	D48	Libhošť	49.6203	18.0737	49.6301	18.1422	Příbor	10	0.25	4

Table 1. *Cont.*

Phase	Station ID	Type	Road Nr.	CHS Name	CHS Lon. (°N)	CHS Lat. (°E)	TR Lon. (°N)	TR Lat. (°E)	TR Name	TR Capac. (MVA)	Missing mv Line (km)	Parking Places (Pcs.)
Phase 3	36	comp	I/6	Pomezí	50.0869	12.2652	50.0969	12.3970	Jindřichov	4	0.4	97
	37	comp	I/6	Verušičky	50.1367	13.1848	50.0638	12.9972	Toužim	12	0.1	16
	38	comp	I/6	Lubenec Ležky	50.1199	13.3670	50.2222	13.3924	Podbořany	20	0	5
	39	comp	D3	Dolní Dvořiště	48.6492	14.4526	48.7323	14.5050	Kaplice	17	0.4	15
	40	comp	D3	Mitrovice	49.5331	14.6632	49.4112	14.6898	Tábor	10	0.8	60
	41	comp	I/35	Jičín	50.4503	15.3473	50.4081	15.3424	Nová Paka	20	0	20
	42	comp	I/35	Vysoké Mýto	49.9346	16.1716	49.9848	16.1918	Choceň	10	1	75
	43	comp	I/43	Letovice	49.5626	16.5727	49.4991	16.6405	Boskovice	3	0	0
	44	comp	I/43	Lanškroun	49.9063	16.5994	49.8923	16.4506	Česká Třebová	15	0	2
	45	comp	D55	Lužice	48.8473	17.0718	48.8781	17.1184	Hodonín	23	0.2	22
	46	comp	D55	Kurovice	49.2847	17.5015	49.3151	17.4491	Hulín	1.5	1.5	80
	47	comp	D48	Chotěbuz	49.7619	18.6091	49.7063	18.6198	Ropice	35	0.15	10
	48	dopln	I/26	Folmava	49.3456	12.8499	49.4505	12.9491	Domažlice	1	0.5	60
	49	dopln	I/7	Hora sv. Šebes	50.4955	13.2678	50.4500	13.4205	Chomutov	10	0.6	4
	50	dopln	D7	Velemyšleves	50.3816	13.5659	50.3808	13.5748	Triangle	25	0.2	33
	51	dopln	I/4	Strážný	48.9017	13.7194	49.0554	13.8056	Vimperk	14	0.5	9
	52	dopln	I/4	Rovná	49.2848	13.9569	49.2966	14.1640	Písek	16	0.4	30
	53	dopln	I/4	Příbram	49.7116	14.0968	49.7021	14.0156	Příbram město	6	0.3	20
	54	dopln	I/20	Protivín	49.1983	14.2110	49.1877	14.3822	Křtěnov	30	0.7	7
	55	dopln	I/10	Malá Skála	50.6458	15.2077	50.7069	15.0898	Jeřmanice	35	0.5	5
	56	dopln	I/53	Znojmo	48.8537	16.1089	48.8382	16.1691	Hodonice	25	0	5
	57	dopln	I/57	Krnov	50.0933	17.7219	50.0822	17.6813	Krnov	15	0	20

3.2. *Results of Power and Spatial Needs (Step 2)*

3.2.1. Scenarios

The calculation using this methodology was performed for the following scenarios:

- AFIR_EK: The scenario is determined by the requirements formulated in the original proposal of AFIR by the European Commission [7] for CHS in the years 2025, 2030, and 2035 (max charging power 350 kW).
- AFIR_EP: The scenario is determined by the requirements of AFIR broadened by a proposal of the European Parliament for CHS in the years 2025, 2027, 2030, and 2032 (max charging power 700 kW; see report to [8] from February 2022).
- Industry baseline: The scenario is detailed in [16,17].
- EV-Leaders: The scenario is detailed in [16,17].
- Road-2-Zero: The scenario is detailed in [16,17].

Based on the data in [17,20,21] and expertly corrected based on [9], the shares of BEV and public charging are determined for the Industry-baseline, EV-Leaders, and Road-2-Zero scenarios for the observed stages and vehicle categories. The vehicle categories are described in detail in [10]. The shares of BEV (ratio of traffic realized by BEV to the total traffic) are presented in Table 2; the shares of public charging were expertly estimated to be 40% on average (distinguished for each vehicle category).

Table 2. The shares of BEV (ratio of traffic realized by BEV to the total traffic) for individual vehicle categories and years.

Scenario	Vehicle Category According to [10]	Vehicle Class According to ECE	Year 2025	Year 2030	Year 2035
Industry baseline	SN	N2	0.3	3.0	9.7
	SNP	N2 + O	0.3	3.0	9.7
	TN	N3	0.4	3.7	11.7
	TNP	N3 + O	0.7	6.2	19.9
	NSN	N3 + O	0.7	6.2	19.9
EV-Leaders	SN	N2	1.9	7.0	23.9
	SNP	N2 + O	1.9	7.0	23.9
	TN	N3	1.7	7.5	25.6
	TNP	N3 + O	1.0	9.5	32.5
	NSN	N3 + O	1.0	9.5	32.5
Road-2-Zero	SN	N2	3.9	10.5	32.4
	SNP	N2 + O	3.9	10.5	32.4
	TN	N3	3.5	11.0	33.9
	TNP	N3 + O	1.7	12.9	39.9
	NSN	N3 + O	1.7	12.9	39.9

The distinction between AFIR_EK and AFIR_EP lies in the proposed maximum output of a single charging station. However, the final compromise version [6] accepted in 2023 requires an individual power output of only 350 kW. In general, it denotes the Regulation for the deployment of alternative fuels infrastructure, and it sets mandatory deployment targets for electric recharging and hydrogen refueling infrastructure for the road sector, for shore-side electricity supply in maritime and inland waterway ports, and for electricity supply to stationary aircraft. The significant part of heavy-duty vehicles states: "Recharging stations dedicated to heavy-duty vehicles with a minimum output of 350 kW need to be deployed every 60 km along the TEN-T core network, and every 100 km on the larger TEN-T comprehensive network from 2025 onwards, with complete network coverage to be achieved by 2030. In addition, recharging stations must be installed at safe and secure parking areas for overnight recharging as well as in urban nodes for delivery vehicles" [21].

3.2.2. Parameters for Charging Stations

For each charging station and for all stations in our model example in the Czech Republic, the following is determined:

- *CHS* Cap: Maximum monthly reserved capacity of the lines (MVA),
- Energy: Annual energy consumption (GWh),
- Techl./Park. Area: Required area of the charging technology and parking (m^2),
- MissPark: Total missing parking area for all charging stations (m^2),
- MissDistr: Power deficiency of the distribution network (MVA),
- Stat. Batt: Nominal energy of the station batteries of all charging stations (MWh),
- ChPts: Number of charging points for individual power levels.

The following result sets are listed in Table 3:

- AFIR_EK for power levels of charging stations at 170 and 350 kW,
- AFIR_EP for power levels of charging stations at 170 and 700 kW,
- Phase 1, according to the methodology for power levels of charging stations at 170 and 350 kW,
- Phase 2, according to the methodology for power levels of charging stations at 170 and 350 kW,
- Phase 3, according to the methodology for power levels of charging stations at 170 and 350 kW,
- Phase 3, according to the methodology for power levels of charging stations at 170, 350, and 700 kW,
- Verification calculation with 100% share of BEVs and 100% share of public charging for power levels from the set {170, 350} or {170, 350, 700} kW.

Table 3. Summary of the main attributes of the proposed charging stations for phases 1 to 3. Shortcuts used in the table heading are listed above within Section 3.2.2, as well as detailed descriptions of the result set. Other shortcuts used in the table: InB—Industry Baseline scenario, EVL— EV-Leaders scenario, R2Z—Road to zero scenario.

Result Set	Scenario	Traffic Feasibility	CHS Cap (MVA)	Energy (GWh/ann)	Techl. Area (m^2)	Park. Area (m^2)	MissPark (m^2)	MissDistr (MVA)	Stat. Batt. (MWh)	ChPts 170 kW	ChPts 350 kW	ChPts 700 kW	ChPts Total
AFIR_EK	2025	100.000	15.4	0.0	280	4315	0	0.0	3.4	77	11	0	88
	2030	100.000	95.9	0.0	724	24,189	4865	5.0	20.2	468	57	0	525
	2035	100.000	119.0	0.0	745	29,743	10,832	7.5	29.5	578	68	0	646
AFIR_EP	2025	100.000	22.0	0.0	287	3305	0	0.0	8.4	44	0	22	66
	2027	100.000	68.0	0.0	720	9455	1928	0.0	25.3	136	0	68	204
	2030	100.000	137.0	0.0	800	21,068	3947	8.0	50.5	343	0	114	457
	2032	100.000	170.0	0.0	850	26,622	9088	12.0	67.3	442	0	136	578
Phase 1	InB	97.914	10.3	16.7	280	4641	0	0.0	0.0	22	34	0	56
	EVL	97.477	15.6	32.1	280	7572	0	0.0	0.0	40	54	0	94
	R2Z	97.518	22.3	58.8	280	10,847	0	0.0	1.6	57	80	0	137
Phase 2	InB	99.683	54.8	152.8	700	28,148	4479	1.6	3.8	129	224	0	353
	EVL	99.674	78.5	247.4	700	38,416	7849	4.8	4.1	179	305	0	484
	R2Z	99.672	101.5	341.7	733	49,716	11,232	8.2	17.5	236	393	0	629
Phase 3	InB	99.998	149.9	490.9	1168	73,062	18,004	15.3	11.8	327	596	0	923
	EVL	99.997	240.3	844.9	1324	110,470	39,634	37.0	38.4	490	913	0	1403
	R2Z	99.997	294.8	1053.7	1501	132,620	56,929	52.1	76.2	579	1113	0	1692
Phase 3	InB_700 kW	99.997	159.5	490.9	1170	72,885	17,685	16.9	12.4	199	414	300	913
	EVL_700 kW	99.997	251.7	845.0	1336	106,245	38,115	39.6	40.1	304	606	428	1338
	R2Z_700 kW	99.997	305.0	1053.7	1515	123,103	50,121	56.4	78.5	348	706	502	1556
Verif	-	99.998	1736.6	6445.7	6992	655,525	913,340	1185.9	440.3	3961	5257	0	9218
100%	700 kW	99.998	1758.8	6446.0	7073	549,733	710,376	1211.4	443.2	1730	3509	1967	7206

3.3. Graphical Results

This section contains selected results in graphical form.

3.3.1. Power Distribution Demands

Power distribution demands of AFIR_EK scenario depicted in Figure 10. In the Figure 11 is depicted power distribution demands of Phase 3, and Figure 12 shows power distribution demands of verification scenario with 100% share of electric trucks. Comparison of different scenarios is shown in the Figure 13.

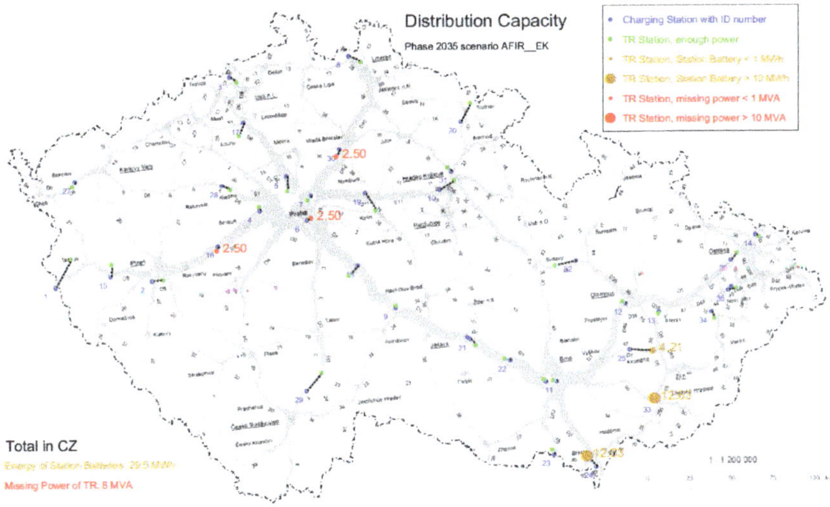

Figure 10. Power distribution demands of AFIR_EK scenario.

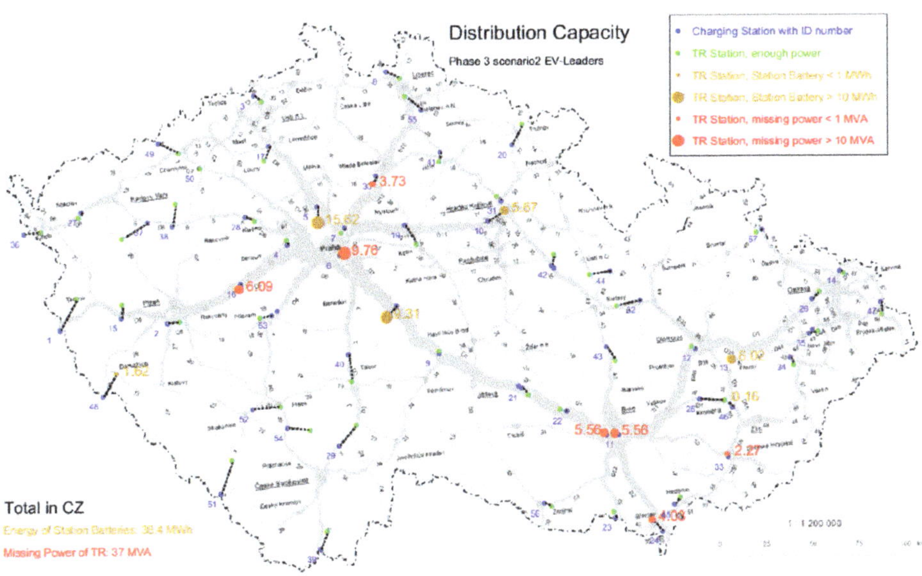

Figure 11. Power distribution demands of Phase 3 EV_Leaders scenario according to the author's methodology.

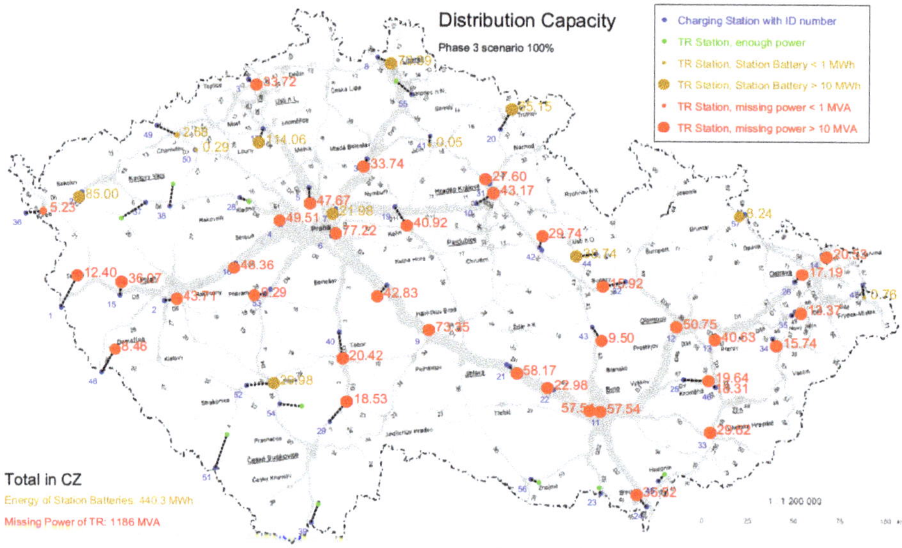

Figure 12. Power distribution demands of verification scenario of 100% share of electric trucks, according to the author's methodology.

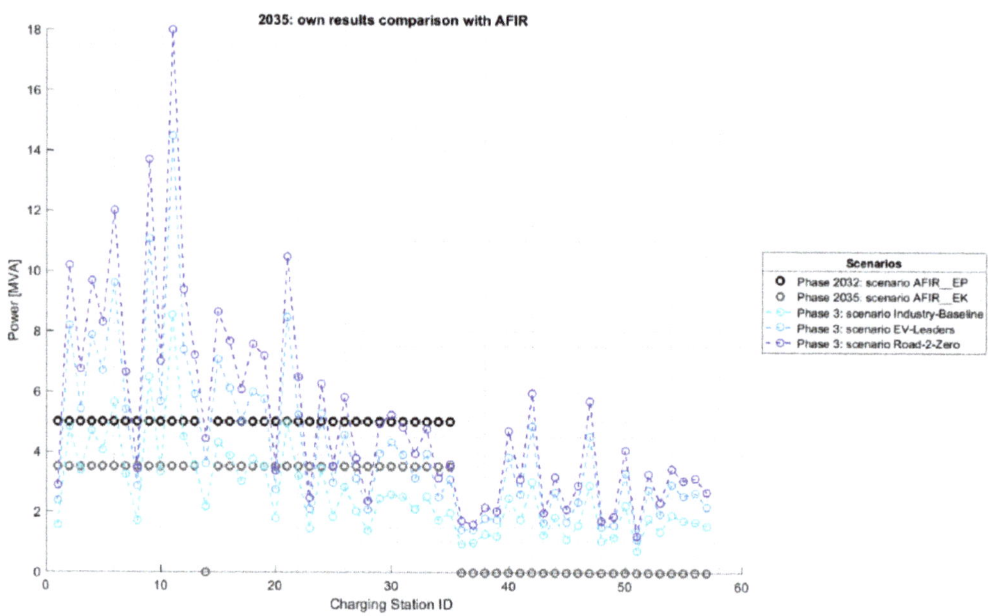

Figure 13. Power distribution demands comparison of different scenarios according to the author's methodology with AFIR requirements.

3.3.2. Spatial Demands for Parking

Following charts show parking space demands for different scenario Figure 14 shows demand for AFIR_EK scenario, Figure 15 shows demand for Phace3 and Figure 16 show demand for of verification scenario of 100% share of electric trucks.

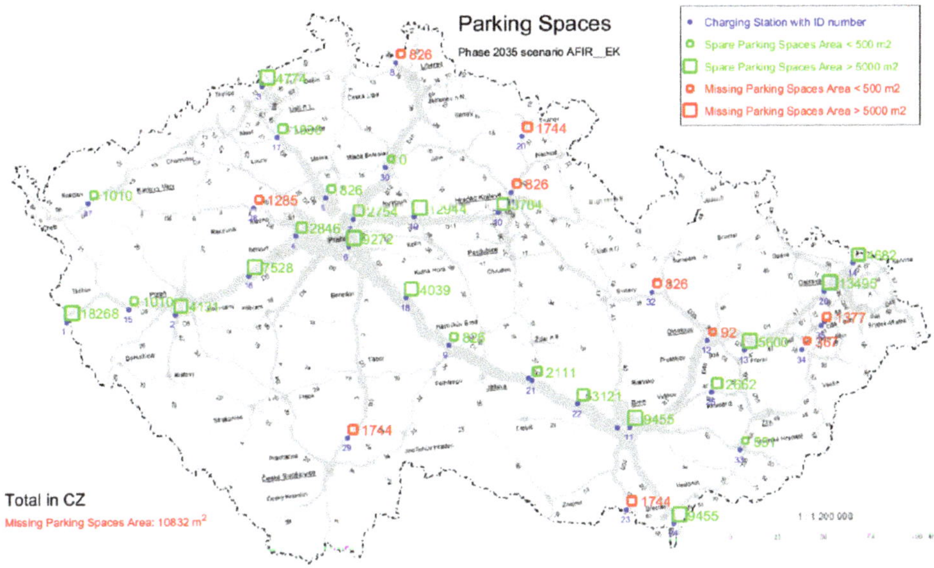

Figure 14. Parking space demands of AFIR_EK scenario.

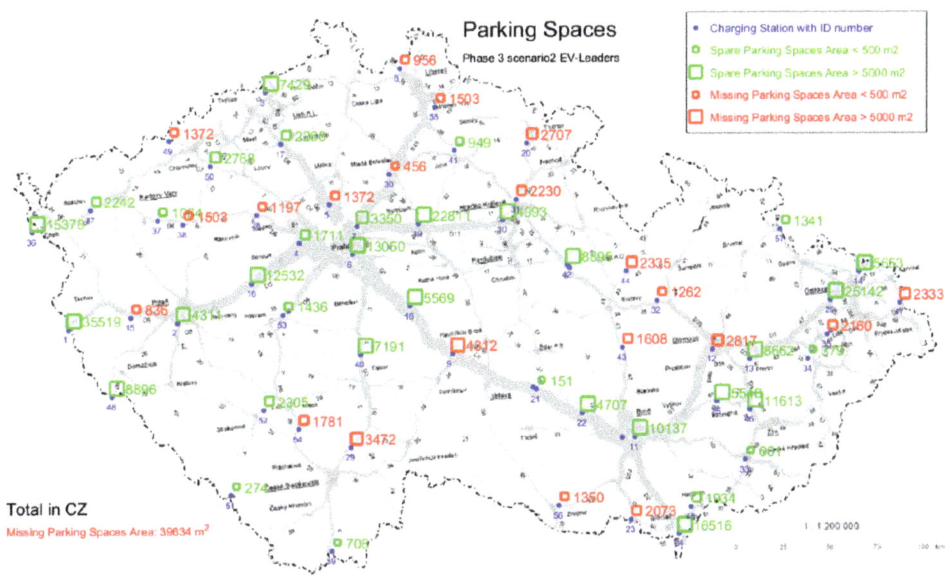

Figure 15. Parking space demands of Phase 3 EV-Leaders scenario according to the author's methodology.

Figure 16. Parking space demands of verification scenario of 100% share of electric trucks, according to the author's methodology.

3.4. Comment on Numerical Results

3.4.1. Scenario AFIR_EK

In the year 2025, the distribution capacity is sufficient, and in one out of the total of 14 CHS, it would be necessary to install a station battery with an energy capacity of 3.4 MWh. Sufficient parking areas are available.

In the year 2030, the distribution capacity would need to be increased by a total of 5 MVA for 2 CHS out of the total of 35. In 3 additional CHS, station batteries with a total energy capacity of 20.2 MWh would need to be installed. Parking areas would need to be increased by a total of 4865 m^2 in five CHS.

In the year 2035, the distribution capacity would need to be increased by a total of 8 MVA for 3 CHS out of the total of 35. In 3 additional CHS, station batteries with a total energy capacity of 29.5 MWh would need to be installed. Parking areas would need to be increased by a total of 10,832 m^2 in ten CHS.

3.4.2. Scenario AFIR_EP

In the year 2025, the distribution capacity is sufficient, and in one out of the total of 14 CHS, it would be necessary to install a station battery with an energy capacity of 8.4 MWh. Sufficient parking areas are available.

In the year 2027, station batteries with a total energy capacity of 25.3 MWh would need to be installed in three CHS. Parking areas would need to be increased by a total of 1928 m^2 in five CHS.

In the year 2030, the distribution capacity would need to be increased by a total of 8 MVA for 2 CHS out of the total of 35. In 3 additional CHS, station batteries with a total energy capacity of 50.5 MWh would need to be installed. Parking areas would need to be increased by a total of 3947 m^2 in five CHS.

In the year 2032, the distribution capacity would need to be increased by a total of 12 MVA for 3 CHS out of the total of 35. In 3 additional CHS, station batteries with a total energy capacity of 67.3 MWh would need to be installed. Parking areas would need to be increased by a total of 9088 m^2 in nine CHS.

3.4.3. Scenario INDUSTRY BASELINE According to the Methodology

In the year 2025, the distribution capacity is sufficient, and no CHS out of the total of 14 would require the installation of a station battery. Sufficient parking areas are available.

In the year 2030, the distribution capacity would need to be increased by a total of 2 MVA for 1 CHS out of the total of 35. In 2 additional CHS, station batteries with a total energy capacity of 3.8 MWh would need to be installed. Parking areas would need to be increased by a total of 4,79 m^2 in seven CHS.

In the year 2035, the distribution capacity would need to be increased by a total of 15 MVA for 4 CHS out of the total of 57. In 3 additional CHS, station batteries with a total energy capacity of 11.8 MWh would need to be installed. Parking areas would need to be increased by a total of 18,004 m^2 in 18 CHS.

3.4.4. Scenario EV-Leaders According to the Methodology

In the year 2025, the distribution capacity is sufficient, and in one out of the total of 14 CHS, it would be necessary to install a station battery with an energy capacity of 0.01 MWh. Sufficient parking areas are available.

In the year 2030, the distribution capacity would need to be increased by a total of 5 MVA for 2 CHS out of the total of 35. In 3 additional CHS, station batteries with a total energy capacity of 4.1 MWh would need to be installed. Parking areas would need to be increased by a total of 7849 m^2 in 8 CHS.

In the year 2035, the distribution capacity would need to be increased by a total of 37 MVA for 6 CHS out of the total of 57. In 6 additional CHS, station batteries with a total energy capacity of 38.4 MWh would need to be installed. Parking areas would need to be increased by a total of 39,634 m^2 in 21 CHS.

3.4.5. Scenario Road-2-Zero According to the Methodology

In the year 2025, the distribution capacity is sufficient, and in one out of the total of 14 CHS, it would be necessary to install a station battery with an energy capacity of 1.6 MWh. Sufficient parking areas are available.

In the year 2030, the distribution capacity would need to be increased by a total of 8 MVA for 3 CHS out of the total of 35. In 3 additional CHS, station batteries with a total energy capacity of 17.5 MWh would need to be installed. Parking areas would need to be increased by a total of 11,232 m^2 in nine CHS.

In the year 2035, the distribution capacity would need to be increased by a total of 52 MVA for 7 CHS out of the total of 57. In 10 additional CHS, station batteries with a total energy capacity of 76.2 MWh would need to be installed. Parking areas would need to be increased by a total of 56,929 m^2 in 24 CHS.

4. Conclusions

The article presents a methodology for the placement of charging stations specifically designed for heavy-duty vehicles. The methodology provides detailed information regarding the spatial distribution and power requirements of the stations. It can be applied in any area where the necessary inputs are available. The study case presented in this article focuses on the specific context of the Czech Republic, reflecting the reality of the country.

Results of the methodology presented in this paper are as detailed as possible, but still, they are conceptual results, not covering such details as, e.g., which part of available distribution power capacity is reserved for which purposes. Only the value of distribution capacity in the area neighboring the respective charging station is evaluated.

Based on the results analyzed in detail in Section 3.4 Comment on Numerical Results, the most feasible considered scenario is AFIR_EK. In the year 2035, the distribution capacity would need to be increased by a total of 8 MVA for 3 charging stations out of the total of 35. In 3 additional charging stations, station batteries with a total energy capacity of 29.5 MWh would need to be installed. Parking areas would need to be increased by a total of 10,832 m^2 in ten charging stations.

AFIR_EP is the more demanding scenario with the necessary increasing the distribution capacity by a total of 12 MVA for 3 charging stations out of the total of 35 necessary station batteries with a total energy of 67.3 MWh and parking places deficit of a total 9088 m^2, in the year 2032.

Results of the author's methodology presented in this paper compare three main scenarios: Industry Baseline, EV-Leaders, and Road-2-Zero in the year 2025, 2030, and 2035. The Industry Baseline scenario is feasible without any modification of distribution capacity or parking areas in 2025, but will require increasing to 2 MVA distribution capacity, 3.8 MWh station batteries, and 4479 m^2 of parking areas in 2030, and increasing to 15 MVA distribution capacity, 11.8 MWh station batteries, and 18,004 m^2 of parking areas in 2035.

The EV-Leaders scenario will require building 0.01 MWh station batteries in 2025, increasing to 5 MVA distribution capacity, 4.1 MWh station batteries, and 7849 m^2 of parking areas in 2030, and increasing to 37 MVA distribution capacity, 38.4 MWh station batteries and 39,634 m^2 of parking areas in 2035.

The most demanding Road-2-Zero scenario will require building 1.6 MWh station batteries in 2025, an increase to 8 MVA distribution capacity, 17.5 MWh station batteries, and 11,232 m^2 of parking areas in 2030, and increasing to 52 MVA distribution capacity, 76.2 MWh station batteries, and 56,929 m^2 of parking areas in 2035.

Author Contributions: Conceptualization, J.H., K.F., N.K., J.S. and H.O.; methodology, J.S., H.O., H.M. and J.H.; software, H.O.; validation, J.H., J.S. and K.F.; formal analysis, J.H.; investigation, N.K.; resources, H.B.; data curation, H.O. and J.S.; writing—original draft preparation, J.H.; writing—review and editing, J.H.; visualization, J.S.; supervision, J.H.; project administration, H.B.; funding acquisition, H.B. All authors have read and agreed to the published version of the manuscript.

Funding: This research has been supported by the grant TA ČR Théta Dynamické dobíjení TK 05010044, CTU in Prague.

Data Availability Statement: Data are contained within the article and referenced materials.

Acknowledgments: Part of this research has been supported by the grant TA ČR Théta 2 TK02010106 Čistá mobilita a její perspektiva v nákladní silniční dopravě. CTU in Prague.

Conflicts of Interest: The authors declare no conflict of interest.

References

1. Mortimer, B.J.; Hecht, C.; Goldbeck, R.; Sauer, D.U.; De Doncker, R.W. Electric Vehicle Public Charging Infrastructure Planning Using Real-World Charging Data. *World Electr. Veh. J.* **2022**, *13*, 94. [CrossRef]
2. Unterluggauer, T.; Rich, J.; Andersen, P.B.; Hashemi, S. Electric vehicle charging infrastructure planning for integrated transportation and power distribution networks: A review. *ETransportation* **2022**, *12*, 100163. [CrossRef]
3. Balcioglu, G. (Ed.) *Charging Infrastructure Electromobility. DKE–German Commission for Electrical, Electronic & Information Technologies in DIN and VDE*; DKE–VDE Association for Electrical, Electronic & Information Technologies: Offenbach am Main, Germany, 2022.
4. Speth, D.; Plötz, P.; Funke, S.; Vallarella, E. Public fast charging infrastructure for battery electric trucks–a model-based network for Germany. *Environ. Res. Infrastruct. Sustain.* **2022**, *2*, 025004. [CrossRef]
5. Speth, D.; Sauter, V.; Plötz, P. Where to Charge Electric Trucks in Europe–Modelling a Charging Infrastructure Network. In Proceedings of the 35th International Electric Vehicle Symposium and Exhibition (EVS35), Oslo, Norway, 11–15 June 2022. Available online: https://publica-rest.fraunhofer.de/server/api/core/bitstreams/4940547d-9398-40ae-90a3-a2d35201b895/content (accessed on 10 July 2023).
6. Regulation of the European Parliament and of the Council on the Deployment of Alternative Fuels Infrastructure, and Repealing Directive 2014/94/EU of the European Parliament and of the Council, Brussels, 13.7.2023, 2021/0223(COD) (AFIR). Available online: https://eur-lex.europa.eu/legal-content/en/TXT/?uri=CELEX:52021PC0559 (accessed on 15 October 2023).
7. Proposal for a Regulation of the European Parliament and of the Council on the Deployment of Alternative Fuels Infrastructure, and Repealing Directive 2014/94/EU of the European Parliament and of the Council, Brussels, 14.7.2021, COM(2021) 0559 Final, European Commission (AFIR–Proposal of EC). Available online: https://eur-lex.europa.eu/legal-content/EN/TXT/PDF/?uri=CELEX:52021DC0550&from=EN (accessed on 23 December 2023).

8. Draft European Parliament Legislative Resolution on the proposal for a Regulation of the European Parliament and of the Council on the Deployment of Alternative Fuels Infrastructure, and Repealing Directive 2014/94/EU of the European Parliament and of the Council, Brussels, 19.10.2022, COM(2021) 0559–C9-0331/2021–2021/2023(COD), European Parliament (AFIR–Proposal of EP for Trilogue Negotations). Available online: https://eur-lex.europa.eu/legal-content/EN/TXT/HTML/?uri=OJ:C:2022:209:FULL&from=EN (accessed on 23 December 2023).
9. Ministry of Transport ČR. National Traffic Census-Celostátní Sčítání Dopravy 2016, ŘSD. Available online: http://scitani2016.rsd.cz/ (accessed on 10 July 2023).
10. Hykš, O. Tabulky dopravních výkonů v atrakčních obvodech jednotlivých nabíjecích stanic, Output of the project grant TAČR Théta 2 TK02010106 Čistá mobilita a její perspektiva.
11. Ministry of Transport ČR, Document TP 189, Stanovení Intenzit Dopravy na Pozemních Komunikacích, Technické Podmínky. 2018. Available online: https://pjpk.rsd.cz/data/USR_001_2_8_TP/TP_189_2018_final.pdf (accessed on 6 October 2023).
12. Hykš, O. Měrná spotřeba pro jednotlivé kategorie vozidel, model SUMO, Mgr. Oldřich Hykš, Output of the project grant TAČR Théta 2 TK02010106 Čistá mobilita a její perspektiva.
13. Croce, A.I.; Musolino, G.; Rindone, C.; Vitetta, A. Traffic and Energy Consumption Modelling of Electric Vehicles: Parameter Updating from Floating and Probe Vehicle Data. *Energies* **2022**, *15*, 82. [CrossRef]
14. Ministry of Transport ČR. Document TP 171, Vlečné křivky pro ověřování průjezdnosti směrových prvků pozemních komunikací, Technické podmínky, 2005.
15. *ČSN 73 6056*; Odstavné a Parkovací Plochy Silničních Vozidel. MDT: Praha, Czech Republic, 1987.
16. Mathieu, L. *Unlocking Electric Trucking in the EU: Recharging in Cities*; Transport & Environment: Brussels, Belgium, 2020. Available online: https://www.smarttransport.org.uk/whitepapers/reports-1/unlocking-the-electric-trucking-in-the-eu-recharging-in-cities (accessed on 10 July 2023).
17. Suzan, S.; Mathieu, L. *Unlocking Electric Trucking in the EU: Recharging along Highways, Electrification of Long-Haul Trucks*; Transport & Environment: Brussels, Belgium, 2021; Volume 2.
18. ČHMÚ, Database of Emtorologicla Reports, Regular Meteorological Meassurements Praha. Available online: https://www.chmi.cz/historicka-data/pocasi/praha-klementinum# (accessed on 10 October 2023).
19. Bínová, H.; Hykš, O.; Hospodka, J.; Kekula, F.; Neubergová, K.; Sadil, J. Document Metodika pro volbu koncepce nabíjecí stanice pro nákladní silniční dopravu, verze 4-0, 2022, Output of the project grant TAČR Théta 2 TK02010106 Čistá mobilita a její perspektiva.
20. Ahmadian, A.; Mohammadi-ivatloo, B.; Elkamel, A. *Electric Vehicles in Energy Systems: Modelling, Integration, Analysis, and Optimization*; Springer Nature: Cham, Switzerland, 2020; ISBN 978-3-030-34448-1.
21. Mathieu, L. *Recharge EU Trucks: Time to Act!* Transport & Environment: Brussels, Belgium, 2020. Available online: https://evmarketsreports.com/recharge-eu-trucks-time-to-act/ (accessed on 10 July 2023).

MDPI AG
Grosspeteranlage 5
4052 Basel
Switzerland
Tel.: +41 61 683 77 34

Infrastructures Editorial Office
E-mail: infrastructures@mdpi.com
www.mdpi.com/journal/infrastructures

www.ingramcontent.com/pod-product-compliance
Lightning Source LLC
Chambersburg PA
CBHW061821210926
51850CB00006B/772